FUNDAMENTALS of
Basic Technical Mathematics

FUNDAMENTALS of
Basic
Technical Mathematics

Sixth Edition

C. Thomas Olivo
Thomas P. Olivo

Delmar Publishers Inc.

ABOUT THE AUTHORS

Dr. C. Thomas Olivo is a foremost authority and leader in authorship of related technical subjects texts in the fields of mathematics, related physical science, and blueprint reading and sketching. He is recognized nationally for contributions to curriculum research and instructional materials development technology and for his work in establishing a system and a national institute for occupational competency assessment of students/trainees, teachers, and industry personnel.

Dr. Olivo worked in industry as a master craftsperson before beginning his career in teaching within secondary schools, post-secondary community colleges and technical institutes, undergraduate studies, and graduate university programs. He served in successively responsible positions as state supervisor, Bureau Chief (Curriculum and Instructional Materials Development), and New York State Education Department Director of Industrial Education. Dr. Olivo pioneered movements for establishing area technical schools (BOCES) and post-secondary technical institutes and community colleges. He served as distinguished professor, College of Education, Temple University, and is known internationally for services in establishing technical education programs and institutions in many countries.

Thomas P. Olivo is a successful teacher of technology and related technical subjects. His experiences range from teaching in secondary schools, to post-secondary institutions, to four-year colleges. Additional services relate to work as assistant professor in technical teacher education at Long Beach State University College. He made significant contributions to the state and nationwide system and products developed in the Curriculum and Instructional Materials Laboratory at Clemson University, where he served as specialist, and during service as a New York State Supervisor of technical education.

Mr. Olivo's writings include coauthorship of textbooks in technical mathematics, related science, and others, and authorship of blueprint reading and technical sketching. Mr. Olivo has worked in manufacturing and in the building construction industry.

For information, address Delmar Publishers Inc.
2 Computer Drive West, Box 15-015
Albany, New York 12212-5015

Printed in the United States of America
Published simultaneously in Canada
by Nelson Canada,
A Division of The Thomson Corporation

10 9 8 7 6 5 4 3 2 1

Cover design by Spiral Design/Lauren Payne
Cover photo by Wm. J. Dederick PHOTOMICROGRAPHY

Library of Congress Cataloging in Publication Data

Olivo, C. Thomas.
    Fundamentals of basic technical mathematics/C. Thomas Olivo, Thomas P. Olivo.—6th ed.
        p.   cm.
    Rev. ed. of: Basic vocational-technical mathematics. 5th ed. © 1985.
    Includes index.
    ISBN 0-8273-4958-0 (textbook)
    1. Mathematics.   I. Olivo, Thomas P.   II. Olivo, C. Thomas. Basic vocational-technical mathematics.   III. Title.
    QA39.2.045   1992
    510—dc20   92-7828
                                                                     CIP

PREFACE

The prime objective of *Fundamentals of Basic Technical Mathematics* is to provide rich, refreshing, functional, instructional materials that assist individuals to develop mathematical competency skills that meet consumer/citizenship needs and occupational requirements. Mathematical concepts, principles, rules, and worked-out examples are applied to meaningful exercises and practical problems that are broadly representative of a number of occupational fields.

ORGANIZATION AND SCOPE

The contents of the book are derived from analyses of many branches of mathematics and from still other studies of representative occupations to establish computational skill needs. The sequencing of units and the methods incorporated within the units are based on proven, effective teaching/learning methods.

Each *Unit,* each *Section,* and each *Part* of this book is founded on extensive occupational studies. The contents, organization, and methods used answer the question: "What kind and how much technical mathematics does an individual need to successfully master in order to prepare for a satisfactory career, to advance within an occupation, and to prepare for purposeful citizenship?"

This new edition retains the solid organizational approach of providing instructional materials in mathematics within four major *Parts* that contain eight *Sections*. Each section is organized in a sequential series of units. Each unit begins with *Objectives* that quickly identify mathematical competencies to be mastered. *Concepts* and *Principles* are clearly described; *Rules* are presented with worked-out *Examples* to demonstrate the steps that are involved.

REVIEW AND SELF TESTS AND ACHIEVEMENT REVIEWS: EXERCISES AND PROBLEMS

At the end of each unit, the *Review and Self Test* provides a wealth of testing materials that may be used for *pretesting* and/or *post testing* to measure the extent to which mathematical skills are mastered. *Section Achievement Reviews,* which include all units within each section, provide additional testing materials.

The test items are of two types: (1) *exercises* for general applications of each new principle and (2) *practical problems*. Exercises provide an easy transition from learning to developing the ability to apply each new mathematical skill in solving practical problems.

Fundamentals of Basic Technical Mathematics is designed for classroom instruction, for adult retraining or upgrading in extension courses, for cooperative industry/institutional training, for business/industry on-the-job related and supplemental instruction, for military specialty training programs, and for self-study.

COMPUTATIONS BY CONVENTIONAL MATHEMATICAL PROCESSES AND CALCULATORS

The textbook units are deliberately designed to first develop mastery of each new basic principle with functional applications, according to conventional mathematics rules and procedures. Complementary computational skills are then developed using four-process simple calculators, scientific/engineering calculators, and graphics calculators.

CONTENT FEATURES

- Part One (*Fundamentals of Basic Mathematics*) relates to concepts, principles, rules, and examples that apply to the four basic mathematical processes; direct and computed measurement systems; percentages and averages, and graphic and statistical measurements. The units in this part permit review and diagnostic assessment for students who have had prior instruction.
- Part Two (*Fundamentals of SI Metric Measurement*). The stage is set for working with: SI metric units; metric measuring instruments; base, supplementary, and derived SI metric units; and measurement conversion. Throughout the book emphasis is placed on both metric and customary unit measurements and quantities.
- Part Four (*Mathematics Applied to Consumer and Career Needs*) applies principles of business mathematics primarily related to finances that affect each person as a consumer and wage earner. Mathematics is viewed in terms of career planning and development.
- Appendix. Included as supplementary resource materials are content-referenced *Handbook Tables;* an updated *Glossary of Mathematical Terms; Answers to Odd-Numbered Problems;* and an easy-to-use *Index.*

ACKNOWLEDGMENTS

The authors express appreciation to the L. S. Starret Company, Sharp Electronics Corporation, and Texas Instruments, Incorporated, for providing technical resource materials and industrial illustrations. Acknowledgment is also made to other business/industry leaders, community college, area technical institute, higher education institution, and adult education specialists who made valuable contributions during successive stages of analyses, assessment, development, and testing.

Albany, New York

C. Thomas Olivo
Thomas P. Olivo

ABOUT THE COVER

Cover Photo: PHOTOMICROGRAPH of a crystal of Acetamide
Equipment: Nikon AFM Camera. Microscope Olympus HA 40X magnification

ACETAMIDE (CH_3 CON H_2) is an excellent solvent for several types of organic and inorganic substances. It is also used to render sparingly soluble substances more soluble in water. Additionally, it is used in the production of aspirin.

DEDICATION

Fundamentals of Basic Technical Mathematics is dedicated to Hilda G. Olivo and Connie, Tom, and Judi in recognition of contributions made during one or more phases of production, beginning with research and occupational analyses and continuing through design, development, editing, testing, and product assessment.

CONTENTS

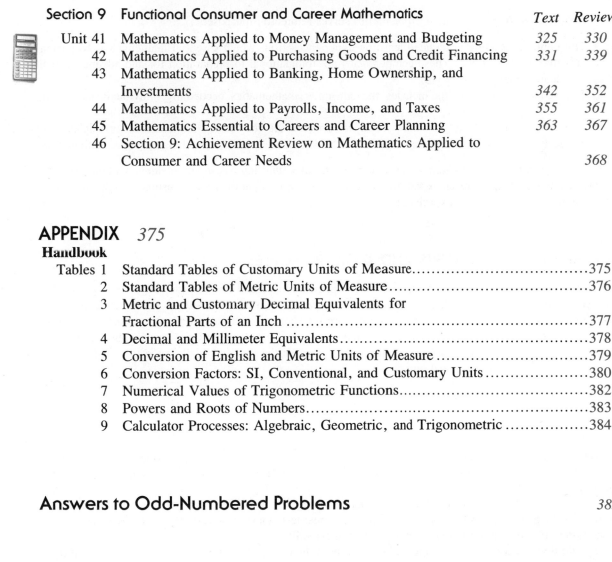

GUIDELINES FOR THE INSTRUCTOR

Fundamentals of Basic Technical Mathematics serves a threefold purpose.

- The contents provide articulated learning experiences from which essential mathematical competencies may be developed. These competencies are foundational to preparation for and advancement in an occupational career.
- The scope of the contents includes the consumer mathematics needed by individuals in meeting everyday needs.
- The instructional units are interlocked to provide all the basic mathematical principles with applications drawn from diverse occupational fields that are needed in teaching technical mathematics.

The scope of the instructional units, the sequence, the rules, the step-by-step procedures, and the extensive exercises and problems are all based on sound analyses of occupational and consumer needs and tested, successful teaching/learning experiences.

MATHEMATICAL CONCEPTS AND BASIC PRINCIPLES

Each new principle or concept in mathematics is presented to fulfill one or more *Objectives of the Unit*. Where applicable, a *Rule* is stated. A practical *Example* is then used to provide an immediate application. Mathematical competency and student progress are then measured by a battery of test items. These are included at the end of each unit as exercises and as practical problems. Within each assignment for each unit, the test items follow a definite sequence that parallels the text and at the same time progress from simple to more complex applications.

PRETESTING AND POST TESTING

Pretesting students for mathematical skills and application competencies may be done by using the test materials that are included in the end-of-unit *Review and Self Test*. These same items may be used for *post testing*.

Pretest results reveal where further instruction is required by an individual or group or determine the point at which instruction begins. After instruction, post tests may be used to determine how successfully the student/trainee has mastered specific mathematical skills.

Additional testing materials are included in the *Achievement Review* for each section. The test items provide a broad comprehensive sampling of mathematical competencies that cut across the topics covered throughout the units within the section.

Finally, the *Instructor's Guide* contains the most comprehensive *Bank of Test Items* available as a supplementary resource. The test items are arranged according to the same Parts, Sections, and Units as the textbook. The Test Bank items may be used to develop a second form of pretest or post test for a unit, section, or part. End-of-course and other comprehensive tests may also be developed by selecting test items from the bank.

STUDENT/TRAINEE PROGRESS RECORDS

A suggested form is included in the *Instructor's Guide* for each student/trainee to self record unit-by-unit progress. The form may be duplicated and made available to each individual. Progress may be recorded with a check mark or (when an assignment is graded) by the score that indicates the mathematical skill competency level of performance.

APPENDIX RESOURCE MATERIALS

Appendix materials are designed as supplementary teaching/learning aids.

- The book is made self-contained by the inclusion of tables, formulas, and other handbook data which a student normally uses.
- The Glossary includes pertinent statements related to mathematical terms, values, quantities, systems, and other technical information. Learning experiences may be built around the Glossary to complement the units.
- Answers to Odd-Numbered Problems permit a quick, easy check on solutions to problems. The *Instructor's Guide* contains the answers to all odd-numbered problems in the *Units;* all problems in the *Section Achievement Reviews,* and all problems in the *Test Bank.*

CORE MATHEMATICS AND COMPLEMENTARY OCCUPATIONAL PROBLEM WORKBOOKS

As stated earlier, *Fundamentals of Basic Technical Mathematics* covers mathematical concepts, principles, and applications to meet occupational needs in a wide diversity of occupations. The units in this book provide the *core of mathematical skill competencies* for a broad spectrum of occupations.

An ancillary series of *Practical Problems in Mathematics* . . . is available for specific occupations. The workbook exercises and problems complement the concepts, principles, rules, formulas, and applications that are covered in detail in *Fundamentals of Basic Technical Mathematics.*

The *Practical Problems in Mathematics* . . . workbooks are designed to extend mathematical competencies to students and other workers in such fields as: business and office occupations, marketing and sales occupations, health and medical services occupations, agricbusiness occupations, trade and technical occupations, and others. The ancillary *Practical Problems in Mathematics* . . . workbooks are publications of Delmar Publishers.

*   *   *   *   *   *

Fundamentals of Basic Technical Mathematics provides a key to the successful development of mathematical competency. This key and competency leads to satisfactory entrance into and progress in careers in major occupational sectors of modern society.

FUNDAMENTALS of
Basic Technical Mathematics

PART ONE

Fundamentals of Basic Mathematics

WHOLE NUMBERS

Unit 1 Addition of Whole Numbers

OBJECTIVES OF THE UNIT
After satisfactorily studying this unit, the student/trainee will be able to
* *Understand the meaning and use of Arabic numbers.*
* *Solve problems requiring the addition of whole numbers.*
* *Check the accuracy of each computation.*
* *Interpret and express whole numbers in the Roman numeral system.*

PRETEST *Use the Review and Self-Test items provided in the Unit Assignment to establish the level of mathematical skills competency and to determine the starting point of instruction.*

A. THE CONCEPT OF WHOLE NUMBERS

The term *whole numbers* refers to complete units where there are no fractional parts left over. Numbers such as 20 and 50; quantities like 144 machine screws, 10 spools, and 57 outlets; and measurements like 75 feet, 125 millimeters, or $875 represent whole numbers because the values do not contain a fraction.

The four basic operations in arithmetic include addition, subtraction, multiplication, and division. Addition is, by far, the most widely used of the four operations. However, even before the basic principles of addition may be applied, the Arabic system of numbers must be understood.

B. THE BASIC ARABIC NUMBER SYSTEM

The Arabic number system is the one that is widely used in this country and in many other parts of the world. This system includes ten digits: 0 1 2 3 4 5 6 7 8 9. These digits may be combined to express any desired number.

The ten numerals make up what is sometimes called the *system of tens* or the *decimal system.* An important fact about this system is that the location or position of a numeral in the written number expresses its value.

The term *digit* is used to identify a particular position. For instance, the first digit (column) in the extreme right position of a number is referred to as the *units digit* or *units column*. The digit in the next position to the left is in the *tens column;* the digit in the third position is in the *hundreds column*. Other examples of the place names of commonly used digits are shown in Figure 1-1. Importantly, the combination of numerals that appear in particular digit positions in a written number expresses its value.

A whole number like 231 is a simple way of saying 200 + 30 + 1. The 231 means that every numeral in the *units* column has its value multiplied by 1; every numeral in the *tens* column, by 10; and every numeral in the *hundreds* column, by 100. The whole number is a shorthand way of representing the sum of the individual place values of the numerals.

Digit Place Names (Whole Numbers)							
Billions	Millions	Hundred-Thousands	Ten-Thousands	Thousands	Hundreds	Tens	Units
●	●	●	●	●	●	●	●

FIGURE 1-1

Numbers may be written in words as three hundred thirty-six; in an expanded form 300 + 30 + 6; or as the numeral, 336.

C. ADDING WHOLE NUMBERS

Addition is simply the process of adding all the numbers in each column in a problem. The answer is called the *sum*.

RULE FOR ADDING WHOLE NUMBERS

- Write one number under another so the unit numerals are in the units columns, tens are in the tens column, and hundreds are in the hundreds column.
- Add all the numbers in the units column.
- Write the result by placing the unit number in the units column.
  Note. Where the sum of a column is greater than 9, put the second number on the left in the tens column.
- Add all the numbers in the tens column. Write the last numeral in this sum in the tens column.
  Note. If this sum is greater than 9, put the second numeral on the left in the hundreds column.
- Continue to add each column of numbers. Find the sum of all columns by adding the numbers in the separate answers in each column.

EXAMPLE: Add 2765 + 972 + 857 + 1724.

Step 1 Arrange numbers in columns.

Step 2 Add all numbers in *units* column.

$$5 + 2 + 7 + 4 = 18$$

Step 3 Add all numbers in *tens* column.

$$6 + 7 + 5 + 2 = 20$$

Step 4 Add all numbers in *hundreds* column.

$$7 + 9 + 8 + 7 = 31$$

Step 5 Add all numbers in *thousands* column.

$$2 + 1 = 3$$

Step 6 Add the sums in each column.

RULE FOR ADDING WHOLE NUMBERS (SIMPLIFIED FORM)

- Write the numbers one under another with each digit in the proper column.
- Add all the numbers in the units column.
- Write the last numeral on the right in the units column.
- *Carry* the remaining numeral (mentally) on the left to the tens column and add it with the rest of the numbers in this column.
- Continue the same process with the remaining columns.

EXAMPLE: Add 2765 + 972 + 857 + 1724.

Add Downward

Step 1 Arrange numbers in columns.

Step 2 Add numbers in *units* column.

$$5 + 2 + 7 + 4 = 18$$

Place 8 in the *units* column. Carry the 1 (mentally) to the *tens* column.

Note. Add all columns downward from top to bottom.

Step 3 Add the numbers in the *tens* column and add the 1 carried over (mentally).

$$(6 + 7 + 5 + 2) + (1) = 21$$

Place 1 in the *tens* column and carry the 2 over to the *hundreds* column.

Step 4 Perform the same addition steps with the *hundreds* and *thousands* columns.

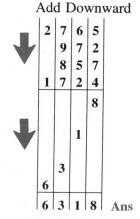

RULE FOR CHECKING ACCURACY

- Find the sum of the numbers in the units column by adding upward (or in the reverse direction of adding).
- Continue to add the numbers in the tens, then hundreds, then thousands columns.
  Note. If the sum of the numbers in the units column is more than 9, carry the number of tens to the tens column and add to the numbers in that column. Follow this same practice with the other columns.
- Compare the result (sum) with the sum obtained by adding in the reverse order.
  Note. If there is an error, repeat the addition step in the reverse order.

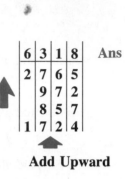

D. HORIZONTAL ADDITION

Many business and industry charts and records and personal forms are designed with horizontal columns for recording values that are to be totaled. The series of values may be added by using either vertical addition or horizontal addition.

RULE FOR THE HORIZONTAL ADDITION OF WHOLE NUMBERS

- Check each number in the series of numbers in the horizontal column that is to be added. Clearly identify each number value in the units, tens, hundreds, thousands, etc. places (digits).
- Start with the numerical value in the units place (digit) of the first number.
- Move to the next number to the right. Add the numerical value in the units place to the first units value.
- Move to the third number to the right. Add each successive units value to the sum of the preceding values in the units place.
- Continue with the numerical values in the series of numbers that are in the tens place; then hundreds, thousands, etc.
- Arrange the numerical values in units, tens, hundreds, thousands, etc. to form the answer.

EXAMPLE: Add **1,925 + 872 + 97 + 1,209**

Step 1 Start with the number value in the units place of the first number.

Step 2 Add this value to the number value in the units place of the second number in the horizontal column.

Step 3 Continue to add each number value in the units place of each successive number to the sum of the preceding numbers.

$$5 + 2 + 7 + 9 = 23$$

└Units place sum

└Carryover to tens place

Step 4 Record the final units place (digit) sum of **3**.

Step 5 Start with the carryover tens place value of **2**. Add the 2 to the numerical value in the tens digit of the first number and each successive number.

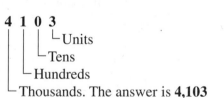

$$1,925 + 872 + 97 + 1,209 = 20$$

└Tens place sum

└Carryover to hundreds place

Step 6 Start with the carryover hundreds place value of 2 and add all number values in the hundreds digit of the first number and each successive number.

Step 7 Combine the sums obtained for the units, tens, hundreds, etc. digits.

4 1 0 3
└Units
└Tens
└Hundreds
└Thousands. The answer is **4,103**

ASSIGNMENT UNIT 1   REVIEW AND SELF TEST

PRETEST *The Review and Self-Test items that follow may be used as a pretest. Pretests are designed to measure a student's beginning level of mathematical skills competency and to determine the starting point of instruction.*

POSTTEST *The Review and Self-Test items may also be applied as a post test. Post tests are planned to establish the student's level of mathematical skills competency after instruction.*

A. The Concept of Whole Numbers (General Applications)

1. Write each number in Arabic figures.
   a. Fifty-nine
   b. One hundred twenty
   c. One thousand eight
   d. Twelve thousand nine hundred eighty-seven

2. Write each number in words.
   a. 12   b. 27   c. 140   d. 926   e. 1,700   f. 7,937

3. Write the numbers in each combination in simplified form.
   a. 50 + 7
   b. 80 + 10 + 9
   c. 100 + 3 + 2
   d. 400 + 50 + 7
   e. 1,000 + 600 + 50
   f. 8,000 + 50 + 7

4. Write each number in expanded form.
   a. 61   b. 103   c. 422   d. 1,006   e. 4,027 f. 6,931

B. Adding Whole Numbers (General Applications)

1. Add each column of numbers vertically. Check all answers.

A	B	C	D	E	F
17	35	67	96	900	808
20	35	24	48	65	15

G	H	I	J	K	L
925	474	72	83	110	9,765
96	888	94	94	87	4,372
		53	40	215	8,988
			12	468	1,009

2. Find the sum of the number combinations (a through f) by adding horizontally. Check each answer.
   a. 12 + 37   c. 77 + 69   e. 548 + 941 + 960
   b. 56 + 25   d. 962 + 829 + 13   f. 827 + 633 + 569

C. Adding Whole Numbers (Practical Problems), Vertical or Horizontal Addition

1. Determine the overall length (A) of the tapered pin.

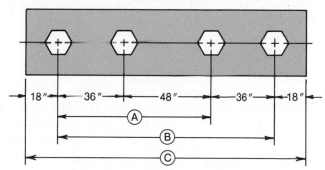

2. The tail lights of a vehicle draw 5 amperes of current; the headlights, 11 amperes; the car heater, 9 amperes; and the ignition coil, 5 amperes. What is the total number of amperes that is drawn?

3. The following amounts of fabrics are required for decorating: 23 yards, 19 yards, 15 yards, and 8 yards. Determine the total yardage of fabric that is needed.

4. Six buildings are to be wired by electricians. The number of outlets that must be installed is 65, 75, 69, 81, 57, and 76, respectively. Find the total number of outlets that must be roughed-in.

5. Determine dimensions Ⓐ, Ⓑ, and Ⓒ for the structural steel beam.

6. A floor coverer laid 675 tiles in a half day; 1,425 the second day; and 1,054 the third day. How many tiles were laid in the $2\frac{1}{2}$ days? Label the answer and check.

7. A plumber made the pipe connections as illustrated. What was the total length of pipe used? (The end-to-end measurement of each pipe is given in inches.)

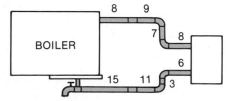

8. Three shipments of 1-inch native pine are received by a contractor: 7,556; 8,750; and 9,898 board feet, respectively. What is the total number of board feet of lumber delivered?

9. The monthly production of motors for refrigerators was as follows: January 29,220; February 32,416; March 37,240; April 39,374; May 45,666; June 52,487; July 36,458; August 35,000; September 32,250; October 51,750; November 62,475; December 50,525. Determine the total output of motors for the year.

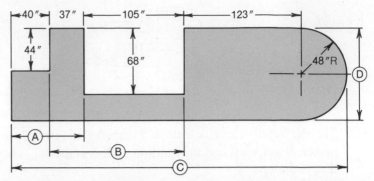

10. Calculate dimensions Ⓐ, Ⓑ, Ⓒ, and Ⓓ for the aluminum template.

11. Find the total length of wire needed by a technician to connect the following lengths of wires: 15 feet, 32 feet, 96 feet, 142 feet, 68 feet, and 84 feet.

12. Determine the total resistance (R_T) of the series electrical circuit by adding the separate resistances ($R_1 + R_2 + R_3 + R_4 + R_5$).

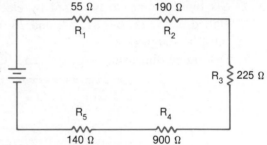

13. The kilowatt hours of electrical energy consumed monthly for six months were 1,412; 1,839; 27,000; 29,787; 32,496; and 1,934. Determine the total kilowatts of energy used.

14. A construction project required A, B, C, and D quantities of diesel fuel for four groups of equipment. (a) Find the number of gallons of diesel fuel used by each group. (b) Determine the total amount of fuel used by the four groups.

Equipment (A)	Equipment (B)	Equipment (C)	Equipment (D)
1,210	4,605	3,202	1,369
989	5,925	3,998	2,787
1,868	9,879	4,687	4,695
		3,896	5,489

15. Examine the main floor plan of the house. Determine the total number of square feet in the master bedroom, living and kitchen/dining rooms, and deck.

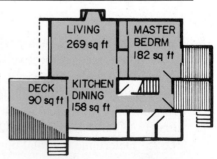

Unit 2 Subtraction of Whole Numbers

OBJECTIVES OF THE UNIT
After satisfactorily studying this unit, the student/trainee will be able to
- *Exchange number values between digits when subtracting whole numbers.*
- *Understand and apply the rule for subtracting whole numbers.*
- *Solve general and practical problems requiring subtraction.*
- *Check answers to subtraction problems.*

PRETEST   *Use the Review and Self-Test items provided in the Unit Assignment to establish the level of mathematical skills competency and to determine the starting point of instruction.*

Subtraction is the process of determining the difference between two numbers or quantities. The number from which another number is to be taken (subtracted) is known as the *minuend*. The number to be subtracted is the *subtrahend*. The result of the process is called the *difference* or *remainder*.

RULE FOR SUBTRACTING WHOLE NUMBERS

- Check the digit values of the larger number (minuend) and the smaller number (subtrahend).
- Subtract the unit digit value in the smaller number from the unit digit value in the larger number.
- Continue subtraction with the numbers in the tens digit, hundreds digit, etc., of the minuend and subtrahend.
- Write out the results of subtraction in each digit place. Label the answer.

EXAMPLE 1: Subtract 346 from 988 *(vertically)*.

Step 1   Write the larger number as the *minuend*.

Step 2   Place the digits in the *subtrahend* in proper columns.

Step 3   Start with the units column and take 6 away from 8. Record the difference (2) in the units column.

Step 4   Continue in the same manner with the tens and hundreds columns.

Step 5   The answer (642) is the difference between the two numbers.

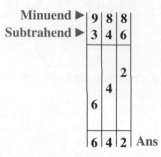

Many times the digits in one or more columns of the subtrahend are larger than the corresponding digits in the minuend. In such cases, numbers are *exchanged* from the minuend. The principle of *exchanging* is based on redistributing numbers. For instance, if 254 is to be subtracted from 723, the digits in both the units and tens places of the subtrahend are larger than those in the minuend.

EXAMPLE 2: *Horizontally* subtract **1,764** from **6,025.**

Step 1   Rearrange the two numbers (minuend and subtrahend) for ease in subtracting.

$$6,025 \; - \; 1,764$$

Step 2   Subtract the unit digit value in the smaller number from the unit digit value in the larger number.

$$6,025 \; - \; 1,764 \; = \; 1$$

Step 3   Continue to subtract the value of the tens digit in the smaller number from the tens digit value in the larger number.

$$6,025 \; - \; 1,764 \; = \; 6$$

Note that (1) is exchanged from the hundreds digit in order to carry on the subtraction.

Step 4   Proceed to subtract the value of each successive digit.

Step 5   Combine the differences obtained for the answer.

$$6,025 \; - \; 1,764 \; = \; 4,261 \text{ Ans.}$$

RULE FOR EXCHANGING NUMBERS IN SUBTRACTION

- Consider 723 as being equal to 700 + 20 + 3.
- Exchange one ten from the tens column for 10 units. Add these 10 units to the units column. The 723 is now equal to 700 + 10 + 13.
- Subtract the 4 in the subtrahend from 13.
- Exchange 100 from the hundreds column and add it to the 10 in the tens column. Thus, the 723 is now equal to 600 + 110 + 13.
  Note. The 5 in the tens column of the subtrahend means 5 times 10 or 50.
- Subtract the 50 in the minuend from the 110 in the subtrahend.
- Subtract the 2 in the hundreds column from the 6 in the same column of the minuend; 600 − 200 = 400.
- Simplify the result. 400 + 60 + 9 = 469   **Ans**

RULE FOR CHECKING THE SUBTRACTION OF WHOLE NUMBERS

- Arrange the numerals in each digit in the subtrahend and those in the answer in columns.
- Add the subtrahend and the difference. When the difference is correct, the sum of the difference and the subtrahend is equal to the minuend.
- Recheck if the answers are not equal. First check the addition, which is the easiest step. If the answers still do not agree, rework the original problem.

ASSIGNMENT UNIT 2   REVIEW AND SELF TEST

PRETEST   *The Review and Self-Test items that follow may be used as a pretest. Pretests are designed to measure a student's beginning level of mathematical skills competency and to determine the starting point of instruction.*

POSTTEST   *The Review and Self-Test items may also be applied as a post test. Post tests are planned to establish the student's level of mathematical skills competency after instruction.*

A. Subtracting Whole Numbers and Checking (General Applications)

1. Subtract each pair of numbers. Check each answer.

A	B	C	D	E	F
78	87	45	98	286	364
34	26	29	59	142	158

G	H	I	J	K	L
753	946	473	707	1,642	2,537
225	168	289	198	456	1,659

2. Perform each operation as indicated. Check each answer.

A	246 − 134	E	3,015 − 2,127
B	727 − 415	F	6,007 − 5,188
C	965 − 847	G	4,112 + 705 + 1,293 − 2,097
D	1,752 − 1,263	H	15,625 + 16,596 + 8,989 − 7,349

3. Determine the difference between each set of numbers and check each answer.
   a. 19,264 and 11,156   c. 10,065 feet and 9,047 feet
   b. 8,537 and 6,759   d. 20,003 miles and 13,365 miles
4. Subtract and check each answer.
   a. 51,219 from 63,422   c. 7,603 acres from 9,502 acres
   b. 9,655 from 13,004   d. 17,092 square miles from 25,001 square miles

B. Subtracting Whole Numbers (Practical Problems)

Note. Check all answers.
1. A contractor has 5,500 board feet of oak flooring. If 2,625 board feet are used on one house, how much flooring is left?
2. A customer's service bill for electricity shows that a total of 1,235 kilowatt hours was used. Of this total, 367 kilowatt hours were used for lighting service and the balance for domestic hot water. How many kilowatt hours were used for hot water?
3. Determine the number of miles traveled for each of five weeks from the odometer readings shown.

Week	1	2	3	4	5
Reading (Start)	32,119	32,899	33,988	35,976	37,065
Reading (End)	32,899	33,988	35,976	37,065	39,001

4. A container (drum) holds 55 gallons of a chemical. In a one-month period, these quantities were used: 5 gallons, 10 gallons, 8 gallons, 7 gallons, 16 gallons, and 8 gallons. How much of the chemical is left?

5. Subtract the following electrical quantities.

a. 297 milliamperes
   −148 milliamperes

b. 632 watts
   −419 watts

c. 54,500 ohms
   −29,750 ohms

6. The cubic yards of gas energy used in five projects for two processes (A) and (B) are given in the table.

Projects	1	2	3	4	5
Process (A)	99	365	1,277	3,018	41,605
Process (B)	87	246	1,098	1,129	32,719

a. Find the increased number of cubic yards of energy required for process A over process B for projects 1 through 5.
b. Determine the total cubic yards of energy required for process A.
c. Determine the total cubic yards for process B.
d. Find the difference between the amounts of gas energy required for process A and process B for the total of projects 1 through 5.
e. Show how the answer to item d. may be checked by a second method.

7. A time and motion study of three workers shows the production reported in the table for processes A through E.

Workers	Production for Processes				
	A	B	C	D	E
1	22	131	2,027	1,169	2,235
2	36	257	3,249	2,479	1,357
3	15	215	2,116	3,368	1,799

a. Identify the fastest production worker for each process A through E.
b. Make a table and indicate the difference in production between (1) the fastest worker and the second fastest worker and (2) the fastest worker and slowest worker.

8. Complete the weekly garment sales chart. Compute the missing quantities for garment types A through D.

Type	Garments Received	Daily Sales - Week Ending June 20						Quantity on Hand June 22
		M	T	W	Th	F	S	
A	178	21	17	9	14	25	31	
B	2,347	345	196	187	294	468	613	
C		93	79	67	105	198	94	217
D		216	237	259	316	419	177	597

9. a. Determine the difference in calories of food quantities A and B for each item (1, 2, 3, and 4).
   b. Find the total food values for A and B.
   c. Indicate the total difference in calories between quantities A and B.

Item	Quantity Food Values in Calories		Difference in Calories
	A	B	
1	46	228	
2	2,280	1,682	
3	7,357	5,469	
4	9,874	10,042	
Total			

10. Compute dimensions (A), (B), (C), and (D). Check each answer.

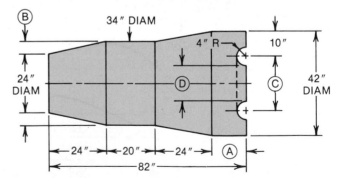

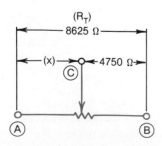

11. Find the ohms of resistance *(X)* between lugs (A) and (B) and (C) of the variable resistor. X = maximum resistance (R_T) − ohms of resistance (B − C).

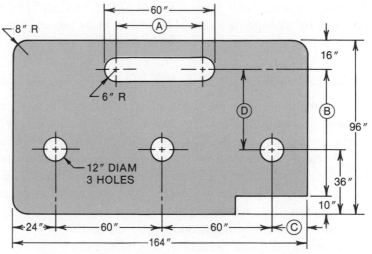

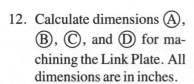

12. Calculate dimensions Ⓐ, Ⓑ, Ⓒ, and Ⓓ for machining the Link Plate. All dimensions are in inches.

Unit 3 Multiplication of Whole Numbers

PRETEST *Use the Review and Self-Test items provided in the Unit Assignment to establish the level of mathematical skills competency and to determine the starting point of instruction.*

Multiplication is a simplified method of adding a quantity a given number of times. For example, instead of writing the number 27 nine times and adding a long column of numbers, 27 may be multiplied by 9. The multiplication process saves time and simplifies the problem. The answer or result obtained by multiplication is called the *product*.

A. ARRANGING NUMBERS FOR MULTIPLICATION

When one number is multiplied by another number, the product is the same, regardless of how the numbers are arranged. For instance, the product of 22 × 11 is the same as 11 × 22. In either case, the number to be multiplied is called the *multiplicand* and the other number, the *multiplier*. The multiplication process is simplified when the smaller number is the multiplier and the larger is

$$\begin{array}{r} 22 \blacktriangleleft \textbf{Multiplicand} \\ \underline{\times 11} \blacktriangleleft \textbf{Multiplier} \\ 484 \blacktriangleleft \textbf{Product} \end{array}$$

the multiplicand. Each digit in the multiplier and multiplicand (and those digits in the number obtained by multiplying) should be placed in its proper column. This simple practice saves time and makes greater accuracy possible.

B. EXPLANATION OF THE MULTIPLICATION PROCESS

In the multiplication process, every number in the multiplicand is multiplied by every number of the multiplier. For example, when the number 47 is multiplied by 26, the numbers in the units column of both multiplicand and multiplier are multiplied first.

Because the product of 7 times 6 is greater than 9, the second digit is mentally *carried over* to the tens column. In this instance, the 42 means 2 units and 4 tens. The 2 is written in the units column and the 4 is carried over and added to the result in the tens column.

Continue by multiplying the number in the tens column of the multiplicand (4) by the same multiplier (6). To this product of 6×4 add the 4 that is carried over.

Write the 8 in the tens column. Since no other numbers in the multiplicand are to be multiplied by 6, the 2 is written in the hundreds column.

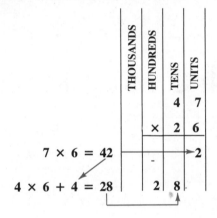

The same multiplication processes are carried out until every number in the multiplicand is multiplied by every number in the multiplier. When the number in the multiplier is in the tens place, the first digit in the product is written in this column. Thus, the next step is to multiply each number in the multiplicand by the number in the tens column of the multiplier.

Actually, the 7 in the multiplicand is multiplied by 20. If this were done, 0 would be written in the units column, 4 in the tens column, and 1 in the hundreds column.

Next, the 4 in the multiplicand represents 4 tens or 40. If the 40 is multiplied by the same 20 in the multiplier, the product is 800. The last step is to add $282 + 140 + 800 = 1222$.

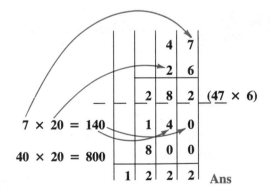

Shortening the Multiplication Process

The multiplication process is shortened and simplified by multiplying by the actual number in the tens column of the multiplier rather than by ten times that number. Then enter the first right digit of the product in the tens column and carry over the remainder to the hundreds column. When multiplying by a number in the hundreds column, put the first right digit in the hundreds column.

C. MULTIPLYING WHOLE NUMBERS

Regardless of the number of digits in either the multiplicand or multiplier, the multiplication process is the same.

RULE FOR MULTIPLYING WHOLE NUMBERS

- Write the larger of the two numbers as the multiplicand; write the smaller as the multiplier.
  Note. Place the numerals in both multiplicand and multiplier under each other in columns: tens in tens column, hundreds in hundreds column.
- Multiply the numbers in the units column of both multiplicand and multiplier. Write the units result in the units column and carry over the tens.
- Multiply the number in the tens column of the multiplicand by the number in the units column of the multiplier. Add the tens remainder to this product.
- Write the first digit on the right in the result in the tens column.
- Carry over any numerals representing hundreds and add to the next result.
  Note. If no other numbers are to be multiplied, write each digit in the result in the proper column.
- Continue to multiply every number in the multiplicand by every number in the multiplier.
- Add the results in each column.

EXAMPLE: Multiply 156 by 78.

Step 1   Write numbers in columns.

Step 2   Multiply every number in the multiplicand by 8.

$6 \times 8 = 48$

$5 \times 8 + 4 = 44$

$1 \times 8 + 4 = 12$

Step 3   Multiply every number in the multiplicand by the number in the tens place of the multiplier (7).

$6 \times 7 = 42$

$5 \times 7 + 4 = 39$

$1 \times 7 + 3 = 10$

Step 4   Add the numbers in each column. The result, 12,168, is the product of 156 × 78.

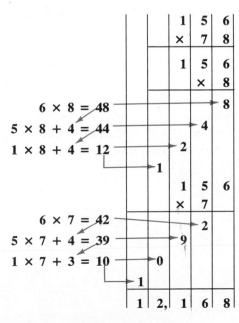

Ans

D. CHECKING THE MULTIPLICATION PROCESS

Two methods of checking multiplication problems are commonly used.

Method I (Checking Each Step)

- Multiply the numbers in the units column of both multiplicand and multiplier.
- Continue until all numbers in the multiplicand are multiplied by all the numbers in the multiplier.
- Check the product against the original product. If the products are not the same, check the steps again.

Method II (Reworking)

- Write the multiplicand as the multiplier; write the multiplier as the multiplicand.
- Multiply each number. Check the final product against the original product.

Regardless of which method is used, the checking process takes almost as much time and effort as the original problem. One of the principal places for error is in failing to remember the numeral that is to be carried over from one column to another. This difficulty may be overcome by writing the number to be carried lightly over the number to be multiplied next. For instance, if 97 is to be multiplied by 7, the problem may be worked out as shown in the example.

The multiplication process is simplified further by learning all the combinations of numbers in a multiplication table from 1 to 10. The product of any two of these numbers may then be given quickly without computation.

$7 \times 7 = 49$

$9 \times 7 + 4 = 67$

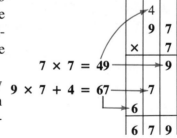

ASSIGNMENT UNIT 3   REVIEW AND SELF TEST

PRETEST
The Review and Self-Test items that follow may be used as a pretest. Pretests are designed to measure a student's beginning level of mathematical skills competency and to determine the starting point of instruction.

POSTTEST
The Review and Self-Test items may also be applied as a post test. Post tests are planned to establish the student's level of mathematical skills competency after instruction.

A. Multiplying Whole Numbers (General Applications)

1. Multiply each set of numbers mentally.

A	B	C	D	E
14	22	78	240	121
6	7	9	8	7

2. Multiply each set of numbers and check each product.

A	B	C	D	E
212 × 6	343 × 5	508 × 9	689 × 6	987 × 7
F	**G**	**H**	**I**	**J**
411	627	303	879	687
×14	×26	×97	×78	×90
K	**L**	**M**	**N**	**O**
8,165	6,057	5,009	7,987	97,009
× 72	×324	×620	×869	×308

B. Multiplying Whole Numbers (Practical Problems)

1. A crew of 17 worked on a construction job 153 days of 8 hours each without a lost-time accident. How many accident-free hours did the crew work?
2. A bricklayer lays an average of 145 bricks an hour. At this rate, how many bricks can be laid in 37 hours?
3. A mason purchased 223 cubic yards of ready-mixed concrete at $36.00 a cubic yard and 38 cubic yards of sand at $9.00 per cubic yard. What was the total cost of materials?
4. The cost of a certain size brass elbow is $7.00 per pound. Determine the cost of 147 pounds of elbows.
5. Compute center distance measurements Ⓐ, Ⓑ, Ⓒ, and Ⓓ for the Fixture Plate.

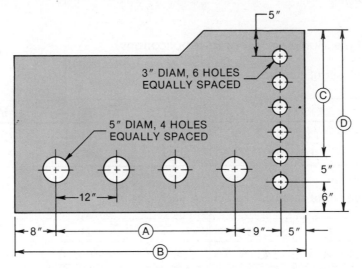

6. It takes 760 shingles per square (100 square feet) when laid 5 inches to weather. How many shingles will be needed to cover 37 squares with the same weathering?
7. A bundle of white cedar shingles contains 250 shingles. How many shingles are in 378 bundles?
8. Two magnets are wound. The first has 57 layers of 98 turns each; the second magnet has 38 layers of 179 turns each. Give the total number of turns in both coils.
9. An electrical contractor purchased 196 conduit boxes at 89 cents each and 87 of another type for $1.35 each. Give the total cost of these materials.
10. A car has three 34-candlepower bulbs, two 26-candlepower, four 9-candlepower, and three 15-candlepower. Find the total candlepower.

11. A truck averages 42 miles an hour for 6 hours daily for 19 days. Another truck averages 37 miles an hour for 6 hours daily for 23 days. A third truck averages 39 miles an hour for 6 hours daily for 22 days. Determine the total mileage of the three trucks.

12. A power plant consumes an average of 17 gallons of fuel per hour. Determine the consumption (a) for a 24-hour day, (b) each week, and (c) for a 31-day month.

13. The hourly production for four different parts (A, B, C, and D) produced on three machines (1, 2, and 3) is recorded in the chart.

Machine	Part A	Part B	Part C	Part D
1	4	10	100	234
2	6	11	120	247
3	9	17	132	379
Unit Cost	3	5	12	27

  a. How many of each part are produced on each machine in an 8-hour shift?
  b. What is the weekly production of each part on each machine for two shifts, each working a 36 hour week?
  c. Determine the weekly cost for each part for the two shifts using the unit cost per part.

14. Compute the amount of material that is needed in the total order for
  a. Hem and seam allowance
  b. Cover length
  c. Completed covers

COVER SPECIFICATIONS

MATERIAL: UNBLEACHED MUSLIN
FINISHED LENGTH: 43"
HEM SEAM ALLOWANCE: 4"
QUANTITY: 725 COVERS

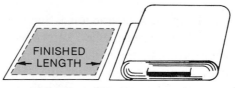

FINISHED
LENGTH

HEM AND SEAM ALLOWANCE

15. Each cylinder of an engine has a displacement of 37 cubic inches. Compute the total displacement of a 6-cylinder engine.

16. Find the voltage in simple circuits Ⓐ and Ⓑ. The voltage (V) equals the current (measured in amperes, A) multiplied by the resistance (R) (measured in ohms, Ω).

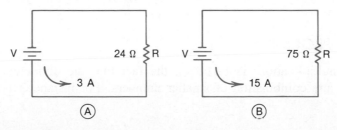

Unit 4 Division of Whole Numbers

OBJECTIVES OF THE UNIT

After satisfactorily studying this unit, the student/trainee will be able to
- *Interpret the arithmetical process of dividing whole numbers.*
- *Divide whole numbers in general and practical problems.*
- *Check answers to problems involving the division of whole numbers.*

PRETEST *Use the Review and Self-Test items provided in the Unit Assignment to establish the level of mathematical skills competency and to determine the starting point of instruction.*

Division is a simplified method of subtracting a quantity a given number of times. For example, if 23 is subtracted from 92, the subtraction process is repeated four times until there is no remainder.

- Subtract 23 from 92 $92 - 23 = 69$
- Subtract 23 from 69 $69 - 23 = 46$
- Subtract 23 from 46 $46 - 23 = 23$
- Subtract 23 from 23 $23 - 23 = \ \ 0$

This method of repeated subtraction is long and becomes more involved as the numbers get larger. In its place, a shorter method known as *division* is used to save time and effort.

In any division problem, the number to be divided is called the *dividend*. The number by which the dividend is divided is the *divisor*. The number that indicates how many times the divisor may be subtracted from the divi-

$$\text{Divisor} \blacktriangleright \quad 23\overline{)92} \quad \blacktriangleleft \text{Dividend}$$

(with 4 marked ◀ Quotient above, and 92 underneath)

dend is the *quotient*. When the divisor cannot be subtracted from the dividend an even number of times, the number left over is referred to as the *remainder*. The two signs or symbols that are commonly used to denote the process of division are ($\div$) and ($\overline{)}$).

A. THE DIVISION PROCESS

Division, like multiplication, is based on the fact that the dividend may be written in expanded form with any combination of smaller numbers. For instance, the number 525 may be thought of as consisting of 500 + 25, or 350 + 175, or any other combination that adds up to 525. The combination depends largely on the divisor. The first step is to break the dividend into a combination of numbers into which the divisor will divide evenly.

If 525 is to be divided by 35, the division process is simplified when the 525 is considered as 350 + 175.

$$35\overline{)525} = 35\overline{)350} + 35\overline{)175} = (10 + 5) = 15 \text{ Ans}$$

In this case, 525 is divided an even number of times by 35. If the number were to be divided by 25, instead of 35, the 525 may be considered in expanded form to be equal to 500 + 25.

$$25\overline{)525} = 25\overline{)500} + 25\overline{)25} = (20 + 1) = 21 \text{ Ans}$$

In these two examples, the dividend in expanded form is made up of different combinations of numbers depending on the divisor.

B. DIVIDING WHOLE NUMBERS

Although this is the principle on which division is based, the actual process is simplified by following a few basic steps.

RULE FOR DIVIDING WHOLE NUMBERS

* Write the number to be divided as the dividend within the division frame; write the divisor on the outside.
* Determine how many times the numerals in the first few digits of the dividend may be divided by the divisor.
* Multiply the divisor by the *trial quotient*. The numeral in the units digit of the divisor is multiplied first, then the tens. The product is placed under the dividend.
  Note. If the *trial quotient* is larger than it should be, the product will be greater than the dividend. When this happens, change the quotient to the next lower number.
* Subtract the product from the dividend.
* Bring down the numeral in the next place in the dividend. If the remainder cannot be divided by the divisor, bring down the next digit in the dividend.
* Repeat the division process until all the digits in the dividend are used.
  Note. When the divisor does not divide evenly into the dividend, the number resulting from the last subtraction is the *remainder*.
* Express the quotient in terms of the quantities that are being divided.

$$\text{Divisor }\overline{)\text{ Dividend}}$$

Trial Quotient

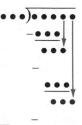

EXAMPLE: *Case 1.* (Without a remainder.) Divide 1984 by 64.

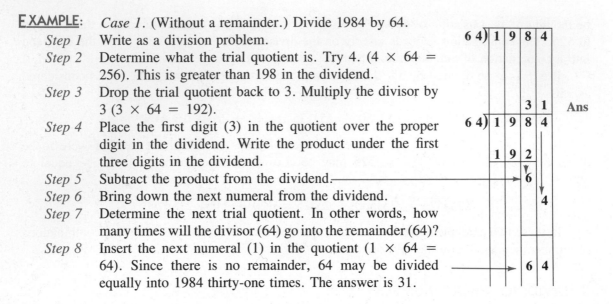

Step 1   Write as a division problem.

Step 2   Determine what the trial quotient is. Try 4. (4 × 64 = 256). This is greater than 198 in the dividend.

Step 3   Drop the trial quotient back to 3. Multiply the divisor by 3 (3 × 64 = 192).

Step 4   Place the first digit (3) in the quotient over the proper digit in the dividend. Write the product under the first three digits in the dividend.

Step 5   Subtract the product from the dividend.

Step 6   Bring down the next numeral from the dividend.

Step 7   Determine the next trial quotient. In other words, how many times will the divisor (64) go into the remainder (64)?

Step 8   Insert the next numeral (1) in the quotient (1 × 64 = 64). Since there is no remainder, 64 may be divided equally into 1984 thirty-one times. The answer is 31.

The division process is simplified if the expanded form of a number is constantly considered. In this example, the first 3 in the quotient is actually 30. Multiplying the divisor (64) by 30 = 1920. Subtracting 1920 from the original 1984 leaves 64. The second digit in the units column is 1, so the quotient is equal to 30 + 1 or 31.

EXAMPLE: *Case 2.* (With a remainder.) Divide 3900 by 47.

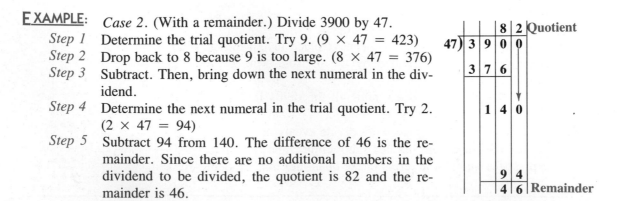

Step 1   Determine the trial quotient. Try 9. (9 × 47 = 423)

Step 2   Drop back to 8 because 9 is too large. (8 × 47 = 376)

Step 3   Subtract. Then, bring down the next numeral in the dividend.

Step 4   Determine the next numeral in the trial quotient. Try 2. (2 × 47 = 94)

Step 5   Subtract 94 from 140. The difference of 46 is the remainder. Since there are no additional numbers in the dividend to be divided, the quotient is 82 and the remainder is 46.

C. CHECKING THE DIVISION OF WHOLE NUMBERS

RULE FOR CHECKING DIVISION

- Multiply the quotient by the divisor.
- Add the remainder to the product. The sum is equal to the dividend when the division is correct.

  Note. If the two quantities are not equal, check the steps again in the checking process. If necessary, rework the original steps in division.

EXAMPLE: Check the correctness of $47\overline{)3,900}$. The given answer is 82 with a remainder of 46.

Step 1   Multiply the quotient by the divisor.

Step 2   Add the remainder (46).

Step 3   Check the sum (3,900) with the original dividend (3,900). Since both agree, the answer is correct.

$$
\begin{array}{r}
\text{Quotient} \\
\text{Divisor} \\
82 \times 47 = 3{,}854 \\
+\ 46 \quad \text{Remainder} \\
\hline
3{,}900 \quad \text{Check}
\end{array}
$$

ASSIGNMENT UNIT 4   REVIEW AND SELF TEST

PRETEST   *The Review and Self-Test items that follow may be used as a pretest. Pretests are designed to measure a student's beginning level of mathematical skills competency and to determine the starting point of instruction.*

POSTTEST   *The Review and Self-Test items may also be applied as a post test. Post tests are planned to establish the student's level of mathematical skills competency after instruction.*

A. Dividing Whole Numbers and Checking (General Applications)

1. Divide each pair of numbers. Check each quotient.

A	B	C	D	E
6)126	4)120	7)147	3)135	7)182

F	G	H	I	J
11)110	12)132	27)810	53)742	92)2,024

2. Perform the operation indicated. Where there is a remainder, mark it (Rem). Check each answer.
   a. Divide 1,250 by 25   c. Divide 1,782 by 162   e. 9,002 ÷ 45
   b. Divide 1,638 by 39   d. 14,091 ÷ 33   f. 80,208 ÷ 121

B. Dividing Whole Numbers (Practical Problems)

1. A mason plastered an area of 425 square yards in 5 days. What was the average number of square yards plastered each day?
2. A contractor agreed to furnish and pour 27 cubic yards of concrete for $1,323.00. What is the cost per cubic yard?
3. A worker caulks 14 windows and uses 14 tubes of caulking compound. How many tubes are used on the average for each window?
4. a. Compute the center-center measurements for slots (A) and holes (B) and (C).
   b. Check each measurement by the multiplication process.

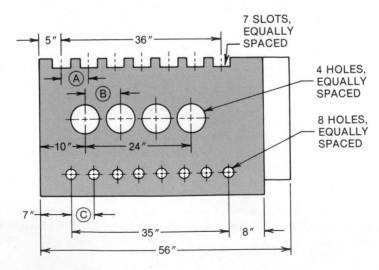

5. A tank holds 4,851 cubic inches of coolant. How many gallons of liquid are needed to fill the tank? (Each gallon contains 231 cubic inches.)

6. How many columns spaced 8'-0" on centers are required for a girder 72'-0" long? (Both ends of the girder are supported on foundation walls.)

7. Determine the rise (A) of each step from the drawing of the stringer.

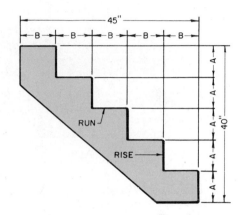

8. Determine the run (B) of each step.

9. A total load of 23,256 watts is distributed equally over 18 branch circuits. Find the load per circuit in watts.

10. What is the average number of feet of wire per outlet used on a job that takes 1,896 feet of insulated wire for 82 outlets?

11. How much money per person can a catering service allot if there will be 350 guests at a total cost of $2,100?

12. A florist has 912 flowers on hand. How many bouquets can be made if there are 6 flowers in each bouquet?

13. A hospital patient is allowed a total of 29,500 calories during a 20-day period. What is the patient's daily allotment?

14. A 24-foot structural steel I-beam weighs 2,136 pounds. Determine the weight per linear foot.

15. The total current in a parallel circuit is 728 milliamperes. There are four branches (A, B, C, and D) into which the current splits equally. Determine and label the amount of current in each branch.

Unit 5 Section 1: Achievement Review on Whole Numbers

OBJECTIVES OF THE UNIT

This achievement review serves as an overall test for Section 1. The unit is designed to measure the student's/trainee's ability to
- *Write whole numbers in simplified or expanded form.*
- *Solve general and practical problems involving addition, subtraction, multiplication, and division of whole numbers.*
- *Find the solutions to problems involving any combination of addition, subtraction, multiplication, and division of whole numbers.*
- *Check each answer.*

SECTION PRETEST/ POSTTEST

The Review and Self-Test items that follow relate to each Unit within this Section. The test items may be used as a Unit-by-Unit pretest and/or Section post test.

UNIT 1. THE CONCEPT OF WHOLE NUMBERS

1. Write each number in expanded form.
   a. 305   b. 620   c. 735   d. 1,025   e. 7,572
2. Write each total in simplified form.
   a. 70 + 5   c. 700 + 20 + 5   e. 6,000 + 200 + 60
   b. 100 + 10   d. 2,000 + 15   f. 10,000 + 700 + 50 + 6
3. Each word or phrase in column I expresses one of the four basic mathematical processes listed in column II. Match each process with the correct term or phrase. Write the A, S, M, or D symbol where it applies in column I.

	Column I				Column II
a.	Times		**f.**	Product of	Addition(A)
b.	Minus		**g.**	Difference between	Subtraction (S)
c.	Plus			Added to	Multiplication (M)
d.	Divided by		**i.**	Sum of	Division (D)
e.	Increased by		**j.**	Quotient of	

ADDITION OF WHOLE NUMBERS

1. Building materials for the repair of a barn include masonry $968, electrical $356, hardware $134, painting $336, and lumber $1,376. Find the total cost of the materials.

2. A school has sixteen electrical circuits with lights in the circuits that total 2,360; 1,648; 1,235; 660; 978; 2,296; 1,586; 1,975; 462; 855; 343; 592; 2,325; 847; 2,015; and 1,238 watts, respectively. Determine the total number of watts consumed when all the circuits are used to capacity.

3. Add each column of numbers vertically. Then add each row across.

201	9,716	7,061	5,012
76	253	2,562	9,109
18	2,009	4,928	6,007
35	5,265	7,465	8,160
257	374	968	5,789

4. Calculate dimensions Ⓐ through Ⓔ from the Die Plate drawing.

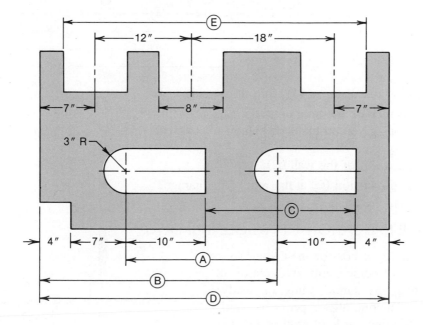

UNIT 2. SUBTRACTION OF WHOLE NUMBERS

1. Subtract the quantities A through F. State each answer in terms of the unit of measure specified in each case.

A	B	C
12,695 square miles − 2,797 square miles	9,085 tons −6,187 tons	3,012 yards −1,029 yards

D	E	F
4,855 board feet −2,068 board feet	6,537 watts −4,659 watts	28,007 barrels −19,018 barrels

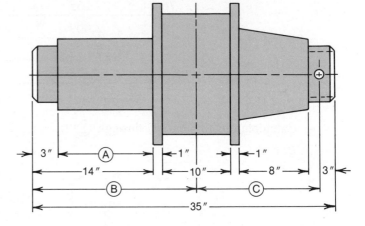

2. Calculate dimensions Ⓐ, Ⓑ, and Ⓒ for the Turned Shaft.

3. Determine the distance Ⓐ from the outside wall to the center of the side door opening from the dimensions given on the floor plan of the two-car garage.
4. Find the width of the wall Ⓑ.
5. What is dimension Ⓒ at the rear of the garage?

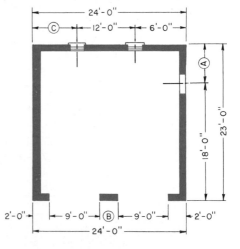

UNIT 3. MULTIPLICATION OF WHOLE NUMBERS

1. Find the total cost of these materials: 65 bags of cement mix at a cost of $8 per bag, 52 cubic yards of ready-mixed concrete at $42 per cubic yard, and 25 cubic yards of sand at $12 per cubic yard.

2. A coil magnet has 56 layers of magnet wire wound around a core. If there are 165 turns of wire per layer, how many turns of wire are there in the coil?

3. Determine the total number of watts in an electrical lighting circuit with this load: twelve 150-watt lamps, three 100-watt lamps, nine 60-watt lamps, and eleven 15-watt lamps.

4. A book contains 277 pages. The type matter on each page measures 5 inches wide by 7 inches long and there are 6 lines of type per inch. Determine how many 5-inch lines must be typeset.

5. Compute center distances Ⓐ, Ⓑ, Ⓒ, and Ⓓ.

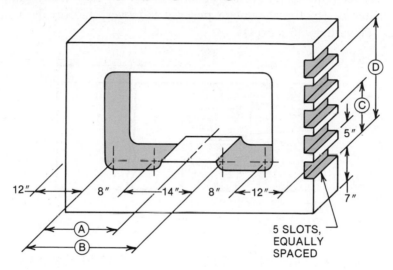

UNIT 4. DIVISION OF WHOLE NUMBERS

1. A total load of 22,931 watts is distributed equally over 23 branch circuits. Give the load per circuit in watts.

2. A 192-page book is to be printed. There are 1,500 copies required, using a paper stock that weighs 178 pounds per thousand sheets and costs $2.00 per pound. From each sheet 32 pages can be printed. Determine the cost of paper stock.

3. In six consecutive months the following quantities of stamped metal parts were heat treated: 462,925; 378,916; 417,829; 382,885; 415,297; and 450,628 pounds. Determine the six-month hourly average per person if the department employs 16 persons on a nine-hour shift, 12 on an eight-hour shift, and 4 on a seven-hour shift. In the six-month period, each person worked 130 days.

4. Find the average hourly production of parts A through E from the monthly (156 hours) production schedule.

Parts Manufactured				
A	B	C	D	E
Average Monthly (156 Hours) Production				
1,248	2,964	21,372	202,332	5,584,332

5. A pattern layout requires 51 inches of cloth. Find (a) how many garments can be cut from the material, (b) the amount of material left over, and (c) the number of yards in 324 inches. (1 yard = 36 inches)

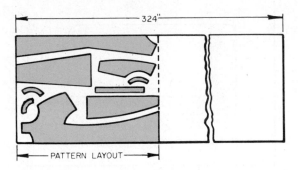

COMBINATIONS OF ADDITION, SUBTRACTION, MULTIPLICATION, OR DIVISION OF WHOLE NUMBERS

1. Establish whether there was a profit or loss on sales A through G. Find the amount of profit (P) or loss (L).

Item	Sales	Cost of Item	Overhead Expense	Profit (P) or Loss (L)
A	$98,000	$75,000	$22,500	
B	15,750	11,429	3,000	
C	9,342	6,210	3,350	
D	56,735	38,750	17,000	
E	126,960	97,873	18,250	
F	7,942	4,975	2,635	
G	11,697	7,539	2,973	

2. Calculate dimensions Ⓐ through Ⓔ for machining the Serrated Plate. Use the basic mathematical processes needed to find each dimension.

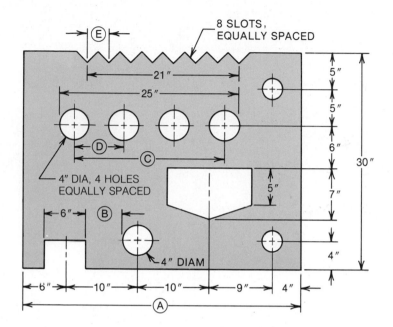

3. Stock control records show an inventory of 9,241 integrated electronic circuits at the close of a four-month period. The following quantities were shipped each month: February, 928; March, 1,641; April, 3,077; and May, 2,104.

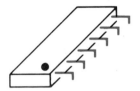

Provide the following information.

a. The total number shipped.

b. The average of the number of circuits shipped each month during the period.

c. The number of integrated circuits remaining in stock at the close of the period.

d. The anticipated shipments for a twelve-month period based on the average per month for the first four-month period.

COMMON FRACTIONS

Unit 6 The Concept of Common Fractions

OBJECTIVES OF THE UNIT

After satisfactorily studying this unit, the student/trainee will be able to
* *Understand the terms and functions of common fractions in relation to shop, laboratory, and everyday problems.*
* *Transfer common fraction values to equivalent measurements on steel tapes, rules, and other measuring tools.*
* *Reduce proper and improper fractions to their lowest terms.*

PRETEST *Use the Review and Self-Test items provided in the Unit Assignment to establish the level of mathematical skills competency and to determine the starting point of instruction.*

Craftspersons are required constantly to take measurements, do layout work, and perform hand and machine operations. These jobs require the use of mathematics to compute missing or needed dimensions.

The use of whole numbers alone is not sufficient to obtain this information. All computations involve either whole numbers or fractions or a combination of both. Numbers, whether they are whole or just parts of a whole, must be added, subtracted, multiplied, and divided.

A. INTERPRETING COMMON FRACTIONS

A *fraction* is a part of a whole quantity. For example, a triangle, square, or circle is divided into two equal parts (Figure 6-1). One part is involved in an operation, as shown by the shaded portion of each illustration. The fractional part of the whole triangle, square, or circle is one-half, as shown at (A), (B), and (C).

(A) 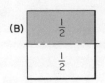   (B)   (C)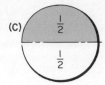

FIGURE 6-1

On shop prints and sketches and in mathematical computations the one-half is usually written $\frac{1}{2}$ and is called a *common fraction*. This fraction shows the number of equal parts of a unit that are taken. If the fraction $\frac{3}{4}$ appears on a drawing, it means the unit one ● is divided into four equal parts ⊕ and that three of the four parts are taken ——→ ◖ .

Common fractions are used daily when taking measurements with line-graduated measuring tools and instruments. Rules, levels, tapes, and other measuring tools are commonly graduated in fourths, eighths, sixteenths, and thirty-seconds of an inch. For greater precision, steel rules are graduated in sixty-fourths of an inch. In the printing trades, measurements are expressed as fine as seventy-seconds of an inch.

Where the measurements are given as fractions, they indicate that the inch has been divided into an equal number of parts. The object must measure a stated number of these parts.

The steel tape, a common measuring tool, may be used to show how the inch looks when it is divided into four, eight, sixteen, thirty-two, and sixty-four equal parts.

1. The inch contains four-fourths. Each equal part is expressed as $\frac{1}{4}$ of the whole.

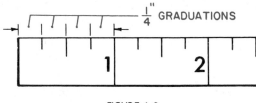

FIGURE 6-2

2. The inch contains eight-eighths. Each equal part is expressed as $\frac{1}{8}$ of the whole.

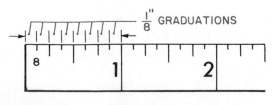

FIGURE 6-3

3. The inch contains sixteen-sixteenths. Each equal part is expressed as $\frac{1}{16}$ of the whole.

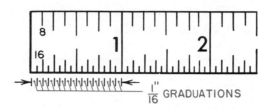

FIGURE 6-4

4. The inch contains thirty-two thirty-seconds. Each equal part is expressed as $\frac{1}{32}$ of the whole.

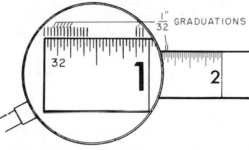

FIGURE 6-5

5. The inch contains sixty-four sixty-fourths. Each equal part is expressed as $\frac{1}{64}$ of the whole.

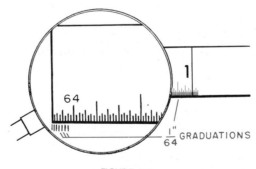

FIGURE 6-6

B. DEFINING PARTS OF FRACTIONS

If on the eighth scale of the steel tape three of the eight equal parts (into which the inch is divided) are needed for a measurement, the fraction would be shown as $\frac{3}{8}$. The terms of this fraction are a *numerator*, which appears over a horizontal line, and a *denominator*, which appears under the line.

Numerator ▶ $\dfrac{3}{8}$ ◀ Denominator

RULE
_____

- The denominator (which is always written below the line) indicates the equal number of parts into which the unit is divided.

_____

RULE
_____

- The numerator (which is always written above the line) indicates the number of equal parts of the denominator that is taken.

_____

C. REDUCING FRACTIONS

A required measurement that is the result of adding, subtracting, multiplying, or dividing fractions is not always expressed as simply as possible. Measurements can be taken or read with greater facility when the fraction is given in its *lowest terms*.

Reducing Common Fractions

The mathematical expression *proper fraction* is used to indicate a fraction whose numerator is smaller than its denominator. Quantities such as

$$\frac{1}{4}, \frac{3}{8}, \frac{29}{64}, \text{ and } \frac{5}{8}$$

are examples of proper fractions.

RULE
_____

- The value of a fraction is not changed when both the numerator and denominator are multiplied or divided by the same number. This number, which can be used to divide both the numerator and denominator of a fraction without a remainder, is called a *common factor*.

_____

RULE FOR REDUCING A COMMON FRACTION TO ITS LOWEST TERMS

- Divide the numerator and denominator by the same number.
  Note. When both the numerator and the denominator cannot be divided further by the same number, the fraction is expressed in its lowest terms.

E XAMPLE: Reduce $\frac{8}{16}$ to its lowest terms.

Step 1   Select a number (common factor) that will divide evenly into both the numerator and denominator.

$$\frac{8}{16} = \frac{4}{8} \qquad\qquad \frac{4}{8} = \frac{2}{4} = \frac{1}{2}$$

Step 2   Continue this division until the numerator and denominator can no longer be evenly divided by the same number.
Note. When it is apparent that both numerator and denominator can be divided by a larger number, for example, 8 in the case of $\frac{8}{16}$, the intermediate steps are omitted.

Reducing Improper Fractions

A fraction whose numerator is greater than its denominator is called an *improper fraction*. Examples of improper fractions are

$$\frac{3}{2}, \frac{25}{16}, \text{ and } \frac{71}{32}.$$

RULE FOR REDUCING AN IMPROPER FRACTION TO ITS LOWEST TERMS

- Divide the numerator (above the line) by the denominator (below the line).
- Reduce the resulting fraction to its lowest terms.

E XAMPLE: Reduce $\frac{20}{16}$ to its lowest terms.

Step 1   Divide 20 by 16. $\qquad\qquad\qquad\qquad\qquad\qquad \frac{20}{16} = 1\frac{4}{16}$

Step 2   Reduce $\frac{4}{16}$ to lowest terms. $\qquad\qquad\qquad\quad \frac{4}{16} = \frac{1}{4}$

Step 3   Answer $\qquad\qquad\qquad\qquad\qquad\qquad\qquad \frac{20}{16} = 1\frac{1}{4}$ **Ans**

ASSIGNMENT UNIT 6   REVIEW AND SELF TEST

PRETEST   *The Review and Self-Test items that follow may be used as a pretest. Pretests are designed to measure a student's beginning level of mathematical skills competency and to determine the starting point of instruction.*

POSTTEST   *The Review and Self-Test items may also be applied as a post test. Post tests are planned to establish the student's level of mathematical skills competency after instruction.*

A. Interpretation of Common Fractions (General Applications)

1. The squares and circles are divided equally into 6, 8, 16, 32, or 64 parts. Visualize (by actual counting of spaces, if necessary) the part of the square or circle represented by fractions (b) to (z).

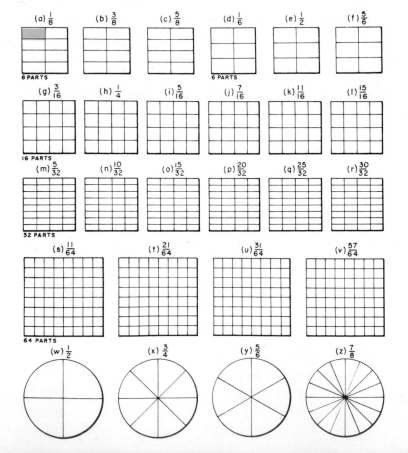

B. Reading Fractional Values (Practical Problems)

1. Draw a table as illustrated. Include the letters in the boxes, the arrows, and all words. Arrange the following twist drills in the table in sizes ranging from the smallest to the largest.

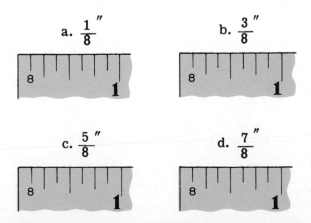

2. Locate the $\frac{1''}{4}$ and $\frac{3''}{4}$ graduations on a steel tape or rule and indicate with an ''x.'' *Note:* If a rule is not available for problems 2 and 3, place a transparent sheet over the illustrations and trace the rules.

a. $\frac{1}{4}''$    b. $\frac{3}{4}''$

3. Locate common fractions (a) through (d) on the eighth scale of a rule. Place an ''x'' at each location.

a. $\frac{1}{8}''$

b. $\frac{3}{8}''$

c. $\frac{5}{8}''$

d. $\frac{7}{8}''$

C. Reduction of Fractions (Practical Problems)

1. Reduce common fractions (a) through (f) to their lowest terms. Then locate each fraction on a rule.

EXAMPLE:

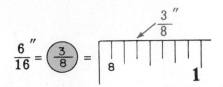

$$\frac{6}{16}'' = \left(\frac{3}{8}\right) =$$

a. $\frac{14}{32}''$

d. $\frac{44}{64}''$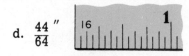

b. $\frac{48}{64}''$

e. $\frac{10}{64}''$

c. $\frac{10}{16}''$

f. $\frac{18}{128}''$

2. Reduce each improper fraction to its lowest terms and locate each measurement on a rule.

   a. $\frac{5''}{2}$ b. $\frac{25''}{4}$ c. $\frac{35''}{8}$ d. $\frac{42''}{32}$ e. $\frac{85''}{64}$

3. Locate each common fraction on the sixteenth scale of a rule for problems (a) through (f).

   a. $\frac{15''}{16}$ b. $\frac{13''}{16}$ c. $\frac{7''}{16}$ d. $\frac{9''}{16}$ e. $\frac{3''}{16}$ f. $\frac{11''}{16}$

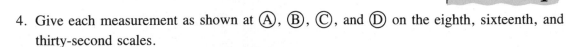

4. Give each measurement as shown at Ⓐ, Ⓑ, Ⓒ, and Ⓓ on the eighth, sixteenth, and thirty-second scales.

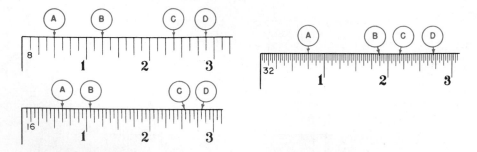

5. Draw lines to the lengths given in (a) through (f).

   a. $3''$ b. $4\frac{1}{2}''$ c. $5\frac{1}{4}''$ d. $3\frac{1}{8}''$ e. $4\frac{3}{4}''$ f. $4\frac{7}{8}''$

6. Give the measurements of the stepped parts shown at Ⓐ and the extensions Ⓑ, Ⓒ, Ⓓ, and Ⓔ.

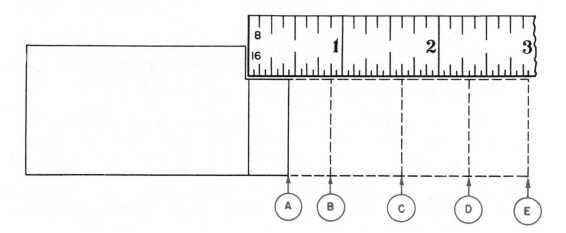

7. Reduce each dimension Ⓐ through Ⓙ to its lowest terms as a common fraction or as an improper fraction.

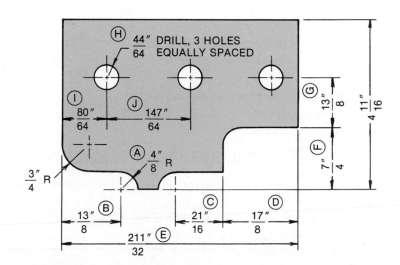

Unit 7 Addition of Fractions

OBJECTIVES OF THE UNIT

After satisfactorily studying this unit, the student/trainee will be able to
- *Reduce common fractions to lowest common denominators.*
- *Add common fractions.*
- *Add combinations of whole numbers, mixed numbers, and common fractions.*

PRETEST   *Use the Review and Self-Test items provided in the Unit Assignment to establish the level of mathematical skills competency and to determine the starting point of instruction.*

The mechanic must often determine overall sizes by adding dimensions that are given on a drawing or written into specifications. The answers to these problems require the addition of whole numbers, common fractions, and combinations of whole numbers and fractions. Such combinations are referred to as *mixed numbers*.

To add combinations of whole numbers and fractions, the denominators of each fraction must be the same number. The smallest number that can be divided by all the denominators is called the *lowest common denominator*.

A. DETERMINING THE LOWEST COMMON DENOMINATOR (LCD)

The simplest method of determining the lowest common denominator (LCD) is used when it is evident that all of the denominators divide evenly into a given number. This method is covered in this unit.

On the drawing of the special pin (Figure 7-1) all of the given sizes must be added to compute the overall dimension (A). Before these measurements (2, $1\frac{31}{32}$, $\frac{3}{8}$, and $\frac{3}{64}$) can be added, the lowest common denominator must be determined.

The lowest number into which each of the denominators (32, 8, and 64) will divide evenly is 64. The 64 is called the *lowest common denominator*.

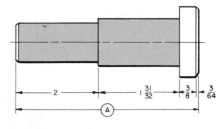

FIGURE 7-1

RULE FOR REDUCING FRACTIONS TO THE LOWEST COMMON DENOMINATOR

- Divide the number selected as the lowest common denominator by the denominator of each given fraction.
- Multiply both the numerator and denominator by this quotient.

EXAMPLE: The lowest common denominator (LCD) for the pin (Figure 7-1) dimension is 64.

Denominator

$$\downarrow \quad 2 \leftarrow \text{Quotient}$$
$$32\overline{)64}$$

Step 1 Divide the LCD by the denominator of first fraction.

Step 2 Multiply the numerator (31) and denominator (32) by the quotient (2).

$$\frac{31 \times 2}{32 \times 2} = \frac{62}{64}$$

Step 3 Continue the same process with the other fractions.

$$\frac{8}{8\overline{)64}} \qquad \frac{3 \times 8}{8 \times 8} = \frac{24}{64}$$

$$\frac{1}{64\overline{)64}} \qquad \frac{3 \times 1}{64 \times 1} = \frac{3}{64}$$

B. ADDING COMMON FRACTIONS

RULE FOR ADDING FRACTIONS

- Change to fractions having a least common denominator.
- Add the numerators.
- Write the sum over the common denominator.
- Reduce the result to its lowest terms.

EXAMPLE: Add $\frac{31}{32}$, $\frac{3}{8}$, and $\frac{3}{64}$.

$$\frac{31 \times 2}{32 \times 2} = \frac{62}{64}$$
$$\frac{3 \times 8}{8 \times 8} = \frac{24}{64}$$
$$\frac{3 \times 1}{64 \times 1} = \frac{3}{64}$$

Step 1 Write fractions in vertical column.

Step 2 Change fractions to same denominator (64).

Step 3 Add numerators.

$$62 + 24 + 3 = 89$$

Step 4 Place result (89) over lowest common denominator (64).

$$\frac{89}{64}$$

Step 5 Reduce to lowest terms.

$$\frac{89}{64} = 1\frac{25}{64} \quad \text{Ans}$$

C. ADDING WHOLE NUMBERS, COMMON FRACTIONS, AND MIXED NUMBERS

A mixed number consists of two parts: (1) a whole number and (2) a fraction. Numbers such as $1\frac{1}{2}$, $256\frac{3}{8}$, and $1,927\frac{3}{5}$ are called *mixed numbers*.

RULE FOR ADDING WHOLE NUMBERS, MIXED NUMBERS, AND COMMON FRACTIONS

- Add the whole numbers.
- Add the fractions.
- Add the two sums.
- Reduce the result to lowest terms.

EXAMPLE: Determine overall dimension Ⓐ of the shaft in Figure 7-2.

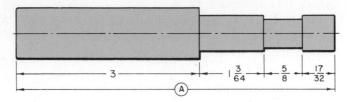

FIGURE 7-2

Step 1	Write all dimensions in vertical columns.	$3 \quad = 3$
Step 2	Change fractions to same denominator (64).	$1\frac{3}{64} \times \frac{1}{1} = 1\frac{3}{64}$
		$\frac{5}{8} \times \frac{8}{8} = \frac{40}{64}$
		$\frac{17}{32} \times \frac{2}{2} = \frac{34}{64}$
Step 3	Add numerators.	$3 + 40 + 34 = 77$
Step 4	Place result (77) over lowest common denominator (64).	$\frac{77}{64}$
Step 5	Reduce $\frac{77}{64}$ to lowest terms.	$1\frac{13}{64}$
Step 6	Add column of whole numbers.	$(3 + 1 = 4)$
Step 7	Add sum of whole numbers (4) to sum of common fractions $\left(1\frac{13}{64}\right)$.	$4 + 1\frac{13}{64} = 5\frac{13}{64}''$ **overall dimension of shaft**

ASSIGNMENT UNIT 7   REVIEW AND SELF TEST

PRETEST *The Review and Self-Test items that follow may be used as a pretest. Pretests are designed to measure a student's beginning level of mathematical skills competency and to determine the starting point of instruction.*

POSTTEST *The Review and Self-Test items may also be applied as a post test. Post tests are planned to establish the student's level of mathematical skills competency after instruction.*

A. Addition of Common Fractions (General Applications)

1. $\frac{1}{6} + \frac{5}{6}$

2. $\frac{1}{8} + \frac{5}{8}$

3. $\frac{1}{4} + \frac{4}{8}$

4. $\frac{1}{3} + \frac{1}{6}$

5. $\frac{1}{2} + \frac{3}{8}$

6. $\frac{1}{6} + \frac{1}{6} + \frac{5}{6}$

7. $\frac{5}{8} + \frac{3}{4} + \frac{3}{8}$

8. $\frac{1}{32} + \frac{7}{8} + \frac{3}{16} + \frac{5}{32}$

9. $\frac{61}{64} + \frac{13}{16} + \frac{5}{8} + \frac{23}{64}$

B. Addition of Common Fractions and Mixed Numbers (General Applications)

1. $121 + 7\frac{5}{12}$

2. $10\frac{9}{16} + 4\frac{1}{8}$

3. $23\frac{5}{8} + 10\frac{5}{8}$

4. $1\frac{17}{64} + 1\frac{13}{64} + \frac{9}{32}$

5. $4\frac{3}{16} + 10\frac{21}{64} + 1\frac{5}{16} + \frac{3}{32}$

6. $3\frac{5}{32} + 2\frac{13}{64} + 1\frac{13}{32} + 3\frac{1}{16} + \frac{1}{4}$

C. Addition of Whole Numbers, Mixed Numbers, and Common Fractions (Practical Problems)

1. Determine dimensions of (A), (B), (C), (D), and (E) (in inches).

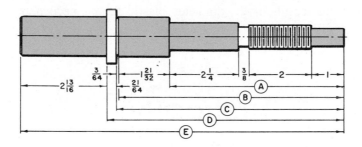

2. Calculate dimensions (A) through (E) for the stand.

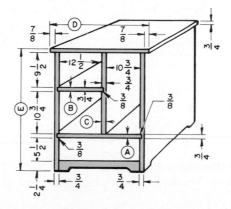

3. Determine the overall lengths of springs A through D.

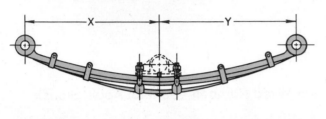

Spring	Distance		Overall Length
	X	Y	
A	$17\frac{1}{2}''$	$15\frac{3}{4}''$	
B	$16\frac{3}{4}''$	$15\frac{7}{8}''$	
C	$17\frac{3}{32}''$	$16\frac{1}{8}''$	
D	$18\frac{21}{32}''$	$17\frac{3}{4}''$	

4. Compute the total horsepower for (a) motors 1, 2, and 3; (b) motors 4, 5, and 6; and (c) all 6 motors. The rated horsepower is indicated.

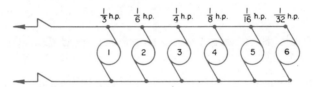

5. Find the total length of $2\frac{1}{2}''$ diameter plastic tubing needed on the job. The required lengths are $35\frac{1}{4}''$, $12\frac{1}{2}''$, $26\frac{3}{8}''$, $8\frac{5}{16}''$, and $19\frac{21}{32}''$. Reduce the answer to lowest terms.

Unit 8 Subtraction of Fractions

OBJECTIVES OF THE UNIT
After satisfactorily studying this unit, the student/trainee will be able to
* *Solve general and practical problems that require the subtraction of proper fractions.*
* *Carry on the addition and subtraction of fractions in the same problem.*

PRETEST   *Use the Review and Self-Test items provided in the Unit Assignment to establish the level of mathematical skills competency and to determine the starting point of instruction.*

The value of a missing dimension must often be determined by subtracting whole numbers, fractions, and mixed numbers. Fractions cannot be subtracted unless they first have the same common denominator.

A. SUBTRACTING PROPER FRACTIONS

RULE FOR SUBTRACTING FRACTIONS

- Express all fractions using the lowest (least) common denominator.
- Subtract the numerators.
- Write the difference over the lowest (least) common denominator.
- Express the resulting fraction in lowest terms.

EXAMPLE: Subtract $\frac{7}{32}$ from $\frac{11}{16}$.

Step 1   Determine the lowest (least) common denominator.

Step 2   Write the fractions in terms of lowest (least) common denomi- $\frac{7}{32} = \frac{7}{32}$ $\frac{11}{16} = \frac{22}{32}$
nator.

$$22 - 7 = 15 \downarrow$$

Step 3   Subtract numerators.

Step 4   Place the numerator result over the lowest (least) common $\frac{15}{32}$ Ans $\downarrow$
denominator.

B. SUBTRACTING A FRACTION FROM A WHOLE NUMBER

RULE FOR SUBTRACTING A FRACTION FROM A WHOLE NUMBER

- Take one unit from the whole number. Change it to a fraction having the same denominator as the fraction that is to be subtracted.
- Subtract the numerators of the original fraction from the one unit that was changed to its fractional value.
- Express the resulting fraction in lowest terms.
- Place the whole number next to the fraction.

EXAMPLE: Subtract $\frac{21}{32}$ from 9.

Step 1   Take one unit from the whole number.

$$9 - 1 = 8$$
$$1 = \frac{32}{32}$$

Step 2   Change the one unit to a fractional equivalent having the same denominator as the fraction to be
subtracted.

$$8\frac{32}{32}$$
$$-\frac{21}{32}$$

Step 3   Arrange the fractions in one column; the whole
number in another.

Step 4   Subtract the numerators.

$$\frac{32}{32} - \frac{21}{32} = \frac{11}{32}$$

Step 5   Place the whole number (8) next to the fraction $\frac{11}{32}$
to get the answer.

$$8\frac{11}{32} \quad \text{Ans}$$

C. SUBTRACTING A MIXED NUMBER FROM A WHOLE NUMBER

RULE FOR SUBTRACTING A MIXED NUMBER FROM A WHOLE NUMBER

- Borrow one unit from the whole number and express it as a fraction that has the same denominator as the mixed number.
- Subtract the fraction part of the mixed number from the fraction part of the whole number.
- Subtract the whole numbers and reduce the resulting mixed number to its lowest terms.

EXAMPLE: Subtract $1\frac{35}{64}$ from 6.

$$\begin{aligned} \text{Step 1} \quad & 6 = 5\frac{64}{64} \\ \text{Step 2} \quad & 1\frac{35}{64} = 1\frac{35}{64} \\ \text{Step 3} \quad & \qquad\quad 4\frac{29}{64} \quad \text{Ans} \end{aligned}$$

D. SUBTRACTING MIXED NUMBERS FROM MIXED NUMBERS

RULE FOR SUBTRACTING MIXED NUMBERS

- Express the fractional part of each mixed number using the least common denominator.
- Borrow one unit, when necessary, to make up a fraction larger than the one to be subtracted.
- Subtract the fractions first and the whole numbers next. Express the result in its lowest terms.

EXAMPLE: Subtract $2\frac{7}{16}$ from $5\frac{11}{32}$.

$$\begin{aligned} \text{Step 1} \quad & 5\frac{11}{32} = 4 + \frac{32}{32} + \frac{11}{32} = 4\frac{43}{32} \\ \text{Step 2} \quad & 2\frac{7}{16} = \qquad\qquad\qquad = 2\frac{14}{32} \\ \text{Step 3} \quad & \qquad\qquad\qquad\qquad\qquad 2\frac{29}{32} \quad \text{Ans} \end{aligned}$$

E. COMBINING ADDITION AND SUBTRACTION OF FRACTIONS

RULE FOR ADDING AND SUBTRACTING FRACTIONS IN THE SAME PROBLEM

- Change all fractions to the least (lowest) common denominator.
- Add or subtract the numerators as required.
- Express the result in lowest terms.

EXAMPLE: Add $1\frac{9}{16} + 3\frac{5}{8} + 2\frac{1}{4}$ and from the sum subtract $2\frac{13}{16}$.

$$1\frac{9}{16} + 3\frac{5}{8} + 2\frac{1}{4} - 2\frac{13}{16} =$$

Step 1 $1\frac{9}{16} + 3\frac{10}{16} + 2\frac{4}{16} - 2\frac{13}{16} =$

Step 2
$$1\frac{9}{16}$$
$$+3\frac{10}{16}$$
$$+2\frac{4}{16}$$

Step 3
$$6\frac{23}{16}$$
$$-2\frac{13}{16}$$

Step 4
$$4\frac{10}{16} = 4\frac{5}{8} \quad \text{Ans}$$

ASSIGNMENT UNIT 8 REVIEW AND SELF TEST

PRETEST *The Review and Self-Test items that follow may be used as a pretest. Pretests are designed to measure a student's beginning level of mathematical skills competency and to determine the starting point of instruction.*

POSTTEST *The Review and Self-Test items may also be applied as a post test. Post tests are planned to establish the student's level of mathematical skills competency after instruction.*

A. Subtraction of Proper Fractions (General Applications)

Subtract.

1. $\frac{5}{8}$
   $-\frac{4}{8}$

2. $\frac{5}{6}$
   $-\frac{1}{6}$

3. $\frac{13}{16}$
   $-\frac{5}{16}$

4. $\frac{9}{32}$
   $-\frac{7}{32}$

5. $\frac{1}{2}$
   $-\frac{1}{4}$

6. $\frac{1}{8}$ from $\frac{5}{8}$

7. $\frac{5}{32}$ from $\frac{9}{16}$

8. $\frac{9}{64}$ from $\frac{23}{32}$

9. $\frac{3}{8} + \frac{1}{8}$ from $\frac{9}{16}$

10. $\frac{37}{64} + \frac{21}{64}$ from $\frac{63}{64}$

B. Subtraction of Fractions from a Whole Number (General Applications)

Subtract.

1. 4
   $-\frac{3}{4}$

2. 7
   $-\frac{15}{16}$

3. 32
   $-\frac{13}{32}$

4. 175
   $-\frac{4}{5}$

5. 72
   $-\frac{61}{64}$

C. Subtraction of a Mixed Number from a Whole Number (General Applications)

Subtract.

1. 2
   $-1\frac{1}{3}$

2. 3
   $-1\frac{3}{8}$

3. 27
   $-1\frac{5}{16}$

4. 142
   $-6\frac{21}{32}$

5. 372
   $-21\frac{5}{64}$

6. $1\frac{21}{32}$ from 3

7. $3\frac{57}{64}$ from 4

8. $3\frac{5}{32} + 1\frac{9}{32}$ from 5

9. $2\frac{53}{64} + 1\frac{1}{64} + 3\frac{1}{4}$ from 8

10. $3\frac{7}{16} + 2\frac{25}{64} + 16\frac{17}{32}$ from 42

D. Subtraction of Mixed Numbers from Mixed Numbers (General Applications)

Subtract.

1. $1\frac{3}{5}$
    $-1\frac{1}{5}$

2. $7\frac{5}{6}$
    $-2\frac{1}{6}$

3. $18\frac{7}{8}$
    $-9\frac{3}{8}$

4. $35\frac{5}{8}$
    $-8\frac{1}{2}$

5. $172\frac{21}{64}$
    $-22\frac{5}{32}$

6. $1\frac{19}{32}$ from $4\frac{29}{32}$

7. $2\frac{31}{32}$ from $5\frac{15}{32}$

8. $4\frac{7}{64}$ from $8\frac{1}{32}$

9. $2\frac{9}{16} + 1\frac{1}{8}$ from $5\frac{51}{64}$

10. $7\frac{1}{64} + 2\frac{31}{32} + 1\frac{1}{4}$ from $12\frac{3}{32}$

E. Addition or Subtraction Processes (Practical Problems)

1. Determine the inside diameter of each size bushing. All dimensions are in inches.

Bushing	Outside Diameter	Single Wall Thickness (T)
A	1	$\frac{1}{16}$
B	$2\frac{1}{8}$	$\frac{3}{16}$
C	$2\frac{15}{32}$	$\frac{9}{64}$
D	$3\frac{1}{64}$	$\frac{15}{32}$
E	$1\frac{9}{64}$	$\frac{9}{32}$

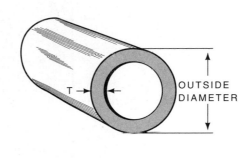

2. Determine dimensions Ⓐ, Ⓑ, Ⓒ, Ⓓ, and Ⓔ. All dimensions are in inches.

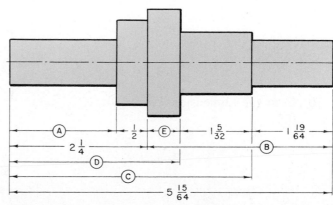

3. How much material will be left from an 18-inch length of strip rule after four pieces are cut off in the following lengths: $2\frac{1}{2}$, $1\frac{1}{4}$, $3\frac{1}{2}$, and $3\frac{3}{4}$ inches.

4. A motor base is to be set up $4\frac{1}{2}$ inches from the floor. Two wooden blocks must be used. If one block is $1\frac{5}{8}$ inches thick, how thick must the second block be?

5. A concrete sidewalk $4\frac{1}{4}$ inches thick consists of a base course and a finish course. The base course thickness is $3\frac{5}{8}$ inches. What is the thickness of the finish course?

6. Three pieces of 3-inch lead pipe are cut from a piece $35\frac{1}{2}$ inches long. The lengths are $7\frac{1}{4}$ inches, $11\frac{3}{8}$ inches, and $6\frac{1}{2}$ inches. If $\frac{3}{8}$ inch of stock is wasted in cutting, how much pipe is left?

7. Four pieces are cut from a 12-foot 2 × 4. The pieces measure 3 feet $9\frac{1}{2}$ inches, 2 feet $6\frac{1}{4}$ inches, 1 foot $4\frac{1}{2}$ inches, and 2 feet $3\frac{3}{8}$ inches. If a total of $\frac{1}{2}$ inch is allowed for the four cuts, how much of the 2 × 4 is left?

8. A piece of radiator hose is $32\frac{1}{2}$ inches long. Short pieces of the following lengths are cut from it: $6\frac{1}{2}$ inches, $5\frac{1}{4}$ inches, $8\frac{13}{16}$ inches, and $10\frac{9}{16}$ inches. How much hose is left?

9. Determine dimensions Ⓐ, Ⓑ, Ⓒ, and Ⓓ.

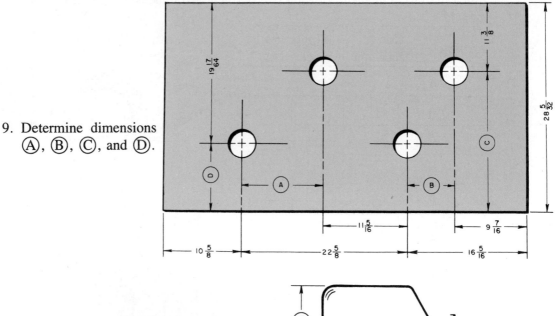

10. Compute dimensions Ⓐ, Ⓑ, Ⓒ, and Ⓓ.

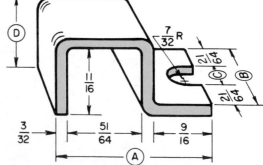

11. The front and rear axles of a light truck are to be aligned. Center distance Ⓐ measures 16'-3$\frac{5}{8}$"; center distance Ⓑ measures 16'-4$\frac{7}{16}$". Determine how much the axles need to be changed to be aligned.

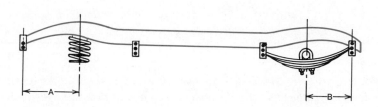

12. Find the voltage Ⓒ in the series circuit according to the voltages indicated for Ⓐ, Ⓑ, and Ⓓ. The source voltage (E_S) is 25$\frac{3}{4}$ volts.

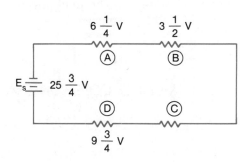

13. Compute the length (L) of the conduit needed according to the dimensions given on the drawing. Express the length in feet and inches.

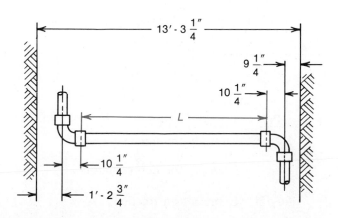

Unit 9 Multiplication of Fractions

OBJECTIVES OF THE UNIT

After satisfactorily studying this unit, the student/trainee will be able to
- *Solve general and practical problems requiring the multiplication of two or more fractions.*
- *Multiply fractions, whole numbers, and mixed numbers.*
- *Simplify the multiplication process by the cancellation method.*

PRETEST *Use the Review and Self-Test items provided in the Unit Assignment to establish the level of mathematical skills competency and to determine the starting point of instruction.*

The multiplication of fractions, like the multiplication of whole numbers, is a simplified method of addition. The multiplication of common fractions, whole numbers, and mixed numbers, typical of those used daily, are covered in this unit.

A. MULTIPLYING COMMON FRACTIONS

RULE FOR MULTIPLYING TWO OR MORE FRACTIONS

- Multiply the numerators.

- Multiply the denominators.

- Write the product of the numerators over the product of the denominators.

- Express the resulting fraction in lowest terms.

EXAMPLE: Multiply $\frac{7}{8}$ by $\frac{3}{4}$.

Step 1	Multiply the numerators.	$(7 \times 3) = 21$
Step 2	Multiply the denominators.	$(8 \times 4) = 32$
Step 3	Express as a fraction.	$\frac{21}{32}$ Ans

B. MULTIPLYING A COMMON FRACTION BY A MIXED NUMBER

RULE FOR MULTIPLYING A COMMON FRACTION BY A MIXED NUMBER

- Express the mixed number as an improper fraction where the numerator is larger than the denominator.
- Multiply the numerator of the improper fraction by the numerator of the common fraction.
- Multiply the denominators.

EXAMPLE: Multiply $4\frac{1}{2}$ by $\frac{1}{8}$.

Step 1 Express the mixed number $\left(4\frac{1}{2}\right)$ as an improper fraction $\left(\frac{9}{2}\right)$.

Step 2 Multiply the numerators.

Step 3 Multiply the denominators.

$$4\frac{1}{2} = \frac{4 \times 2 + 1}{2} = \frac{9}{2}$$

$$(9 \times 1) = 9$$

$$(2 \times 8) = 16$$

$$4\frac{1}{2} \times \frac{1}{8} = \frac{9}{2} \times \frac{1}{8} = \frac{9}{16} \text{ Ans}$$

C. MULTIPLYING FRACTIONS AND WHOLE AND MIXED NUMBERS

RULE FOR MULTIPLYING FRACTIONS, WHOLE NUMBERS, AND MIXED NUMBERS IN ANY COMBINATION

- Express all mixed numbers as improper fractions.
- Place all whole numbers over a denominator of 1.
- Multiply all numerators.
- Multiply all denominators.
- Express resulting product in lowest terms.

EXAMPLE: Multiply $8\frac{3}{8} \times 1\frac{1}{4} \times 2$.

Step 1 Express $\left(8\frac{3}{8}\right)$ as an improper fraction $\left(\frac{67}{8}\right)$.

Express $\left(1\frac{1}{4}\right)$ as an improper fraction $\left(\frac{5}{4}\right)$.

Step 2 Place 2 over denominator of 1.

Step 3 Multiply all numerators.

Step 4 Multiply all denominators.

Step 5 Express $\left(\frac{670}{32}\right)$ in lowest terms.

$$\frac{8 \times 8 + 3}{8} = \frac{67}{8}$$

$$\frac{1 \times 4 + 1}{4} = \frac{5}{4}$$

$$\frac{2}{1}$$

$$(67 \times 5 \times 2) = 670$$

$$(8 \times 4 \times 1) = 32$$

$$\frac{670}{32}$$

$$\frac{670}{32} = 20\frac{30}{32} = 20\frac{15}{16} \text{ Ans}$$

D. CANCELING TO SIMPLIFY THE MULTIPLICATION PROCESS

The multiplication of fractions can be simplified by removing the *common factor*. The common factor is any number by which the numerator and denominator may be evenly divided. The process is called *cancellation*.

RULE FOR EVENLY DIVIDING A NUMERATOR AND DENOMINATOR

- Select a number *(common factor)* by which the numerator and denominator may be evenly divided.
- Divide by the common factor.
- Reduce the result to the lowest terms.

EXAMPLE: Multiply $72 \times 3\frac{5}{8}$.

Step 1   Express the mixed number $\left(3\frac{5}{8}\right)$ as an improper fraction $\left(\frac{29}{8}\right)$.

Step 2   Select a number that is common to any numerator and any denominator (8 in this case).

Step 3   Divide the numerator and denominator by this factor (8).

$$\frac{\overset{9}{\cancel{72}}}{1} \times \frac{29}{\cancel{8}} = 261 \quad \textbf{Ans}$$

E. SHORTCUTS IN MULTIPLYING BY ONE-HALF $\left(\frac{1}{2}\right)$

EXAMPLE: *Case 1.* Find $\frac{1}{2}$ of $\frac{7}{8}$.

Step 1   Multiply the denominator (8) by 2.

Step 2   Use the numerator (7) as it is. The answer is $\frac{7}{16}$.

EXAMPLE: *Case 2.* Find $\frac{1}{2}$ of $2\frac{3}{4}$.

Step 1   Take one-half of 2. $\frac{1}{2} \times 2 = 1$

Step 2   Multiply the denominator (4) of the fraction $\left(\frac{3}{4}\right)$ by 2. $4 \times 2 = 8$

Step 3   Use the same numerator (3). $\frac{3}{8}$

Step 4   Combine the whole number 1 and the fraction $\frac{3}{8}$. The answer is a mixed number $\left(1\frac{3}{8}\right)$. $1 + \frac{3}{8} = 1\frac{3}{8}$ **Ans**

EXAMPLE: *Case 3.* Find $\frac{1}{2}$ of $5\frac{13}{16}$.

Step 1 Take one-half of 4 (the largest number in the $\frac{1}{2} \times 4 = 2$
mixed number that will divide exactly).

Step 2 Express the remainder $\left(1\frac{13}{16}\right)$ as an improper fraction. $1\frac{13}{16} = \frac{29}{16}$

Step 3 Place the numerator (29) over twice the denominator. $\frac{29}{2 \times 16} = \frac{29}{32}$

Step 4 Combine the whole number (2) with the fraction $\left(\frac{29}{32}\right)$. $2 + \frac{29}{32} = 2\frac{29}{32}$ Ans

Special Note. When more than one operation is called for within a problem, multiplication or division operations are completed first. Other operations (addition and subtraction) are then performed. These are done in order from left to right.

ASSIGNMENT UNIT 9   REVIEW AND SELF TEST

PRETEST *The Review and Self-Test items that follow may be used as a pretest. Pretests are designed to measure a student's beginning level of mathematical skills competency and to determine the starting point of instruction.*

POSTTEST *The Review and Self-Test items may also be applied as a post test. Post tests are planned to establish the student's level of mathematical skills competency after instruction.*

A. Multiplication of Proper Fractions (General Applications)

Multiply.

1. $\frac{1}{4}$ by $\frac{1}{2}$

2. $\frac{5}{9}$ by $\frac{1}{8}$

3. $\frac{1}{6}$ by $\frac{7}{12}$

4. $\quad \frac{7}{8}$

$\quad$ by $\frac{1}{4}$

5. $\quad \frac{5}{6}$

$\quad$ by $\frac{5}{12}$

6. $\left(\frac{21}{32} + \frac{3}{32}\right)$

$\quad$ by $\frac{1}{8}$

7. $\left(\frac{5}{8} + \frac{3}{16}\right)$

$\quad$ by $\frac{1}{2}$

8. $\left(\frac{19}{32} + \frac{1}{4}\right)$ by $\frac{3}{8}$

9. $\left(\frac{1}{64} + \frac{17}{32} - \frac{5}{64}\right)$ by $\frac{13}{16}$

10. $\left(\frac{1}{8} + \frac{19}{64} - \frac{3}{16}\right)$ by $\frac{13}{32}$

B. Multiplication of Common Fractions and Mixed Numbers (General Applications)

Multiply.

1. $\frac{1}{6}$ by $1\frac{1}{6}$

2. $1\frac{1}{4}$ by $\frac{1}{2}$

3. $2\frac{3}{8}$ by $\frac{1}{4}$

4. $\frac{5}{8}$ by $6\frac{1}{4}$

5. $\frac{5}{6}$ by $3\frac{7}{12}$

6. $\begin{array}{r} 2\frac{9}{16} \\ \times \frac{3}{8} \\ \hline \end{array}$

7. $\begin{array}{r} \frac{15}{32} \\ \times 2\frac{1}{8} \\ \hline \end{array}$

8. $3\frac{1}{2} + \left(1\frac{3}{4} \times \frac{7}{8}\right)$

9. $1\frac{9}{16} + \left(\frac{5}{32} \times \frac{5}{8}\right)$

10. $\left(17\frac{1}{4} + 3\frac{1}{32} - 2\frac{1}{8}\right) \times \frac{27}{64}$

C. Multiplication of Mixed Numbers (General Applications)

Multiply.

1. $1\frac{1}{3}$ by $2\frac{1}{6}$

2. $1\frac{9}{16}$ by $10\frac{3}{4}$

3. $6\frac{5}{8}$ by $2\frac{7}{32}$

4. $3\frac{5}{6} \times 1\frac{3}{4} \times 3\frac{1}{8} \times 6\frac{1}{2}$

5. $\left(2\frac{53}{64} - 1\frac{9}{32}\right) \times 2\frac{1}{2} \times 3\frac{3}{4}$

D. Shortcuts in Multiplying by One-Half $\left(\frac{1}{2}\right)$ (General Applications)

Find by the shortcut method.

1. $\frac{1}{2}$ of $\frac{1}{4}$

2. $\frac{1}{2}$ of $\frac{7}{16}$

3. $\frac{1}{2}$ of $2\frac{3}{32}$

4. $\frac{1}{2}$ of $3\frac{5}{6}$

E. Multiplication of Whole Numbers, Fractions, and Mixed Numbers (Practical Problems)

1. What lengths of bar stock will be needed to machine the quantity of parts A, B, C, D, and E?

Parts	Quantity	Length	Each Saw Cut
A	10	$\frac{1}{2}''$	$\frac{1}{16}''$
B	12	$1\frac{5}{32}''$	$\frac{3}{32}''$
C	64	$2\frac{21}{64}''$	$\frac{1}{8}''$
D	100	$1\frac{3}{32}''$	$\frac{7}{64}''$
E	75	$1\frac{9}{16}''$	$\frac{1}{8}''$

2. Determine the cost of materials A, B, C, D, and E. Round off answers to two places.

Materials	Length (ft)	Weight (lbs/ft)	Unit Cost (lb)
A	2	$\frac{1}{2}$	$2.16
B	$10\frac{1}{4}$	$\frac{1}{4}$	$2.76
C	$7\frac{5}{12}$	$1\frac{1}{2}$	$3.96\frac{1}{2}$
D	$9\frac{1}{2}$	$3\frac{3}{4}$	$1.98\frac{1}{8}$
E	$11\frac{3}{4}$	$2\frac{1}{16}$	$1.45\frac{11}{16}$

3. Compute dimensions Ⓐ, Ⓑ, Ⓒ, and Ⓓ.

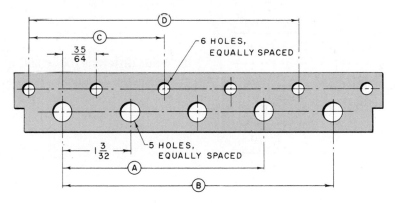

4. Find the special incentive earnings for weeks A, B, C, D, and E and the total earnings from data given in the table. Overtime incentive earnings amounting to time and a half are paid for all hours over 40 per week.

Week	Hours per Week	Rate per Hour
A	40	$.87\frac{1}{2}$
B	44	.87\frac{1}{2}$
C	48	.87\frac{1}{2}$
D	$43\frac{1}{4}$	.99\frac{1}{2}$
E	$43\frac{3}{4}$	.99\frac{1}{2}$

5. a. Find the daily production on parts A and B produced by methods 1, 2, and 3 for a $7\frac{1}{2}$ hour day.

   b. Determine the unit costs for all conditions.

Method	Part A		Part B	
	Hourly Production	Unit Cost	Hourly Production	Unit Cost
1	10	2	21	$2\frac{4}{5}$
2	14	$2\frac{1}{2}$	27	$3\frac{1}{8}$
3	15	$2\frac{3}{4}$	$29\frac{1}{2}$	$4\frac{1}{8}$

Foods	Quantity	Cost per Unit
A	$127\frac{1}{2}$ lbs	$2.16/lb
B	$29\frac{3}{4}$ doz	4.63/doz
C	$63\frac{1}{4}$ qts	.87/qt
D	$17\frac{1}{2}$ yds	7.47/yd

6. Four different foods and quantities are given in the table. Compute the cost of each food.

7. Compute the total length (feet/inches) of insulated cable required to fill the bill of materials for rooms A, B, C, and D.

Room	# Pieces	Length
A	7	10′-6″
B	26	9″
C	8	6′-3″
D	7	9′-9″

Unit 10 Division of Fractions

OBJECTIVES OF THE UNIT
After satisfactorily studying this unit, the student/trainee will be able to
* *Solve general, shop, and laboratory problems involving the division of whole numbers and mixed numbers.*
* *Determine the dimensional and other measurement quantities that involve multiplication and division processes.*

The division of fractions refers to the process of determining how many times one number is contained in another. While the division of fractions is not used as often as the other mathematical processes, the principles are applied constantly.

PRETEST *Use the Review and Self-Test items provided in the Unit Assignment to establish the level of mathematical skills competency and to determine the starting point of instruction.*

A. DIVIDING FRACTIONS

RULE FOR DIVIDING FRACTIONS

* Turn the dividing fraction around so the denominator becomes the numerator and the numerator becomes the denominator. This step is often expressed as "invert the divisor."
* Change the division sign to a multiplication sign and multiply.

EXAMPLE: Divide $\frac{7}{8}$ by $\frac{3}{4}$.

Step 1 Invert the divisor $\left(\frac{3}{4}\right)$ to $\left(\frac{4}{3}\right)$.

Step 2 Cancel the factor (4) common to both numerator and denominator.

$$\frac{7}{\overset{}{\underset{2}{8}}} \times \frac{\overset{1}{4}}{3} =$$

Step 3 Multiply remaining fractions.

$$\frac{7}{2} \times \frac{1}{3} = \frac{7}{6}$$

Step 4 Express in lowest terms.

$$\frac{7}{6} = 1\frac{1}{6} \quad \text{Ans}$$

B. DIVIDING FRACTIONS AND WHOLE NUMBERS

RULE FOR DIVIDING A FRACTION AND A WHOLE NUMBER

- Express the whole number as a fraction whose denominator is 1.
- Invert the divisor.
- Proceed as in the multiplication of fractions.

EXAMPLE: *Case 1*. Divide 20 by $\frac{7}{8}$.

Step 1 Express the whole number (20) as a fraction having an equivalent value.

$$20 = \frac{20}{1}$$

Step 2 Invert the divisor $\left(\frac{7}{8}\right)$ to $\left(\frac{8}{7}\right)$.

Step 3 Multiply and simplify.

$$\frac{20}{1} \times \frac{8}{7} = \frac{20 \times 8}{1 \times 7} = \frac{160}{7} = 22\frac{6}{7} \quad \text{Ans}$$

EXAMPLE: *Case 2*. Divide $\frac{15}{32}$ by 6.

Step 1 Express the divisor (6) as a fraction having an equivalent value $\left(\frac{6}{1}\right)$.

Step 2 Invert the divisor $\left(\frac{6}{1}\right)$ to become $\left(\frac{1}{6}\right)$.

Step 3 Cancel factor (3).

$$\frac{\overset{5}{\cancel{15}}}{32} \times \frac{1}{\underset{2}{\cancel{6}}} =$$

Step 4 Multiply.

$$\frac{5}{32} \times \frac{1}{2} = \frac{5}{64} \quad \text{Ans}$$

C. DIVIDING MIXED NUMBERS

RULE FOR DIVIDING MIXED NUMBERS

- Express the mixed numbers as improper fractions.
- Invert the divisor.
- Multiply the fractions.

EXAMPLE: Divide $1\frac{9}{16}$ by $3\frac{1}{8}$.

Step 1 Express the $1\frac{9}{16}$ as $\frac{25}{16}$, the divisor $\left(3\frac{1}{8}\right)$ as $\frac{25}{8}$.

$$1\frac{9}{16} \div 3\frac{1}{8} =$$

$$\frac{25}{16} \div \frac{25}{8} =$$

Step 2 Invert the divisor $\left(\frac{25}{8}\right)$ to $\frac{8}{25}$.

$$\frac{25}{16} \times \frac{8}{25} =$$

Step 3 Cancel like factors (25) and (8).

$$\frac{\overset{1}{\cancel{25}}}{\underset{2}{\cancel{16}}} \times \frac{\overset{1}{\cancel{8}}}{\underset{1}{\cancel{25}}} =$$

Step 4 Multiply.

$$\frac{1}{2} \times \frac{1}{1} = \frac{1}{2} \quad \text{Ans}$$

D. COMBINING THE MULTIPLICATION AND DIVISION OF FRACTIONS AND WHOLE AND MIXED NUMBERS

RULE FOR SOLVING PROBLEMS REQUIRING THE MULTIPLICATION AND DIVISION OF FRACTIONS

- Express all mixed numbers as improper fractions.
- Invert the divisor or divisors and change the division sign or signs to multiplication sign(s).
- Cancel like factors from numerator and denominator.
- Multiply remaining fractions.
- Express product in lowest terms.

EXAMPLE: $9\frac{1}{4} \div \frac{7}{8} \div \frac{1}{32} \times 1\frac{17}{32} =$

Step 1 Express all mixed numbers as improper fractions.

$$9\frac{1}{4} = \frac{37}{4} \quad 1\frac{17}{32} = \frac{49}{32}$$

Step 2 Invert the divisors $\left(\frac{7}{8}\right)$ and $\left(\frac{1}{32}\right)$ to $\left(\frac{8}{7}\right)$ and $\left(\frac{32}{1}\right)$ and change division signs to multiplication signs.

$$9\frac{1}{4} \div \frac{7}{8} \div \frac{1}{32} \times 1\frac{17}{32} =$$

$$\frac{37}{4} \times \frac{8}{7} \times \frac{32}{1} \times \frac{49}{32} =$$

Step 3 Cancel like factors (7), (4), and (32).

$$\frac{37}{\cancel{4}} \times \frac{\overset{2}{\cancel{8}}}{7} \times \frac{\overset{1}{\cancel{32}}}{1} \times \frac{\overset{7}{\cancel{49}}}{\cancel{32}} =$$

Step 4 Multiply remaining fractions and express result in lowest terms.

$$\frac{37}{1} \times \frac{2}{1} \times \frac{1}{1} \times \frac{7}{1} = \frac{518}{1} = \textbf{518}$$

Ans

ASSIGNMENT UNIT 10   REVIEW AND SELF TEST

A. Division of Common Fractions (General Applications)

Divide.

1. $\frac{3}{4}$ by $\frac{1}{4}$

2. $\frac{1}{2}$ by $\frac{1}{6}$

3. $\frac{1}{4}$ by $\frac{3}{8}$

4. $\frac{5}{8}$ by $\frac{1}{2}$

5. $\frac{3}{16}$ by $\frac{1}{4}$

6. $\frac{5}{16}$ by $\frac{11}{16}$

7. $\frac{7}{12}$ by $\frac{5}{6}$

8. $\frac{9}{64}$ by $\frac{5}{32}$

9. $\left(\frac{13}{16} \times \frac{3}{32}\right)$ by $\frac{53}{64}$

10. $\left(\frac{21}{32} \times \frac{15}{64} \times \frac{1}{4}\right)$ by $\frac{7}{16}$

B. Division of Fractions and Whole Numbers (General Applications)

Divide.

1. 3 by $\frac{1}{3}$

2. 4 by $\frac{1}{2}$

3. 5 by $\frac{5}{8}$

4. $\frac{7}{16}$ by 7

5. $\frac{15}{32}$ by 30

6. $\frac{55}{64}$ by 11

7. 15 by $\frac{9}{16}$

8. 8 by $\frac{7}{16}$

9. $\left(\frac{5}{8} - \frac{1}{4} + \frac{1}{16} + \frac{1}{8}\right)$ by $\frac{3}{16}$

10. $\left(\frac{9}{32} - \frac{5}{64} \times \frac{1}{16} \times \frac{3}{32}\right)$ by $\frac{5}{8}$

C. Division of Mixed Numbers (General Applications)

1. $1\frac{1}{2} \div 1\frac{1}{2}$

2. $3\frac{1}{2} \div 4\frac{1}{4}$

3. $5\frac{3}{8} \div 3\frac{1}{4}$

4. $2\frac{3}{4} \div 4\frac{3}{8}$

5. $12\frac{1}{4} \div 6\frac{5}{16}$

6. $13\frac{3}{16} \div 11\frac{3}{4}$

7. $2\frac{15}{16} \div 6\frac{3}{8}$

8. $12\frac{3}{32} \div 1\frac{7}{64}$

D. Multiplication and Division of Fractions, Whole Numbers, and Mixed Numbers

Reduce all answers to lowest terms.

1. $\frac{1}{2} \times \frac{1}{4} \div \frac{1}{2}$

2. $1\frac{1}{6} \times \frac{5}{12} \div \frac{1}{6}$

3. $12 \times 6\frac{3}{8} \div \frac{7}{8}$

4. $21\frac{1}{4} \times 9\frac{5}{8} \div \frac{3}{4}$

5. $2\frac{9}{32} \div 1\frac{3}{16} \times 2\frac{1}{4}$

6. $16\frac{21}{64} \times 12\frac{3}{8} \div 2\frac{1}{4} \div \frac{7}{8}$

7. $1\frac{1}{4} \times \frac{9}{16} \times \frac{5}{8} \div 3\frac{1}{32}$

8. $\left(17\frac{1}{2} + \frac{3}{8} + \frac{1}{4} - \frac{5}{8}\right) \div 12\frac{1}{8}$

E. Division or Multiplication of Fractions, Whole Numbers, and Mixed Numbers (Practical Problems)

1. How many parts that are $2\frac{1}{2}$ inches can be stamped from a 20-inch strip? No waste allowance is made for trim.

2. How many pieces of stock $\frac{7}{8}$ of an inch long can be cut from a 30-inch bar of drill rod if $\frac{1}{16}$ of an inch is allowed on each piece for cutting?

3. How many pieces $10\frac{5}{16}$ inches long may be cut from a 12-foot length of a 2 × 4? Allow $\frac{3}{16}$ of an inch between cuts for waste.

4. How many billets of cold-drawn steel $3\frac{9}{32}$ inches long can be cut from a bar 48 inches long? Allow $\frac{1}{16}$ of an inch for the saw cut and another $\frac{1}{16}$ of an inch for facing.

5. Determine the number of pieces that can be blanked from a 50-yard roll of brass when each stamping is $4\frac{1}{2}$ inches long and each piece requires an additional $\frac{5}{32}$ of an inch for positioning.

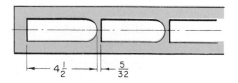

6. A drafting scale of $\frac{1}{4}$ of an inch to 1 foot is used on a drawing of a house. Compute the length of rooms A, B, C, and D.

Room	Scale Measurement (length)	Actual Room Length
A	$9\frac{1}{4}$ in.	
B	$10\frac{1}{4}$ in.	
C	$13\frac{1}{8}$ in.	
D	$14\frac{5}{16}$ in.	

7. The I-beam lintel of a brick doorway opening is 4'-4" long. It weighs $97\frac{1}{2}$ pounds. Find the number of pounds per foot in the I-beam.

8. Determine dimensions Ⓐ and Ⓑ.

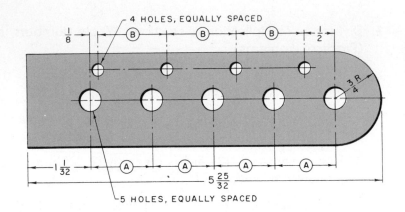

9. In milling flat surfaces, the work is fed against a revolving plain milling cutter with a specified feed.
   a. Find the length of time it will take to mill pieces A, B, C, D, and E for the lengths of work and feed indicated in the table.
   b. Determine the total time for milling.

Cut	Length of Work	Feed per Minute
	(in inches)	
A	4	$\frac{3}{4}$
B	11	$1\frac{3}{8}$
C	$11\frac{1}{2}$	$2\frac{7}{8}$
D	$13\frac{9}{16}$	$3\frac{3}{4}$
E	$6\frac{3}{32}$	$8\frac{1}{8}$

10. The average hourly production of a plastic part, the number of work hours per item per station, and the required production are given.

   a. Find the daily production for items 1, 2, and 3.

   b. Determine the number of days required to produce the units needed of the three items.

Item	Average Hourly Production	Daily Work Hours per Station	Required Units
1	20	14	3220
2	18	15	3720
3	$16\frac{1}{2}$	$20\frac{1}{2}$	5158

11. Calculate the current of a heating element in a simple circuit in amperes. The resistance of the element is $32\frac{3}{4}$ ohms, operating in a 48-volt circuit.
   Note. The current equals the voltage divided by the resistance. Round the answer to the nearest $\frac{1}{2}$ ampere.

12. Find the quantity (cubic feet) of gas that is consumed per hour for heat pumps A, B, and C, rounded to the nearest $\frac{1}{8}$ cubic feet (cu ft).

Heat Pump	Operating Hours	Gas Consumption
A	$8\frac{1}{2}$	$37\frac{1}{8}$ cu ft
B	6 hr., 30 min.	$492\frac{3}{8}$ cu ft
C	72 hr., 15 min.	$108\frac{5}{8}$ CCF ∠

∠ CCF = 100 cubic feet (cu ft)

Section 2
Unit 11 Achievement Review on Common Fractions

OBJECTIVES OF THE UNIT

This achievement review serves as an overall test for Section 2. The unit is designed to measure the student's/trainee's ability to

- *Visualize fractional parts of an object and reduce answers for measurements and other computed quantities to lowest terms.*
- *Apply the appropriate mathematical processes to solve problems that require the addition, subtraction, multiplication, or division of fractions, whole numbers, or mixed numbers.*
- *Check each answer.*

SECTION PRETEST/ POSTTEST

The Review and Self-Test items that follow relate to each Unit within this section. The test items may be used as a Unit-by-Unit pretest and/or Section post test.

UNIT 6. THE CONCEPT OF COMMON FRACTIONS

1. What common fraction is represented by the shaded area of each square?

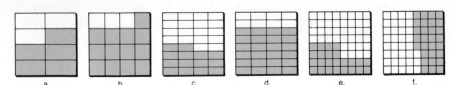

a.   b.   c.   d.   e.   f.

2. Reduce the fractional dimensions of (A), (B), (C), (D), (E), and (F), given in the drawing, to lowest terms. Locate each dimension on a rule.

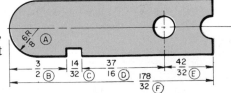

UNIT 7. ADDITION OF FRACTIONS

1. Add the following fractions.

a. $\frac{1}{8} + \frac{1}{4} + \frac{3}{8}$

b. $\frac{3}{32} + \frac{5}{8} + \frac{3}{4}$

c. $\frac{27}{64} + \frac{3}{16} + \frac{3}{32}$

d. $1\frac{1}{16} + \frac{3}{64} + 4\frac{17}{32}$

e. $3\frac{23}{32} + 2\frac{1}{8} + \frac{57}{64} + \frac{1}{4} + \frac{3}{16}$

f. $\frac{1}{9} + \frac{1}{6} + 2\frac{7}{8} + 3\frac{27}{32}$

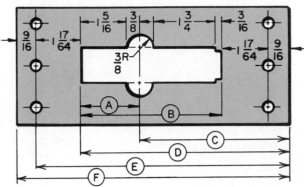

2. Determine the dimensions Ⓐ, Ⓑ, Ⓒ, Ⓓ, Ⓔ, and Ⓕ for the stripper plate.

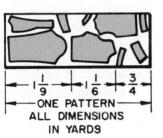

3. A pattern layout for a jacket is illustrated. Determine the length of material needed for one pattern.

ONE PATTERN
ALL DIMENSIONS
IN YARDS

UNIT 8. SUBTRACTION OF FRACTIONS

1. Subtract the following fractions.

a. $\frac{9}{16} - \frac{3}{16}$ d. $122 - 3\frac{3}{8}$ g. $19 - 2\frac{25}{64}$

b. $\frac{19}{64} - \frac{7}{32}$ e. $17 - \frac{55}{64}$ h. $13\frac{9}{16} - 7\frac{9}{32}$

c. $3 - \frac{27}{32}$ f. $203 - 6\frac{9}{16}$ i. $5\frac{37}{64} - 2\frac{1}{4} - 1\frac{7}{64}$

2. Determine dimensions Ⓐ, Ⓑ, Ⓒ, Ⓓ, Ⓔ, and Ⓕ. Reduce all answers to lowest terms.

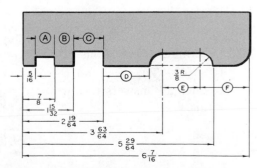

UNIT 9. MULTIPLICATION OF FRACTIONS

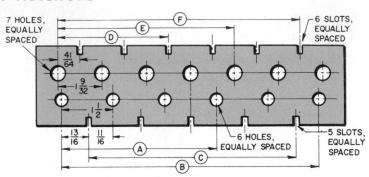

1. Determine dimensions Ⓐ, Ⓑ, Ⓒ, Ⓓ, Ⓔ, and Ⓕ. Reduce all answers to lowest terms.

2. Determine the length of corduroy fabric needed for jackets A, B, and C. State any fractional answer in terms of the nearest next whole number.

Jacket	Material per Jacket (including selvages)	Quantity	Length of Material Needed
A	$2\frac{7}{8}$ yds.	70	
B	$3\frac{1}{4}$ yds.	120	
C	$3\frac{5}{8}$ yds.	325	

UNIT 10. DIVISION OF FRACTIONS

1. Divide the following fractions and reduce all answers to lowest terms.

   a. $\frac{1}{8} \div \frac{3}{4}$ c. $11 \div \frac{5}{6}$ e. $3\frac{3}{32} \div 1\frac{5}{64}$

   b. $\frac{21}{64} \div \frac{7}{16}$ d. $\frac{21}{32} \div 7$ f. $2\frac{3}{16} \div 1\frac{3}{64} \div 4 \div 1\frac{3}{64}$

2. Determine dimensions Ⓐ, Ⓑ, Ⓒ, Ⓓ, Ⓔ, and Ⓕ. Reduce all answers to lowest terms.

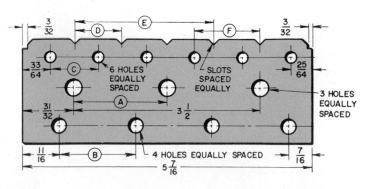

COMBINED PROCESSES WITH COMMON FRACTIONS

1. The resistance of each of three resistors in series is given in ohms (Ω) on the diagram. The current through the circuit is $1\frac{1}{2}$ amperes (A). Determine the following values.
   a. The total resistance (R_T).
   b. The source voltage (E_S).

Note. The total resistance (R_T) equals the sum of the three separate resistors (R_1, R_2, and R_3). The source voltage (E_S) equals the current (I) multiplied by the total resistance (R_T).

2. The current through a diode is $\frac{1}{24}$ ampere. The voltage drop across the diode is $\frac{3}{10}$ volt. Find the resistance reduced to lowest terms.

Note. Resistance is equal to the voltage (E) divided by the current (I).

3. Find the number of board feet of lumber that is required for parts A, B, and C. Round-off the fractional board feet to the nearest whole number.
   Note. The number of board feet equals

$$\frac{\text{Thickness}^{('')} \times \text{Width}^{('')} \times \text{Length}^{('')}}{12} \times \text{Number of pieces}$$

Part	Number Required	Thickness × Width	Length
A	16	2" × 4"	6' − 6"
B	4	2" × 8"	8' − 9"
C	5	1" × 10"	5' − 10"

DECIMAL FRACTIONS

Unit 12 The Concept of Decimal Fractions

OBJECTIVES OF THE UNIT

After satisfactorily studying this unit, the student/trainee will be able to
- *Understand how fractional measurements and other quantities are written in the decimal system.*
- *Read and write equivalent values for whole numbers and fractional parts in the decimal system.*
- *Round off decimals.*

PRETEST   *Use the Review and Self-Test items provided in the Unit Assignment to establish the level of mathematical skills competency and to determine the starting point of instruction.*

Machine, hand, and assembly operations must often be performed to a greater degree of accuracy than a fractional part of an inch. Where this accuracy is required, as in mating or interchangeable parts, precise dimensions are given on specifications, drawings, and sketches. These dimensions are given in thousandths, ten-thousandths, and, for extremely accurate work, in hundred-thousandths and millionths of an inch.

A. INTERPRETING THE DECIMAL SYSTEM

This system, which is based on ten (10), is known as the decimal system. The decimal system has been adopted universally throughout many industries because of the ease and accuracy with which dimensions may be measured and computed. Steel rules, micrometers, indicators, and other precision instruments are available for taking measurements based on the decimal system.

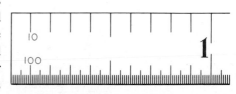

FIGURE 12-1

Describing a Decimal Fraction

A decimal fraction is a fraction. The denominator is 10, 100, 1,000, 10,000, or any other value that is obtained by multiplying 10 by itself a specified number of times. Instead of looking like a common fraction, the decimal fraction is written on one line with a period in front of it. This is possible because the denominator is always one (1) followed by zeros. By placing a period before the number that appears in the numerator, the denominator may be omitted. This period is called a decimal point. For example, the common fraction $\frac{5}{10}$ is written as the decimal .5; $\frac{5}{100}$ is written as .05; and $\frac{5}{1000}$ is written as .005.

Writing Decimal Fractions

Any whole number with a decimal point in front of it is a decimal fraction. The numerator is the number to the right of the decimal point. The denominator is always one (1) with as many zeros after it as there are places in the number to the right of the decimal point.

For example, the fraction $\frac{9}{10}$ may be written as the decimal fraction .9. This means that 9 is the numerator; the denominator is 1 with as many zeros as there are places (or digits) in the number to the right of the decimal point. In this case, there is one place to the right of the decimal point so the denominator is 10. The decimal fraction .9 is, therefore, the same as $\frac{9}{10}$.

To illustrate further

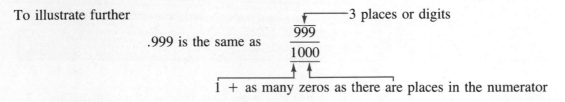

.999 is the same as $\dfrac{999}{1000}$

3 places or digits

1 + as many zeros as there are places in the numerator

B. EXPRESSING DECIMAL VALUES

Writing Whole Numbers and Decimal Fractions

With whole numbers and fractions, the whole number is placed to the left of the decimal point. The decimal fraction appears to the right. Three examples are given to show how different quantities may be expressed.

<u>EXAMPLE</u>:

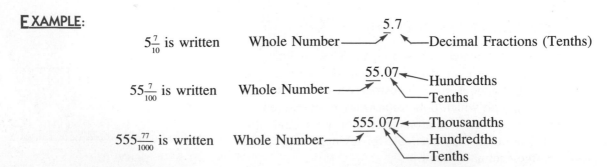

$5\frac{7}{10}$ is written   Whole Number ——— 5.7 ——— Decimal Fractions (Tenths)

$55\frac{7}{100}$ is written   Whole Number ——— 55.07 ——— Hundredths / Tenths

$555\frac{77}{1000}$ is written   Whole Number ——— 555.077 ——— Thousandths / Hundredths / Tenths

Indicating Degree of Accuracy

On drawings and in computations, whole numbers are sometimes expressed in the decimal system with zeros following the decimal point to indicate the degree of precision to which certain dimensions must be held.

E XAMPLE: The quantity 2″ is written 2.00″ if the dimension must be accurate to the second decimal place. If accuracy to the thousandth part of an inch is required, the 2″ is written 2.000″.

A decimal fraction such as .46 is often written (0.46). The zero that is placed before the decimal point emphasizes the fact that the decimal fraction is less than one.

C. READING DECIMALS

A decimal is read like a whole number except that the name of the last column or place to the right of the decimal point is added.

E XAMPLE:

- 0.63 is read sixty-three *hundredths*.
- 0.136 is read one hundred thirty-six *thousandths*.
- 0.5625 is read five thousand six hundred twenty-five *ten-thousandths*.
- 3.5 is read three and five *tenths*.
- 2.15625 is read two and fifteen thousand six hundred twenty-five *hundred-thousandths*.
- 0.0625 is read six hundred twenty-five *ten-thousandths*. This quantity is also commonly expressed as sixty-two and a half *thousandths*.

Simplified Method of Reading Decimals

Dimensions involving whole numbers and decimals are frequently expressed in an abbreviated form.

E XAMPLE:

- A dimension like (7.625) is spoken of as "seven point six two five."
- A dimension like (21.3125) is spoken of as "twenty-one point three one two five."

The use of decimal fractions provides an easy method of solving problems. Accurate computations may be made in addition, subtraction, multiplication, and division of fractions having a denominator of 10, 100, 1000, and the like. The units that follow deal with each one of the four fundamental mathematical operations as applied to decimal fractions.

D. ROUNDING OFF DECIMALS

The degree of precision to which a part is to be machined or finished sometimes determines how accurately the answer to a problem is computed. Many drawings indicate an accuracy in terms of thousandths or ten-thousandths of an inch. However, in computing dimensions, the answers may be accurate to four, five, or more decimal places.

The process of expressing a decimal to the number of decimal places needed for a predetermined degree of accuracy is called *rounding off decimals*.

RULE FOR ROUNDING OFF DECIMALS

- Check the drawing, sketch, or specifications to determine the required degree of accuracy.
- Look at that digit in the decimal place that indicates the required degree of accuracy.
- Increase that digit by 1 if the digit that follows immediately is 5 or more.
- Leave that digit as it is if the digit that follows is less than 5. Drop all other digits that follow.

EXAMPLE: The sum of a column of decimals is .739752. A part must be machined to an accuracy of only three places. Round off the decimal to three places.

Step 1	Write the computed decimal (6 places).	**.739752**
Step 2	Locate the digit that shows the number of .001″. The third digit (9) does this.	**.739**
Step 3	Look at the fourth-place digit to determine whether or not the third digit should remain the same or be increased.	**.7397**
Step 4	Increase the 9 by 1 because the fourth-place digit (7) is greater than (5). The correct answer is .740.	**.740** **Ans**

All of the intermediate steps in the example are given to serve as a guide in rounding off decimals. With actual practice, it is possible to round off a decimal to any desired degree of accuracy by just looking at it.

ASSIGNMENT UNIT 12   REVIEW AND SELF TEST

PRETEST
The Review and Self-Test items that follow may be used as a pretest. Pretests are designed to measure a student's beginning level of mathematical skills competency and to determine the starting point of instruction.

A. Writing Equivalent Decimal Values (General Applications)

1. Examine each circle and square. Determine visually what fractional part of each circle or square is represented by the shaded portions (A through I).

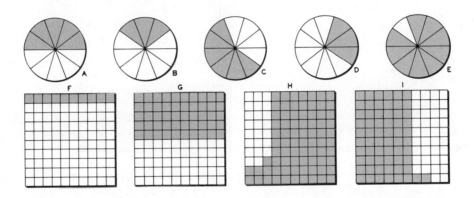

2. Express decimals (a, b, c, d, e, and f) in words.
   a. .3   c. 1.25   e. 5.375
   b. .07   d. 0.3125   f. 27.01563

3. Write the value of each quantity (a, b, c, d, e, and f) as a decimal.
   a. Seven tenths
   b. Sixteen hundredths
   c. Fifteen thousandths
   d. Eleven ten-thousandths
   e. Two thousand one hundred fifty-two ten-thousandths
   f. Three point one eight seven five

4. Write the following fractions as decimal fractions.
   a. $\frac{1}{10}$ e. $\frac{73}{100}$ i. $\frac{3}{10,000}$ m. $\frac{1,000}{10,000}$
   b. $\frac{3}{10}$ f. $\frac{1}{1,000}$ j. $\frac{19}{10,000}$ n. $\frac{793}{100,000}$
   c. $\frac{9}{100}$ g. $\frac{93}{1,000}$ k. $\frac{205}{10,000}$ o. $\frac{1,027}{100,000}$
   d. $\frac{29}{100}$ h. $\frac{157}{1,000}$ l. $\frac{1,923}{10,000}$ p. $\frac{30,019}{100,000}$

5. Write the following mixed numbers as decimals.
   a. $1\frac{1}{10}$ c. $25\frac{91}{100}$ e. $2525\frac{21}{10,000}$
   b. $3\frac{9}{100}$ d. $272\frac{67}{1,000}$ f. $362\frac{2,007}{10,000}$

B. Reading Decimal Measurements and Rounding Off Values (Practical Problems)

1. Secure a rule with a tenth and hundredth scale on it. Locate dimensions (a, b, c, d, and e) on the tenth scale.

   a. $\frac{1}{10}$ d. 1.2

   b. $\frac{3}{10}$ e. 1.5

   c. $\frac{5}{10}$

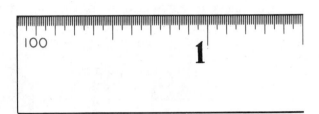

2. Locate dimensions (a, b, c, d, and e) on the hundredth scale.

   a. $\frac{10}{100}$ d. 1.20

   b. $\frac{33}{100}$ e. 1.32

   c. $\frac{77}{100}$

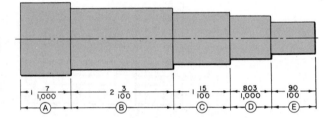

3. Express dimensions (A) through (E) on the drawing as decimals.

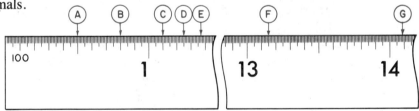

4. Read dimensions (A) through (G) on the rules as illustrated. Express as decimal fractions and decimals.

5. Round off decimal fractions (a) through (i) to two decimal places.

   a. .756   d. 29.409   g. 221.7557
   b. 1.952   e. 2.5644   h. 0.89673
   c. 7.324   f. 18.2707   i. 20.99974

6. Express each electrical measurement as a common fraction or mixed number. Reduce each answer to lowest terms.

   a. A copper wire diameter of 0.090″   c. A voltage of 105.625 volts
   b. A resistance of 22.35 ohms   d. An average unit cost of $.875

7. Convert each quantity to its equivalent decimal value, rounded to three decimal places.

   a. A wire resistance of $9\frac{3}{32}$ ohms   c. An installation time of 4 hours 20 minutes
   b. A wire length of $6\frac{3}{8}''$ d. A power rating of $3\frac{15}{16}$ watts

Unit 13 Addition of Decimals

PRETEST *Use the Review and Self-Test items provided in the Unit Assignment to establish the level of mathematical skills competency and to determine the starting point of instruction.*

On many drawings and sketches, dimensions must be computed that require the addition of two or more decimals.

A typical example is illustrated (Figure 13-1). The distance Ⓐ on the gage may be determined by adding the decimal dimensions 2.20″, 2.76″, and .50″. The addition of these decimals is the same as the addition of regular whole numbers. The exception is that the location of the decimal point must be given.

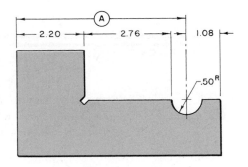

FIGURE 13-1

A. ADDING DECIMALS

RULE FOR ADDING DECIMALS

- Write the given numbers one under the other so that all of the decimal points are aligned in a vertical column.

- Add each column of numbers the same as for regular whole numbers.

- Locate the decimal point in the answer by placing it in the same column in which it appears with each number.

EXAMPLE: Add .875 + 1.2 + 375.007 + 71.1357 + 735.

Step 1 Write the numbers under each other so that all of the decimal points are aligned in a vertical line.

```
  .875
 1.2
375.007
 71.1357
735.
```

Note. Zeros are sometimes added to the numbers so that they all have an equal number of places after the decimal point. This practice may be followed to eliminate errors.

```
  .8750
 1.2000
375.0070
 71.1357
735.0000
```

Step 2 Add each column.

Step 3 Locate the decimal point in the answer in the same column in which it appears with the numbers being added.

1183.2177   Ans

ASSIGNMENT UNIT 13   REVIEW AND SELF TEST

PRETEST *The Review and Self-Test items that follow may be used as a pretest. Pretests are designed to measure a student's beginning level of mathematical skills competency and to determine the starting point of instruction.*

POSTTEST *The Review and Self-Test items may also be applied as a post test. Post tests are planned to establish the student's level of mathematical skills competency after instruction.*

A. Addition of Decimals (General Applications)

1. Add.

a.	b.	c.	d.
.5	9.3	76.8	195.7
.6	17.7	119.32	83.02
8.3	72.4	24.6	9.006

e. .4 + .7 + .4

f. 269.1 + 201.3

g. 0.57 + 29.35 + 1.6

h. 0.872 + 1.54 + 725.093

i. 2.9834 + 0.7256 + 329.7 + 21.0006

j. 0.00850 + 0.93006 + 3,225.06 + 0.0875

2. Add decimals (a) through (e). Then round off each sum correct to three decimal places.

  a. 25.0097
   0.9237
   <u>1.125 </u>

  c. 11.61254 + 0.735 + 1.3 + 625.003125

  d. .7 + 1.707 + 22.0625 + 3.09375 + 0.625

  e. 7.251 + 0.98475 + .03125 + 25.0 + 5.105

  b. .7895
   .6842
   12.7
  <u>231.0924</u>

B. Addition of Decimal Measurements (Practical Problems)

1. Determine dimensions Ⓐ, Ⓑ, Ⓒ, Ⓓ, and Ⓔ for the Die Plate.

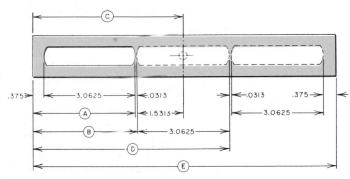

2. Determine baseline dimensions Ⓐ, Ⓑ, Ⓒ, Ⓓ, and Ⓔ.

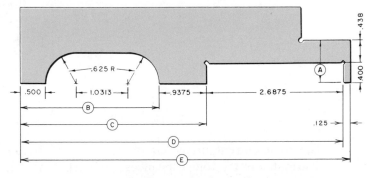

3. Determine overall dimension Ⓐ of the Template. Round off the answer to three decimal places.

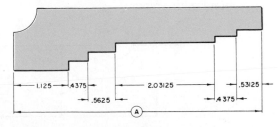

4. Find dimensions Ⓐ through Ⓔ. Start each computation with the dimensions given on the drawing. Round off answers to two decimal places.

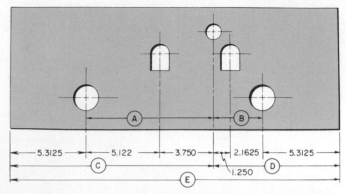

5. A counter top is 1.725″ thick. It is covered with laminated plastic .0625″ thick. Give the total thickness of the top.

6. Find the thickness of the plywood panel as illustrated.

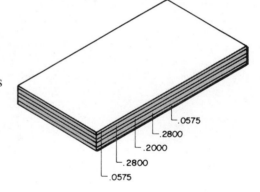

7. The total current in the circuit is measured by an ammeter Ⓐ. The total current is distributed in the four appliances. State what the reading is on the ammeter to the closest hundredth ampere.

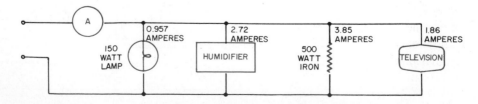

Unit 14 Subtraction of Decimals

PRETEST *Use the Review and Self-Test items provided in the Unit Assignment to establish the level of mathematical skills competency and to determine the starting point of instruction.*

Subtracting one decimal dimension from another is common practice in the factory, store, hospital, and home. The mathematical operation is the same as the subtraction of whole numbers with one exception. The position of the decimal point in the answer is similar to the addition of decimals.

The drawing of the tapered plug gage (Figure 14-1) shows a typical method of dimensioning a taper. The difference in diameters can readily be determined by the subtraction of decimals. The smaller diameter (1.5936″) is subtracted from the larger diameter (1.875″). The answer is referred to as the *difference*.

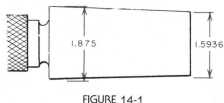

1.875   1.5936

FIGURE 14-1

A. SUBTRACTING DECIMALS

RULE FOR SUBTRACTING DECIMALS

* Write the given numbers so that the decimal points are under each other.

* Subtract each column of numbers the same as for regular whole numbers.

* Locate the decimal point in the answer by placing it under the column where it appeared in the problem.

EXAMPLE: Determine the amount of taper of the tapered plug gage. Give the dimension to the nearest thousandth.

Step 1	Write the two dimensions (1.875″ and 1.5936″) so that the smaller is under the larger. The decimal points are in the same column.	**1.875** **1.5936**
	Note. Zeros may be added to one of the numbers (1.875). It has fewer digits after the decimal point than the other number. The addition of zeros does not change the value.	**1.8750** **1.5936**
		1.8750
Step 2	Subtract the numbers by starting with the last digit on the right.	**1.5936**
Step 3	Locate the decimal point in the answer.	**0.2814**
Step 4	Round off the answer to the required number of places. 0.2814″ rounded off to three places.	**= 0.281** Ans

ASSIGNMENT UNIT 14   REVIEW AND SELF TEST

PRETEST   *The Review and Self-Test items that follow may be used as a pretest. Pretests are designed to measure a student's beginning level of mathematical skills competency and to determine the starting point of instruction.*

POSTTEST   *The Review and Self-Test items may also be applied as a post test. Post tests are planned to establish the student's level of mathematical skills competency after instruction.*

A. Subtracting Decimals (Practical Problems)

1. Find the difference in diameters (taper) of the Taper Plugs A, B, C, D, and E.

Plug	Diameter	
	Large End	**Small End**
A	1.750″	1.500″
B	3.0935″	2.837″
C	4.5312″	3.687″
D	10.1563″	8.3125″
E	7.0781″	5.7812″

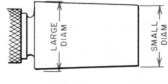

2. Find dimensions Ⓐ, Ⓑ, Ⓒ, Ⓓ, and Ⓔ for the plate.

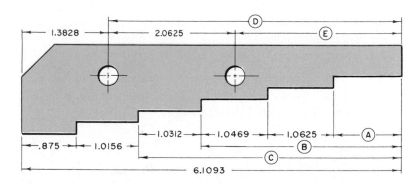

3. Determine dimensions Ⓐ, Ⓑ, Ⓒ, Ⓓ, and Ⓔ. Round off each answer to three decimal places.

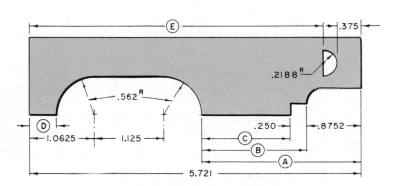

4. Determine dimensions Ⓐ, Ⓑ, Ⓒ, Ⓓ, and Ⓔ correct to two decimal places.

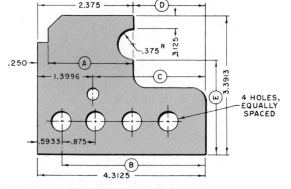

5. Find dimensions Ⓐ, Ⓑ, Ⓒ, Ⓓ, and Ⓔ correct to three decimal places.

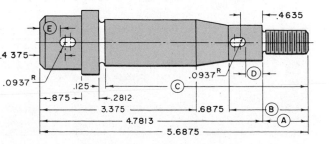

Wire Size	
National Gage Number	Diameter in Inches
10	0.10190
11	0.09074
12	0.08081
13	0.07196
14	0.06408
15	0.05707
16	0.05082

6. The wire diameter for sizes 10 through 16 of copper conductors is given in the table. Indicate the difference in diameter between sizes.
   a. 10 and 16
   b. 10 and 11
   c. 13 and 14
   d. 15 and 16

7. The opening between two parts is three and a half thousandths (0.0035″). A shim 0.00225 is available. Find the additional thickness to fill the opening.

8. A diet permits a daily intake of 1.67 quarts of liquid a day; 0.375 quart and 0.750 quart of liquid are used for two meals. Determine the remaining amount of liquid that may be taken during the day. Round off the answer to two decimal places.

9. Refer to the diagram for the quantity of current (I) flowing through each branch circuit. The total supply current (I_T) in the parallel circuit is 2.625 amperes (A). The current in I_A is 0.938 A and the current in I_B is 0.625 A. Find the current in I_C. Note. $I_C = I_T - I_A - I_B$.

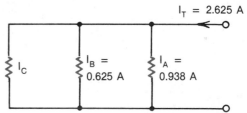

Unit 15 Multiplication of Decimals

OBJECTIVES OF THE UNIT

After satisfactorily studying this unit, the student/trainee will be able to
- *Solve general and practical measurement problems involving the multiplication of decimals.*
- *Convert percent values to decimals and decimals to percent values.*
- *Check solutions obtained by the multiplication of decimals.*

Multiplying decimals is a convenient and simplified method of adding them. One method of multiplying is to take one of the numbers to be multiplied, list it the number of times indicated by the multiplier and then add. This procedure is cumbersome, however, and it is easier with less chance for error if the two numbers are multiplied.

The multiplication process is identical to the multiplication process used for whole numbers. The exception is in pointing off the decimal places in the answer.

A. MULTIPLYING DECIMALS

RULE FOR MULTIPLYING DECIMALS

- Multiply the same as for whole numbers.
- Total the number of decimal places to the right of the decimal point in both of the numbers being multiplied.
- Locate the decimal point in the answer.
  Note. Start at the extreme right digit in the answer. Count off as many places to the left as there are in both the multiplier and multiplicand.

EXAMPLE: Multiply 27.935×7.07.

Step 1   Multiply (27.935) by (7.07).

Step 2   Add the number of digits to the right of the decimal point in both numbers that have been multiplied.

Step 3   Start at the right in the product. Count off the five decimal places to the left. Place the decimal point here. The result is the product of multiplying the two decimals.

$$
\begin{array}{r}
\textbf{27.935} \blacktriangleleft \textbf{(3 decimal places)} \\
\textbf{7.07} \blacktriangleleft \textbf{(2 decimal places)} \\
\hline
\textbf{195545} \\
\textbf{195545} \\
\hline
\textbf{197.50045}\quad \textbf{Ans} \\
\blacktriangle \\
\textbf{(5 decimal places)}
\end{array}
$$

B. CONVERTING PERCENT VALUES TO DECIMALS

The term percent means comparison in terms of so many hundredths of a quantity. In most percentage problems, multiplication is the mathematical process involved. The percent value is often converted to the decimal system for ease in computing and accuracy in pointing off the required number of decimal places. Instead of writing out the word percent each time, the symbol (%) is used.

RULE FOR CHANGING A PERCENT TO A DECIMAL

- Remove the percent (%) sign.

- Place a decimal point two digits to the left of the number for the given percent. *Note*. If the percent is a mixed number, change the fraction to a decimal and place this value after the whole number.

- Use the decimal value for the given percent the same as any other decimals to perform the required mathematical operations.

EXAMPLE: Change 12% to a decimal.

Step 1	Remove percent (%) sign.	**12**
Step 2	Place decimal point two digits to the left.	**.12**
Step 3	12% = .12	**.12** **Ans**

C. CONVERTING DECIMALS TO PERCENT

RULE FOR CHANGING A DECIMAL TO A PERCENT

- Move the decimal point two places to the right.

- Place the percent (%) sign after this number.

EXAMPLE: Change .055 to a percent.

Step 1	Move decimal point two places to the right.	**05.5**
Step 2	Place percent (%) sign after this number.	**5.5%**
Step 3	.055 = 5.5%	**5.5%** **Ans**

Different ways of representing common and decimal fraction and percent values of the same quantities are shown in Figure 15-1.

Fraction	Decimal Equivalent	Percent Value	Expressed in Words
$\frac{1}{100}$	1.0	1%	One one-hundredth
$\frac{10}{100}$	.10	10%	Ten hundredths
$\frac{100}{100}$	1.00	100%	One hundred hundredths
$\frac{150}{100}$	1.50	150%	One hundred fifty hundredths
$\frac{175}{1000}$	.175	17.5%	One hundred seventy-five thousandths

FIGURE 15-1

D. CHECKING

Problems in multiplication may be checked by one of two methods (the same as for whole numbers).

- The multiplier and multiplicand may be interchanged.
- The multiplication process may be repeated, using the same multiplier and multiplicand.

In either case, one of the most important steps is to check the location of the decimal point in the product.

ASSIGNMENT UNIT 15   REVIEW AND SELF TEST

PRETEST   *The Review and Self-Test items that follow may be used as a pretest. Pretests are designed to measure a student's beginning level of mathematical skills competency and to determine the starting point of instruction.*

POSTTEST   *The Review and Self-Test items may also be applied as a post test. Post tests are planned to establish the student's level of mathematical skills competency after instruction.*

A. Multiplication of Decimals

1. Multiply each whole number and decimal fraction. Check each answer.

   a. 9 by .8
   b. 16 by 1.5
   c. 12 by .72
   d. .37 by 100
   e. 1.3×98
   f. $9.5 \times .76$
   g. 11.7×1.82
   h. 11.31×6.14
   i. 92.07×7.392
   j. 1.0313×2.937
   k. 125.002×2.14
   l. 10.063×2.030

2. Multiply these decimal fractions. Check each answer. Then, round off the answers to four decimal places.

   a. 10.0625 × 6.437   b. 1.0937 × 3.0313   c. 1.5 × 3.7 × 5.12

B. Conversion of Percent Values

1. Determine the weight of metals A, B, C, and D, which were alloyed to cast a bronze plate weighing 79.5 pounds. The composition of the bronze is indicated by percent of each alloying metal.

A	B	C	D
Copper	Tin	Zinc	Lead
81%	6.5%	7.25%	5.25%

C. Multiplication of Decimal Values (Practical Problems)

1. Determine the cost of bar stock needed for parts A, B, C, D, and E. Give the answer in terms of dollars and cents for each part and the total cost.

Parts	Weight (Pounds per Foot)	Required Number of Feet	Cost per Pound
A	2.5	3.5	.75
B	2.25	7.5	.63
C	1.25	7.75	.93
D	7.5	27.125	.527
E	.0125	10.25	4.375

2. Find the lengths of insulating strip required to blank out each of the quantity of plates specified in A, B, C, D, and E. Round off all answers to one decimal place. Also determine the total length required.

Part	Quantity	Length of Plate	Allowance for Blanking
A	100	1.5	.25
B	100	2.25	.25
C	75	2.375	.25
D	75	2.625	.125
E	75	2.8906	.0937

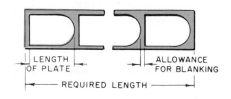

LENGTH OF PLATE   ALLOWANCE FOR BLANKING

REQUIRED LENGTH

3. Determine the distance in inches that a tool travels for each of five cuts (A, B, C, D, and E). Each distance should be rounded off to one decimal place. Also determine the total distance.

Cuts	RPM	Feed per Rev. (in inches)	Time (in minutes)
A	900	.005	1.5
B	424	.008	2.25
C	368	.015	6.75
D	336	.062	5.75
E	128	.062	25.25

4. The table gives the thickness, number of laminations, and meter readings of resistance for each lamination for each of three different core materials.
   a. Determine the thickness of cores 1, 2, and 3.
   b. Find the total resistance of the laminations in the three cores.
   Note. Round off all answers to two decimal places.

Core Material	Thickness per Lamination	Number of Laminations	Resistance per Lamination
1	.003″	200	.1 amp
2	.012″	21	.125 amp
3	.1352″	15	.3157 amp

5. It takes an average of 3.4 hours to lay 100 square feet (one square) of finish flooring. Compute the time required to lay finish flooring in rooms A, B, and C.

Room	Floor Area (Squares)	Flooring Time
A	12.2	
B	13.75	
C	17.87	

6. The unit price, weight per container, and quantity required for four foods are given. Determine the cost of each item correct to two decimal places.

Food	Cost ¢ per Ounce	Ounces per Container	Quantity	Cost
A	.086	12	24 pkgs	
B	.078	18	12 pkgs	
C	.062	15.75	24 cans	
D	.034	15.75	48 cans	

7. An electric (kilowatt hour) meter registers $\boxed{7\ 0\ 2\ 3}$ on June 1. On July 1 the reading is $\boxed{7\ 8\ 6\ 9}$. The difference represents electrical energy used. The cost of electrical energy is $0.07252 per kilowatt hour (kW·h).
   a. Find the cost of the electrical energy.
   b. The bill is reduced by a fuel adjustment of $0.00303 per kW·h. Indicate the amount the bill is reduced.
   c. A sales tax of 0.975% is levied against each adjusted utility bill. Compute the sales tax on the adjusted bill.
   d. Determine the total charges.

8. The inductive reactance (in ohms) in the ac circuit equals $2\pi \cdot I \cdot f$. Substitute the value of 0.032 Hz for *(f)* and 60/s for *(I)*.
   Calculate the inductive reactance in ohms. Round the answer to two decimal places. Use $\pi = 3.1416$.

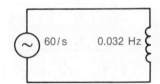

9. Determine the power in watts for the values given in the circuit diagram, correct to two decimal places.
   Note. The power *(W)* equals the voltage (V) multiplied by the current in amperes (A).

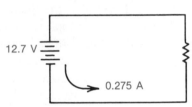

Unit 16 Division of Decimals

OBJECTIVES OF THE UNIT
After satisfactorily studying this unit, the student/trainee will be able to
* *Solve general and practical measurement problems involving the division of decimals.*
* *Translate and express common fractions as decimal fractions and mixed numbers as decimals.*
* *Use decimal equivalent charts and shop methods to select equivalent decimal or fractional measurements.*

PRETEST   *Use the Review and Self-Test items provided in the Unit Assignment to establish the level of mathematical skills competency and to determine the starting point of instruction.*

Division is the simplified process of computing the number of times one number is contained in another. The division of decimals, like all other mathematical operations for decimals, is essentially the same as for whole numbers. An additional consideration is the location of the decimal point in the answer.

A. DIVIDING DECIMALS

RULE FOR DIVIDING DECIMALS

- Place the number to be divided (called *dividend*) inside the division box.
- Place the *divisor* outside.
- Move the decimal point in the divisor to the extreme right. The divisor then becomes a whole number.
- Move the decimal point the same number of places to the right in the dividend. *Note*. Zeros are added in the dividend if it has fewer digits than the divisor.
- Mark the position of the decimal point in the *quotient*. The position is directly above the decimal point in the dividend.
- Divide as for whole numbers. Place each figure in the quotient directly above the digit involved in the dividend.
- Add zeros after the decimal point in the dividend if the dividend cannot be divided exactly by the divisor.
- Continue the division until the quotient has as many places as are required for the answer.

EXAMPLE: *Case 1*. Divide 25.5 by 12.75.

Step 1 Move the decimal point in the divisor to the right (2 places).

Step 2 Move the decimal point in the dividend to the right the same number of places (2). Since there is only one digit after the decimal, add a zero to the dividend.

Step 3 Place the decimal point in the quotient.

Step 4 Divide as for whole numbers.

EXAMPLE: Case 2. Divide 123.573 by 137.4.

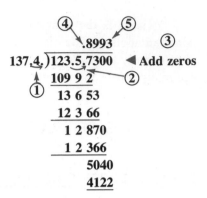

The answer must be correct to three decimal places. Note. The division process is usually carried out to one more than the required number of places in the answer. The last digit may then be rounded off for greater accuracy. In this case, .8993 is rounded off to .899. **Ans**

B. EXPRESSING COMMON FRACTIONS AS DECIMAL FRACTIONS

Dimensions used in machine, bench, or assembly operations are given in terms of common fractions or decimal fractions. Yet, when the actual measurements are taken, it is often necessary to either express a common fraction as a decimal fraction or a decimal fraction as a common fraction. Converting from one system to the other is comparatively easy. It involves, in most cases, the division of numbers with three or fewer digits.

RULE FOR EXPRESSING A COMMON FRACTION AS A DECIMAL FRACTION

• Divide the numerator by the denominator.

EXAMPLE: Express $\frac{5}{16}$ as a decimal fraction.

$$
\begin{array}{r}
.3125 \quad \text{Ans} \\
16\overline{)5.0000} \\
4\ 8 \\
\hline
20 \\
16 \\
\hline
40 \\
32 \\
\hline
80 \\
80 \\
\hline
\end{array}
$$

Step 1   Divide the numerator (5) by the denominator (16).
Step 2   Place a decimal point after the 5.
Step 3   Locate the decimal point in the quotient.
Step 4   Add as many zeros as are needed to obtain a quotient that can be rounded off to the required number of decimal places.
Step 5   Divide. The resulting answer (.3125) is the decimal fraction equivalent of the common fraction $\left(\frac{5}{16}\right)$.

C. EXPRESSING MIXED NUMBERS AS DECIMALS

RULE FOR EXPRESSING A MIXED NUMBER AS ITS DECIMAL EQUIVALENT

- Express the mixed number as an improper fraction.
- Divide the numerator by the denominator.
- Carry out the division to the number of decimal places required for the degree of accuracy involved.

EXAMPLE: Determine the decimal equivalent of $1\frac{3}{64}$ correct to three decimal places. The answer is 1.047.

Step 1 Divide the numbers to four decimal places.

Step 2 Round off the decimal equivalent (1.0468) to three decimal places. The answer is 1.047.

$$1\frac{3}{64} = \frac{67}{64} = 64\overline{)67.0000} \quad \frac{1.0468 = 1.047 \quad \text{Ans}}{}$$

$$\begin{array}{r} \underline{64} \\ 3\ 00 \\ \underline{2\ 56} \\ 440 \\ \underline{384} \\ 560 \\ \underline{512} \\ 48 \end{array}$$

D. EXPRESSING DECIMAL FRACTIONS AS COMMON FRACTIONS

RULE FOR EXPRESSING A DECIMAL FRACTION AS A COMMON FRACTION

- Write the number after the decimal point as the numerator of a common fraction.
- Write the denominator as 1 with the same number of zeros after it as there are digits to the right of the decimal point.
- Express resulting fraction in lowest terms.

EXAMPLE: Express .09375 as a common fraction.

Step 1 Write number after the decimal point as the numerator.   9375 = 9,375

Step 2 Determine denominator.   1 + 5 zeros = 100,000

Step 3 Express fraction in lowest terms. $\dfrac{9,375}{100,000} = \dfrac{3}{32}$ Ans

E. SIMPLIFIED METHODS OF DETERMINING DECIMAL OR FRACTIONAL EQUIVALENTS

Tables of Decimal and Fractional Equivalents (Figure 16-1) are usually found in all trades, drafting rooms, production shops, and toolrooms. These tables are printed in many forms: as enlarged wall charts, as reference tables in handbooks and kits, and as handy-guide plastic cards. The tables are used extensively for determining either the decimal or fractional equivalent of a given value. The simplified use of the charts is preferred to the longhand method of dividing. The charts ensure accuracy and speed up the process of computation.

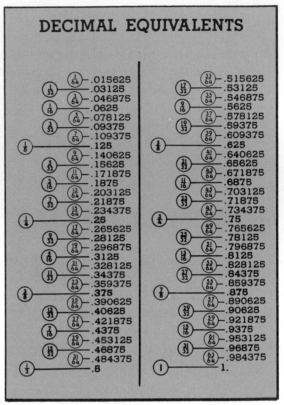

FIGURE 16-1

Selecting Decimal or Fractional Values

The fractions on these charts are given in steps of $\frac{1}{64}''$ in one column with the corresponding decimal equivalents in another. Most tables are carried out to six places. Decimal fractions may be rounded off to any desired degree of accuracy. The decimal equivalent is found by locating the given fraction in the left-hand column. The equivalent value is located in the decimal column. Reverse the practice for finding fractional equivalents of decimals.

Shop Method of Determining Value

Even without the charts, most good mechanics can quickly determine the decimal equivalents of all fractions by increments of $\frac{1}{64}''$. In addition to knowing the decimal equivalents of eighths and quarters, many craftspeople also learn that the decimal equivalent of $\frac{1}{64}''$ is .015625, that of $\frac{1}{32}''$ is .03125", and $\frac{1}{16}''$ is .0625".

By thinking of the required fractions as being equal to or so much smaller or larger than a known decimal value, it is comparatively easy to determine decimal equivalents of fractions. For example, assume that the decimal equivalent of $\frac{29}{32}$ is needed. The mechanic usually thinks of that fraction as a "thirty-second over seven-eighths." It would be equal to .875 plus $\frac{1}{32}$ (.03125) or (.90625). This same line of reasoning provides a shortcut in determining decimal equivalents of most fractions.

ASSIGNMENT UNIT 16 REVIEW AND SELF TEST

PRETEST *The Review and Self-Test items that follow may be used as a pretest. Pretests are designed to measure a student's beginning level of mathematical skills competency and to determine the starting point of instruction.*

POSTTEST *The Review and Self-Test items may also be applied as a post test. Post tests are planned to establish the student's level of mathematical skills competency after instruction.*

A. Division of Decimals (General Applications)

1. Divide.
   a. 11.8 by 100
   b. 23.7 by 1,000
   c. 10 by 2.5
   d. 104.26 ÷ 26
   e. 7.5 ÷ 22.5
   f. 26.0313 ÷ 10.25

2. Divide and round off each answer correct to the number of decimal places indicated.
   a. .875 ÷ 6.25 (one place)
   b. 2.234 ÷ 24.63 (three places)
   c. .4375 ÷ 156.25 (three places)
   d. 145.26 ÷ 13.750 (two places)

B. Reduction of Common Fractions and Mixed Numbers to Decimals (General Applications)

1. Compute the three-place decimal fraction equivalent of each of the following.
   a. $\frac{3}{4}$
   b. $\frac{5}{8}$
   c. $\frac{9}{16}$
   d. $\frac{1}{6}$
   e. $\frac{7}{72}$
   f. $\frac{13}{32}$
   g. $\frac{29}{64}$

C. Determination of Decimal and Fractional Sizes by Table

1. Find the decimal value of each fractional size drill by using a decimal equivalents table.
   a. $\frac{1}{2}''$
   b. $\frac{3}{4}''$
   c. $\frac{3}{8}''$
   d. $\frac{5}{16}''$
   e. $\frac{7}{32}''$
   f. $\frac{19}{64}''$

2. Find the fractional equivalent to each decimal by using a decimal equivalents table.
   a. .250"   b. .875"   c. .5625"   d. .6875"   e. .71875"   f. .046875"

D. Division of Decimals (Practical Problems)

1. Determine the depth of cut for each tooth on broaches A, B, C, D, and E.

Part	Depth to Be Broached (in inches)	
A	.1	10
B	.126	42
C	,255	150
D	.063	30
E	.0924	42

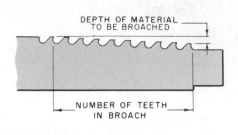

DEPTH OF MATERIAL TO BE BROACHED

NUMBER OF TEETH IN BROACH

2. Determine how many spacing collars are needed for sawing operations A, B, C, D, and E. (All dimensions are given in inches.)

Operation	Thickness of Collars	Spacing Required
A	.5	4.5
B	.25	3.5
C	.125	2.375
D	.1875	1.875
E	.09375	1.125

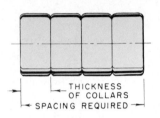

THICKNESS OF COLLARS
SPACING REQUIRED

3. In a production operation, pins A, B, C, D, and E are cut off to the lengths indicated. Determine the whole number of pieces that can be cut from the workable lengths of stock given. (All dimensions are given in inches.)

Pins	Workable Length	Length of Pin	Allowance for Cutoff and Facing
A	10	$\frac{1}{2}$	.120
B	33	$\frac{3}{4}$	.120
C	68	$1\frac{1}{4}$	.125
D	68	$2\frac{1}{8}$	.125
E	11	$\frac{17}{32}$	.094

LENGTH OF STOCK THAT CAN BE USED (WORKABLE LENGTH)
ADDITIONAL STOCK FOR CHUCKING

4. A power sewing machine is set at eight stitches per inch of seam.
   a. Determine the length of one stitch.
   b. Calculate the number of stitches in a seam that is 20.25″ long.

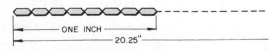

ONE INCH
20.25"

5. Express the decimal quantities of food products A, B, C, and D.

Food Product	Contents	
	Decimal Value	Fractional Equivalent
A	14.437 oz.	
B	25.4 g	
C	175.375 qts.	
D	9.56 lbs.	

6. Six wall receptacles are to be equally spaced starting four feet from each wall to the center line of the first receptacle. The distance between walls is 117′-0″. Determine the center-to-center distance between each receptacle.

Section 3
Unit 17 Achievement Review on Decimal Fractions

OBJECTIVES OF THE UNIT
This achievement review serves as an overall test for Section 3. The unit is designed to measure the student's/trainee's ability to
- *Express fractional dimensions and other quantity measurements as decimal fractions.*
- *Apply the appropriate mathematical processes to solve problems that require the addition, subtraction, multiplication, or division of decimals.*

SECTION PRETEST/ POSTTEST

The Review and Self-Test items that follow relate to each Unit within this section. The test items may be used as a Unit-by-Unit pretest and/or Section post test.

UNIT 12. THE CONCEPT OF DECIMAL FRACTIONS

1. Write dimensions Ⓐ, Ⓑ, Ⓒ, Ⓓ, and Ⓔ as decimals. Locate each dimension on a rule or scale.

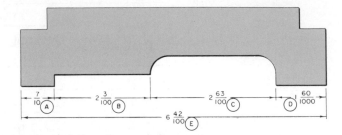

UNIT 13. ADDITION OF DECIMALS

1. Determine distances Ⓐ, Ⓑ, Ⓒ, Ⓓ, and Ⓔ.

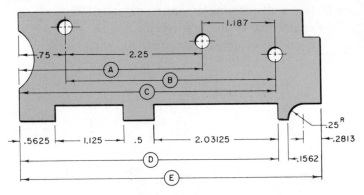

2. Prepare a table similar to the one illustrated and fill in the appropriate columns.
   a. List the values obtained for A, B, C, D, and E in the problem B.1.
   b. Round off each measurement to three decimal places.
   c. Express each decimal as a mixed number.

	Value Obtained	Rounded to Three Places	Expressed as Mixed Number
A			
B			
C			
D			
E			

3. The elevation at the top of the girder is 39.72 feet. Determine the elevation on the surface of the terrazzo floor.

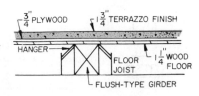

UNIT 14. SUBTRACTION OF DECIMALS

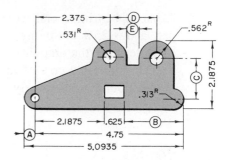

1. Determine dimensions (A), (B), (C), (D), and (E). Recheck the accuracy of each dimension to three places.

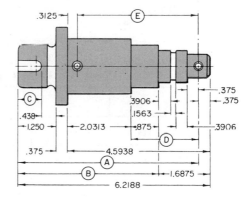

2. Determine dimensions (A), (B), (C), (D), and (E) correct to three decimal places.

UNIT 15. MULTIPLICATION OF DECIMALS

1. Determine the overall dimension of the saw and spacing collar combination for each of five setups given in the table (A, B, C, D, and E). Dimensions must be correct to three decimal places.

Setup	Collars		Saws	
	No.	Width	No.	Width
A	4	.5	3	.1
B	8	.25	7	.1
C	7	.25	6	.06
D	10	.187	9	.03
E	12	.1875	11	.0313

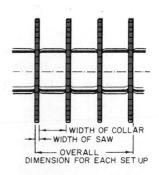

WIDTH OF COLLAR
WIDTH OF SAW
OVERALL
DIMENSION FOR EACH SET UP

2. A bar of bearing metal weighing 142.5 pounds is composed of 79.5% tin, 11.25% copper, and 9.25% antimony.
   a. Determine the amount of tin, copper, and antimony in the bar.
   b. Determine the cost of the copper and tin if these metals sell for $6.976 and $10.512 per pound, respectively.

UNIT 16. ADDITION, MULTIPLICATION, AND DIVISION OF DECIMALS

1. Determine the whole number of parts A, B, C, D, and E that can be machined from the workable lengths of stock given. (Each dimension is in inches.) *Note.* Round off the length of each part to two decimal places before determining the number of parts.

Parts	Length of Part	Allowance for		Spoilage	Workable Length
		Cutting Off	Facing		
A	$\frac{1}{2}$	.100	.010	1%	28
B	$\frac{3}{4}$	.100	.010	2.5%	34
C	1.25	.100	.005	4.5%	$69\frac{1}{2}$
D	2.125	.120	.015	$3\frac{1}{4}\%$	$69\frac{1}{2}$
E	3.187	.125	.015	$5\frac{1}{2}\%$	$69\frac{1}{2}$

2. How many pieces can be stamped from each length of stock for parts A, B, C, D, and E?

Part	Length of Stock (in feet)	Length of Stamping	Allowance on One End
			(in inches)
A	16	$1\frac{1}{10}$	$\frac{2}{10}$
B	$12\frac{1}{2}$	2.25	$\frac{2}{10}$
C	$12\frac{3}{4}$	3.13	.25
D	$25\frac{1}{4}$	3.187	.25
E	$240\frac{3}{4}$	5.125	.275

3. Compute the micrometer readings for the finished diameters of cylinders A, B, and C.

Cylinder		Oversize	Finished
	Diameter	Bore	Diameter
A	$3\frac{1}{2}''$	+0.040"	
B	$4\frac{1}{4}''$	+0.0625"	
C	$5\frac{1}{8}''$	+0.0781"	

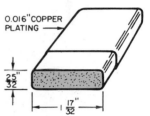

4. A carbon brush has a 0.016" copper plating on all sides. Find the width and thickness of the carbon brush as illustrated.

5. Determine the power consumption *(P)* in watts of the transistor shown in the diagram. The quantity of current *(I)* is 0.125 amperes (A); voltage *(E)* across the power transistor is 8.25 V. *Note.* Power equals current multiplied by the voltage.

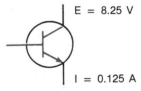

6. Find the voltage *(E)* in a circuit where a transistor draws a current *(I)* of 1.02 amperes and has a power consumption *(P)* of 0.75 watts.
Note. The voltage equals power divided by current.

7. A new piston is 3.875" diameter. The rate of wear (reduction of diameter) is 0.0015" for each 7,500 miles of use.

   a. Determine the total wear at (1) 41,250 miles and again at (2) 76,875 miles. Round the answer to four decimal places.

   b. Use each answer to find the piston diameter at the end of (1) 41,250 miles and (2) 76,875 miles.

MEASUREMENT: DIRECT AND COMPUTED (CUSTOMARY UNITS)

Unit 18 Principles of Linear Measure

OBJECTIVES OF THE UNIT

After satisfactorily studying this unit, the student/trainee will be able to

- *Understand how common fractional and decimal units of linear measurement are applied to rules, tapes, and other line-graduated measuring instruments.*
- *Interpret standard and vernier micrometer readings and relate these to precision measurements in the ranges of 0.001" and 0.0001".*
- *Set up combinations of working gage blocks for extremely accurate linear measurements of ± 0.000 004".*

PRETEST   *Use the Review and Self-Test items provided in the Unit Assignment to establish the level of mathematical skills competency and to determine the starting point of instruction.*

The exchange of goods and services among nations depends on communications. Though languages differ, it is necessary that universally accepted international standards for measurements, technical terms, and data be established. The current worldwide standards movement is toward metrication, using the International System of Units (SI). "Metrication" refers to any program or process of conversion to SI metrics. This means that the continuing development of SI metric units by the International Standards Organization must in turn be adopted by the nations of the world.

The United States is one of the few nations that continues to use a nonmetric system of measurement. Even after the legislative adoption of SI metrics, the system will not be mandated nor will it be the sole system of measurement. Conversion is intended to be voluntary, based on industry-by-industry decision.

The measurements with which each individual must be familiar in order to solve common mathematical problems in business, industry, agriculture, health occupations, and in the home and for other daily activities, include

- Linear measure
- Circular measure

- Area measure
- Volume measure

These measurements are covered in this section by customary units. These units are used to define the quantity of each measurement. The term "customary" refers to the American system of units based on the British system. "Conventional metric" units and the newer SI metrics follow in Part 2. Standardization, in terms of these three major systems, affects the life of each individual.

Direct and Indirect Measurement

Some measurements are taken directly. When measuring tools, weights, instruments, and other line-graduated rules are used and the quantity is read directly, the term "direct measurement" applies. There are other instances where it is impractical or impossible to take direct measurements. In such cases, dimensions are computed, resulting in an *indirect or computed measurement*.

A. CONDITIONS AFFECTING DEGREE OF PRECISION

The British, American, metric, SI metric, and other special systems of measurement all provide for varying degrees of precision. For some applications, a value rounded off to the closest $\frac{1}{100}$ is adequate. In other cases, a dimensional tolerance, accurate to three or more decimal places, is required. Besides dimensional accuracy, precision also is affected by surface finish and temperature. The movement of parts and mechanisms requires differing conditions of finish and measurement precision.

Precision applies equally to indirect measurements. These may relate in the fields of science to light, heat, and sound, and nuclear and other energy sources. Precision relates to all industries: foods, health, construction, banking, manufacturing, and services. These, and all other industries, use measurements.

Measuring Tools and Instruments

The greater the degree of precision, the more precise the measuring instruments must be. Higher precision usually involves higher production expenses because of additional processing costs. Where a rough direct measurement within $\frac{1}{16}''$ is required, the dimension may be measured easily with a ruler. If a drawing shows a tolerance of plus or minus $\frac{1}{32}''$ or $\frac{1}{64}''$, a line-graduated steel rule is a practical measuring tool to use.

A micrometer is needed for tolerances within a plus or minus "one-thousandth" range (± 0.001″). An operation like precision grinding to a tolerance of plus or minus one-ten-thousandth (0.0001″) requires a micrometer having vernier graduations reading in ten-thousandths of an inch. Still other parts require that direct measurements be taken to within limits of two-millionths (0.000002″) of an inch. Gage blocks are used industrially in combination with other measuring instruments to make direct measurements to this limit.

Common rules, measuring instruments, and accessories are described in this section. The different measurement applications range from a precision of $\frac{1}{16}″$ to two-millionths (0.000002″). These experiences are intended to develop skill in applying mathematical principles to direct and computed measurements.

Linear Measurements

Linear measure is the measurement of straight-line distances between two points, lines, or surfaces. In this section, linear measurements are treated in terms of British linear units, which are still the most widely used. The yard is the standard unit of length. The smallest unit of measure in the British system is the inch.

In 1856 England presented the United States with two bronze bars as a standard representation of the yard. The American system of linear measure is based on the British system. The bronze bars are kept for historical significance and not accuracy. The standard for all linear measurements was authorized by law in 1893 as the National Standard of Length.

Table of Linear Measure		
12 inches	=	1 foot
3 feet	=	1 yard
$5\frac{1}{2}$ feet	=	1 rod

Smallest unit of measure = one inch

FIGURE 18-1

A. APPLICATIONS OF RULES, TAPES, AND LINE-GRADUATED TOOLS TO LINEAR MEASUREMENTS

Linear measurements may be made by craftspersons and technicians with solid rules, flexible steel tapes, and other line-graduated measuring instruments. Containers and other vessels are sometimes graduated to permit linear measurements to be taken directly (Figure 18-2).

Consumers generally use a ruler, yardstick, steel tape, or tape measure. When the tape measure is made of fabric, measurements should be checked for accuracy against a more precise measuring tool. Line graduations of sixteenths and eighths are common on consumer measuring tools.

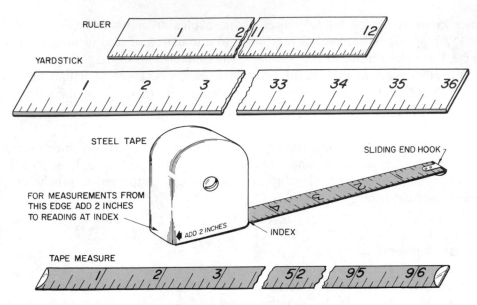

FIGURE 18-2 Consumer linear measuring tools

Standard Unit of Linear Measure

The most commonly used unit of measure is the inch. As a standard unit of linear measure, the inch is subdivided into smaller fractional parts representing either common or decimal fraction equivalents.

The fractional divisions of an inch that are most commonly used on rules represent halves, quarters, eighths, sixteenths, thirty-seconds, and sixty-fourths of an inch (Figure 18-3).

The decimal system is used when smaller units of measure are required. It is common practice in the shop and laboratory to express fractional parts of an inch in decimals, which are called "decimal equivalents." For example, the decimal equivalent of one-fourth $\frac{1''}{4}$ would be two hundred fifty thousandths (.250"). See Figure 18-4.

Measurements up to $\frac{1''}{100}$ may be made directly with a steel rule graduated in fiftieths

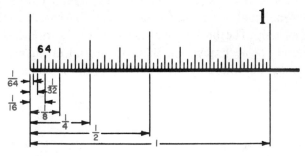

FIGURE 18-3 Enlarged view of fractional parts of an inch

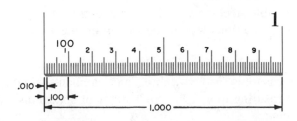

FIGURE 18-4 Enlarged view of decimal fractions

and hundredths of an inch. The use of such rules reduces the possibility of error that results from changing common fractions to decimals.

B. APPLICATION OF THE CALIPER TO LINEAR MEASUREMENTS

The outside and inside caliper, while not a graduated measuring tool, is used in combination with the rule to measure linear distances. The caliper is used to transfer linear measurements from the work to the steel rule.

Ordinarily, the smallest measurement that can be taken with a caliper and rule is $\frac{1}{64}''$ in the case of common fractions and $\frac{1}{100}''$ for decimal fractions. Where measurements in terms of thousandths are to be taken, the caliper size is measured with a micrometer.

C. PRINCIPLES OF MICROMETER MEASUREMENT

The standard micrometer is used to measure parts requiring an accuracy of one thousandth. These readings are obtained by turning a graduated thimble on a graduated barrel (Figure 18-5). The movement of this thimble is at the rate of $\frac{1}{40}''$ per turn (Figure 18-6). The $\frac{1}{40}''$ is determined by the pitch of the screw threads, which are concealed. As the thimble turns, the spindle moves closer to or further away from the anvil. The anvil is a stationary part of the frame.

The barrel of the micrometer has 40 vertical graduations to indicate this movement of (.025″). Each fourth division on the barrel is marked for ease in reading (Figure 18-7).

The thimble of the micrometer is divided into 25 equal parts (Figure 18-8). As each line crosses the horizontal line on the barrel, the space between the anvil and spindle is greater or smaller by $\frac{1}{25}$ of a revolution or (.001″). Each fifth division of the spindle is numbered 5, 10, 15, 20, and 0 or 25. These numbers are in terms of thousandths: .005″, .010″, .015″, .020″, and .025″.

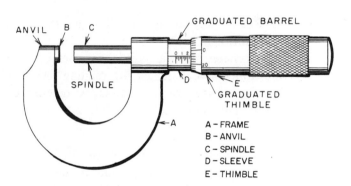

A – FRAME
B – ANVIL
C – SPINDLE
D – SLEEVE
E – THIMBLE

FIGURE 18-5

ONE REVOLUTION OF THIMBLE MOVES THE SPINDLE $\frac{1''}{40}$ OR .025″

FIGURE 18-6

GRADUATIONS ON BARREL OF MICROMETER

ONE DIVISION = $\frac{1''}{40}$ = .025″

FIGURE 18-7

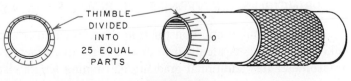

THIMBLE DIVIDED INTO 25 EQUAL PARTS

FIGURE 18-8

RULE FOR READING A MICROMETER

- Note the last vertical line that is visible on the barrel.
- Determine its reading with respect to a numbered graduation.
- Add to this reading the number of the line on the sleeve that crosses the horizontal line on the barrel.

EXAMPLE: Determine the micrometer reading (Figure 18-9).

Step 1 Read last numbered vertical line on barrel.

Step 2 Add .025″ for each additional line that shows (reading on barrel).

Step 3 Add number of graduation on sleeve (reading on sleeve).

Step 4 Read the required reading as the sum of all additions.

.400
.025
.017
442″ **Ans**   FIGURE 18-9

Some micrometer manufacturers recommend that the number of vertical graduations on the barrel be multiplied by 25 to get the reading. Although this may be necessary at first, with practice the value of the graduations can be determined readily. It is also possible to measure accurately to the half-thousandth by splitting the distance between lines on the thimble.

EXAMPLE: The decimal equivalent of $\frac{9''}{16}$ is .5625″. The measurement would indicate a reading of .550″ on the barrel plus .0125″ on the thimble (Figure 18-10).

READING MIDWAY BETWEEN .013 AND .012 = .0125

FIGURE 18-10

The micrometer head includes the graduated barrel, graduated sleeve and spindle, and modified frame. The micrometer head has many applications in addition to the micrometer caliper. On internal work the micrometer principle is applied in a measuring tool called an inside micrometer. Examples of other applications are the depth micrometer and height gage attachment for V-blocks.

Micrometer heads of slightly different construction are used widely for accurate machining, measuring, and inspection processes. Regardless of the construction or size, the micrometer principle remains the same. The total measurement is equal to the sum of the reading on the barrel and sleeve.

D. PRINCIPLES OF VERNIER MICROMETER MEASUREMENT

In appearance and construction the vernier micrometer, with which measurements can be taken to .0001″, is identical to the standard micrometer. The only difference is that the vernier micrometer has additional graduations running lengthwise on the barrel (Figure 18-11).

Readings in one ten-thousandths of an inch are obtained by the vernier principle. Pierre Vernier applied the vernier principle in 1631. The vernier principle is comparatively simple. There are ten graduations on the top of the barrel. These occupy the same space as nine divisions on the thimble.

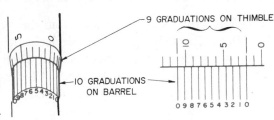

FIGURE 18-11

The difference between the width of one of the nine spaces on the thimble and one of the ten spaces on the barrel is one-tenth of one space.

Since each space on the thimble represents one-thousandth of an inch, the difference between the graduation on the thimble and barrel is one-tenth of one-thousandth or one ten-thousandth ($\frac{1}{10}$ of $\frac{1}{1,000} = \frac{1}{10,000} = .0001''$).

RULE FOR READING A VERNIER MICROMETER

- Read as a standard micrometer graduated in thousandths.
- Add to this reading the number of the line on the vernier scale of the barrel that coincides with a line on the sleeve. This number gives the ten-thousandths to be added.

EXAMPLE: The micrometer reading in the illustration in thousandths is (.281″). Add to this reading the number of the line on the vernier scale that coincides with a line on the thimble. In this case, the first line coincides. This indicates one ten-thousandth. The second line would be two-tenths of one-thousandth, the third line three-tenths. These tenth readings continue until the zero line on the barrel coincides with a line on the thimble. In this position the fourth-place number in the reading is zero.

To read the vernier micrometer in Figure 18-12, correct to four decimal places:

Step 1   Take regular reading.
Step 2   Determine which lines on vernier scale coincide.   **.281**
Step 3   Add regular reading to vernier reading.   **.0001**
                                                          .2811″   Ans

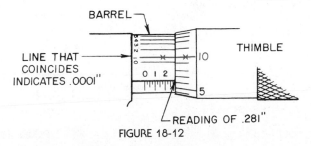

FIGURE 18-12

E. PRINCIPLES OF VERNIER CALIPER MEASUREMENT

The vernier caliper differs from the vernier micrometer in construction and principle of operation. The reading on the vernier caliper is not obtained by any relationship between the pitch of a screw and the movement of a thimble. Instead, the vernier caliper legs are slid into position. They are accurately adjusted for measurement by means of a fine screw that moves a sliding leg on a beam. The measurement is then determined by adding the reading on the beam and a graduation on the vernier scale.

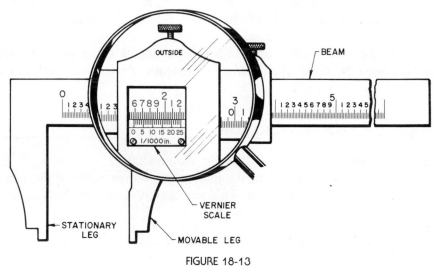

FIGURE 18-13

By varying the design of the stationary or solid leg, the beam can be used for other measuring needs as a height gage and depth gage. In many instances the beam is fitted to the table of a machine and the scale to a stationary part. Accurate linear measurements for machine operations may thus be taken with greater ease and less chance of error than by using the graduated collar on machines (Figure 18-13).

Each inch on the beam of a vernier caliper is divided into 40 equal parts. The distance between each graduation is $\frac{1''}{40}$ or .025″.

The vernier scale has 25 divisions that correspond to 24 divisions on the beam (Figures 18-14 and 18-15). The difference between one division on the scale and one on the beam is $\frac{1}{25}$ of .025″ or .001″.

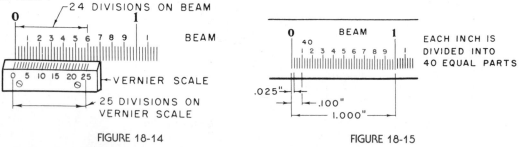

FIGURE 18-14 FIGURE 18-15

If the movable or sliding leg of the caliper (to which the vernier is attached) is moved to the right until the first line on the beam and vernier coincide, the leg will open .001″ (Figure 18-16).

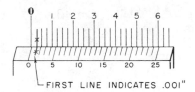

FIGURE 18-16

If this movement continues until the jaw is opened (.025″), the zero line on the vernier scale and the first line on the beam will coincide (Figure 18-17).

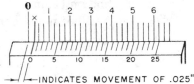

FIGURE 18-17

RULE FOR MEASURING WITH THE VERNIER CALIPER

- Read the graduation on the beam to the left of the zero on the vernier scale.
- Determine the number of the line on the vernier that coincides with a line on the beam.
- Add the reading on the beam to that of the vernier.

EXAMPLE: *Case 1.*

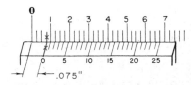

FIGURE 18-18

Step 1	Read graduation on beam to left of zero on vernier scale (Figure 18-18).	**.075**
Step 2	Determine which line on vernier scale coincides with line on beam.	**.000**
Step 3	Add beam and vernier scale readings.	**.075″ Ans**

EXAMPLE: *Case 2.*

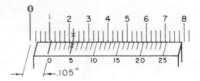

FIGURE 18-19

Step 1	Read graduation on beam to left of zero on vernier scale (Figure 18-19).	**.100**
Step 2	Determine number of line on vernier scale that coincides with line on beam.	**.005**
Step 3	Add beam and vernier scale readings.	**.105″** Ans

EXAMPLE: *Case 3.*

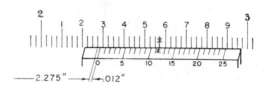

FIGURE 18-20

Step 1	Read graduation on the beam to left of zero on vernier scale in Figure 18-20 (two inches plus .275).	**2.275**
Step 2	Determine number of line on vernier scale that coincides with line of beam.	**.012**
Step 3	Add beam and vernier scale readings.	**2.287″** Ans

F. PRINCIPLES OF PRECISION GAGE BLOCK MEASUREMENT

Gage blocks are hardened rectangular- or square-shaped blocks of steel. Gage blocks have a high-quality surface finish and are dimensionally accurate within millionths of an inch.

The dimensional accuracy of a gage block relates to its *size, degree of flatness, parallelism* between the reference faces, and the *quality of* the *surface finish*. In general, the highest quality newer gage blocks are dimensionally accurate to within ±0.000 002″. This high precision level is required for calibrating instruments and other laboratory and scientific work under temperature-controlled conditions.

Gage blocks that are used for regular layout and precise linear measurements in the shop are dimensionally accurate to within ±0.000 004". Blocks earlier produced are accurate to within ±0.000 008".

Gage blocks are available in different combinations of sizes and numbers in a *set*. Fractional gage blocks range from $\frac{1}{64}$" to 12.000" by sixty-fourths. Other gage block sets are produced in increments (steps) of 0.001", 0.0001", or 0.000 025", depending on the required degree of accuracy and range of linear measurements. Each set provides for a tremendous range. For example, it is possible to make more than 80,000 different linear combinations with a 35-gage block set.

In practice, the combination of blocks to use for a specified dimension is determined by the addition and subtraction of decimals.

RULE FOR DETERMINING GAGE BLOCK COMBINATIONS

- Determine from the required dimension what number is in the ten-thousandths column.
- Select the gage block from the series .1001 to .1009 whose fourth-place number is the same as the required dimension.
- Select any gage block from the .001 series whose third-place (thousandths) dimension ends in the same number of thousandths as the required dimension.

  Note. Make sure that the sum of the two blocks is not greater than the required dimension.
- Select any gage block from the .01 series whose second-place dimension (hundredths) ends in the same number of hundredths as the right dimension.
- Continue to add blocks until the sum of the gages equals the required dimension.

EXAMPLE: Determine the combination of blocks needed for a measurement of 5.9325″ using the 35-piece set indicated below. (See Figures 18-21 and 18-22.)

Series	Thicknesses of English Measurement Gage Blocks in Inches								
.0001 Series	.1001	.1002	.1003	.1004	.1005	.1006	.1007	.1008	.1009
.001 Series	.101	.102	.103	.104	.105	.106	.107	.108	.109
.010 Series	.110	.120	.130	.140	.150	.160	.170	.180	.190
.100 Series	.100	.200	.300	.400	.500				
1.000 Series	1.000	2.000	3.000						

FIGURE 18-21

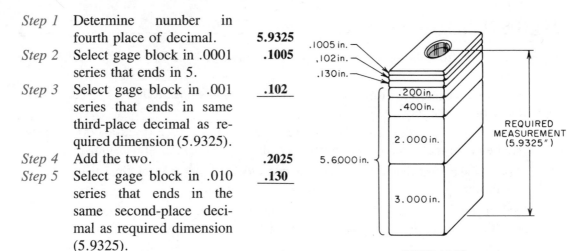

Step 1	Determine number in fourth place of decimal.	**5.9325**	
Step 2	Select gage block in .0001 series that ends in 5.	**.1005**	
Step 3	Select gage block in .001 series that ends in same third-place decimal as required dimension (5.9325).	**.102**	
Step 4	Add the two.	**.2025**	
Step 5	Select gage block in .010 series that ends in the same second-place decimal as required dimension (5.9325).	**.130**	
Step 6	Add the three gages.	**.3325**	

FIGURE 18-22

Step 7 Subtract the decimal value of the gages (.3325″) from the required dimension (5.9325″). = 5.600″

Step 8 Select the smallest number of blocks in the .100 and 1.000 series to obtain this number. 5.600″ = 3.000″ + 2.000″ + .400″ + .200″

Step 9 Check the gage block combination for accuracy. Also, determine whether or not a combination having fewer blocks may be used.

ASSIGNMENT UNIT 18   REVIEW AND SELF TEST

PRETEST   *The Review and Self-Test items that follow may be used as a pretest. Pretests are designed to measure a student's beginning level of mathematical skills competency and to determine the starting point of instruction.*

A. Direct Measurement (Practical Problems)

1. Determine the reading of each measurement.

a.

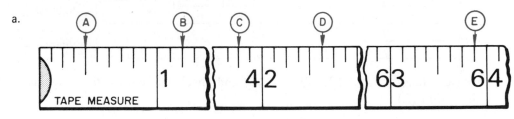

b.

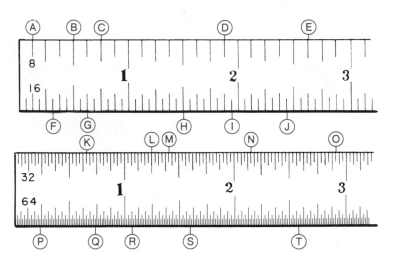

2. Determine the reading of each measurement indicated on the rule, which is graduated in 10ths, 50ths, and 100ths of an inch.

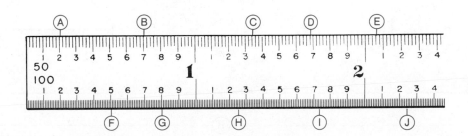

3. Measure the length of lines (a) through (j) to the degree of accuracy indicated in each case.

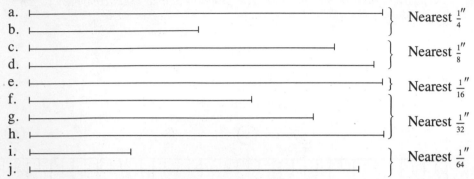

a. $\qquad\qquad\qquad$ } Nearest $\frac{1}{4}''$
b. $\qquad\qquad$

c. $\qquad\qquad\qquad$ } Nearest $\frac{1}{8}''$
d. $\qquad\qquad\qquad$

e. $\qquad\qquad\qquad$ } Nearest $\frac{1}{16}''$
f. $\qquad\qquad$

g. $\qquad\qquad\qquad$ } Nearest $\frac{1}{32}''$
h. $\qquad\qquad\qquad$

i. $\qquad$ } Nearest $\frac{1}{64}''$
j. $\qquad\qquad$

4. Measure lengths Ⓐ, Ⓑ, Ⓒ, and Ⓓ and check the sum of these against the overall dimension Ⓔ.

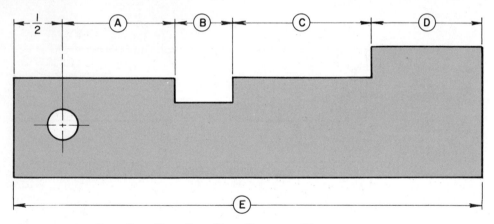

5. Measure lengths Ⓐ, Ⓑ, Ⓒ, Ⓓ, Ⓔ, Ⓕ, and Ⓖ and check the sum against the overall dimension Ⓗ.

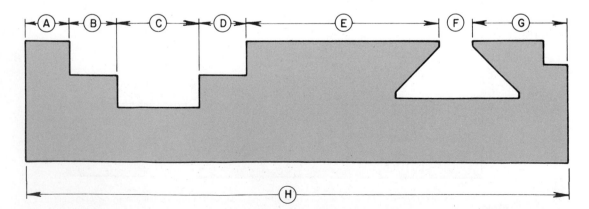

6. Measure the outside diameters of the metal bars Ⓐ, Ⓑ, Ⓒ, Ⓓ, and Ⓔ with a caliper. Transfer the measurement to a rule and record the diameter to the nearest $\frac{1}{50}''$.

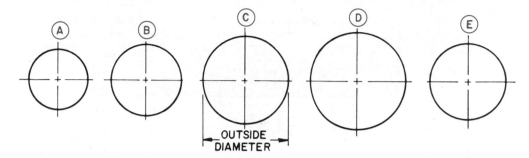

7. Measure the length of lines (a) through (d) to the degree of accuracy indicated in each case.

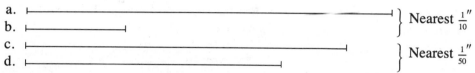

a. } Nearest $\frac{1}{10}''$
b.
c. } Nearest $\frac{1}{50}''$
d.

8. Measure the diameter of bored holes Ⓐ, Ⓑ, Ⓒ, Ⓓ, Ⓔ, and Ⓕ with an inside caliper. Then transfer the measurement to a rule and record the diameter to the nearest $\frac{1}{32}''$.

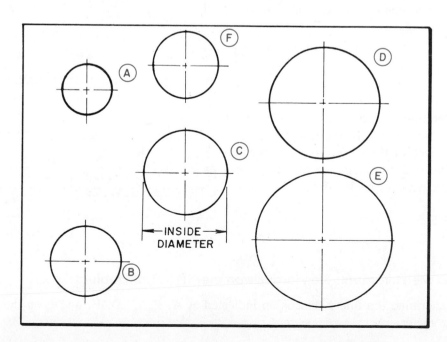

B. Indirect Measurement (Practical Problems)

1. Determine dimensions Ⓐ, Ⓑ, Ⓒ, Ⓓ, Ⓔ, and Ⓕ.

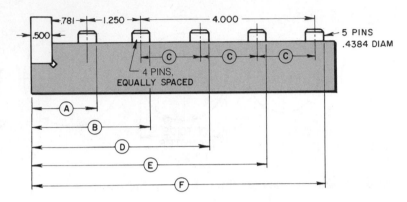

2. Find dimensions Ⓐ through Ⓙ to which the shaft must be rough turned. Allow $\frac{1}{16}''$ on all diameters and $\frac{1}{32}''$ on all faces for finish machining.

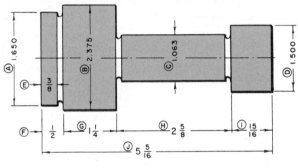

3. Calculate dimensions Ⓐ through Ⓙ to which the part must be finish turned before grinding. Allow .010″ on all diameters and .008″ on all faces for grinding.

4. Determine (a) the outside perimeter of the house and (b) the inside perimeter of each of the six rooms. The perimeter equals the sum of the linear measurements of each side of an object.

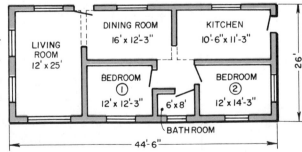

C. Direct Measurement: Standard Micrometer (Practical Problems)

1. Determine the linear dimension indicated at A, B, C, D, E, and F on the standard micrometer.

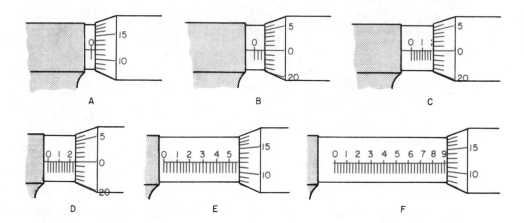

D. Direct Measurement: Vernier Micrometer

1. Determine the linear dimension on the vernier micrometer settings at A, B, and C.

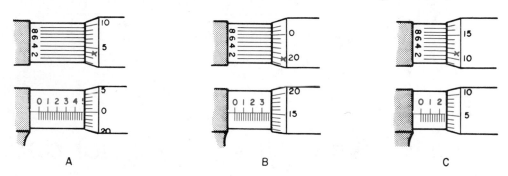

E. Direct Measurement: Vernier Caliper (Practical Problems)

1. Determine the vernier caliper readings A, B, C, and D.

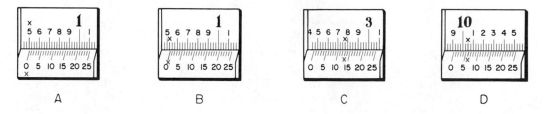

F. Direct and Computed Measurement: Gage Blocks (Practical Problems)

1. The set of 81 gage blocks that may be used to make 120,000 accurate combinations of measurements in steps of .0001″ from .200″ to over 24″ is listed in table form.

Series	Thicknesses of Gage Blocks (in inches)									
.0001 Series	.1001	.1002	.1003	.1004	.1005	.1006	.1007	.1008	.1009	
.001 Series	.101	.102	.103	.104	.105	.106	.107	.108	.109	
	.110	.111	.112	.113	.114	.115	.116	.117	.118	.119
	.120	.121	.122	.123	.124	.125	.126	.127	.128	.129
	.130	.131	.132	.133	.134	.135	.136	.137	.138	.139
	.140	.141	.142	.143	.144	.145	.146	.147	.148	.149
.050 Series	.050	.100	.150	.200	.250	.300	.350	.400	.450	
	.500	.550	.600	.650	.700	.750	.800	.850	.900	.950
1.000 Series	1.000	2.000	3.000	4.000						

Determine the sizes of gage blocks that, when combined, will give the required dimensions A through J.

A	.500
B	.750
C	.875
D	.265

E	.6001
F	.7507
G	1.2493

H	2.2008
I	11.0049
J	11.885

Unit 19 Principles of Angular and Circular Measure

OBJECTIVES OF THE UNIT

After satisfactorily studying this unit, the student/trainee will be able to

- *Interpret parts and features of circles and angles in relation to computing circular and angular measurements.*
- *Express angular measurements in terms of degrees, minutes, and seconds.*
- *Compute the circular length of an arc.*
- *Measure and lay out angles by using a flat semicircular protractor or a more accurate bevel protractor.*

PRETEST   *Use the Review and Self-Test items provided in the Unit Assignment to establish the level of mathematical skills competency and to determine the starting point of instruction.*

The measurement of circles, curved surfaces, cylinders, and angles takes place daily in the home, in business, and in industry. Measuring these rounded surfaces calls for an understanding of the fundamental principles relating to circular and angular measurement.

Some problems require the application of a principle to compute a missing value or dimension. Under actual conditions, measurements are made directly on the job, using tools and instruments that will measure to the required degree of accuracy.

This unit gives step-by-step procedures for computing answers (where basic principles alone are applied) and for direct measurement with tools and instruments.

A. DEVELOPING A CONCEPT OF CIRCULAR AND ANGULAR MEASURE

Defining the Circle and Its Parts

A circle is defined as a closed, curved line on a flat surface. Every point on the closed, curved line is the same distance from a fixed given point called a *center*. The distance around the circle, or the periphery of the circle, is the *circumference*. This circumference is measured either in terms of the standard units of linear measure or in degrees.

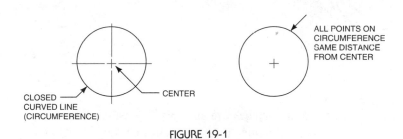

FIGURE 19-1

The *diameter* is the straight line through the center of the circle. Its ends terminate in the circumference. The diameter divides the circle into two equal half circles called *semicircles*. The *radius* is a straight line starting at the center of the circle and terminating in the circumference. The symbol *(R)* or *(r)* is usually used to indicate radius. The term *radial* is in common usage. A line or surface on a circular object is radial if, when extended, it cuts through the center.

In all circles, the circumference is related to the diameter in a specific way. It is approximately $3\frac{1}{7}$ or 3.1416 times the diameter. This value is a constant relationship between the diameter and circumference of the same circle. For convenience it is called "pi" or π.

To find the circumference of a circle, multiply the diameter by the value of π.

On many objects, only a *segment* of a circle is used. The segment refers to the area of that portion of the circle that is being considered. The segment consists of an *arc* on the circumference whose ends are joined by a straight line in the circle.

An *arc* is a curved part of the circumference. The straight line that extends inside the circle to join both ends of the arc is known as a *chord*.

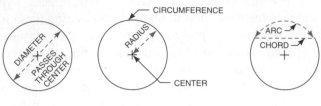

FIGURE 19-2

Two other terms are widely used: *concentric circles* and *eccentric circles*. When two or more circles have a common center, they are *concentric*. Circles that are on the same flat surface and are used in the same part but do not originate from a common center are called *eccentric circles*. On concentric circles the inner circle is sometimes designated by the letters *(ID)* for inside diameter; the outer circle is designated by the letters *(OD)* for outside diameter.

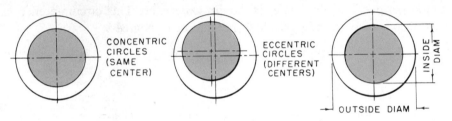

FIGURE 19-3

Defining the Angle and Its Parts

Most measurements of round surfaces are in linear measure. However, many dimensions are expressed in angular measure. Angular measure indicates the size of an opening formed by two lines or surfaces (intersecting lines). The lines or surfaces open as they extend from a common starting point called a *vertex*. The angles formed are measured in *degrees, minutes,* and *seconds*.

The size of an angle depends upon the space between the sides and not on the length of the sides. In speaking of an angle, it may be referred to as $\angle ABC$, or $\angle 1$, or $\angle A$ and the angle is marked or lettered accordingly.

A circle is divided into 360 equal parts that are called degrees. An angle of one degree is formed by drawing two lines from the center to two consecutive points that cut $\frac{1}{360}$ of the circumference.

CIRCLE =
360 DEGREES

$\frac{1}{360}$ OF CIRCLE =
1 DEGREE

1 DEGREE

FIGURE 19-4

Degrees are designated by placing a small symbol to the right and slightly above the number: thus, 4 degrees is 4°. Sometimes degrees are given in tenths. For example, four and three-tenths degrees is written 4.3°.

Each degree is divided into 60 equal parts, each of which is called a *minute*. A degree therefore equals 60 minutes. Minutes are indicated by placing a single line in the same relative position as the degree sign. Thus, 15 minutes is written 15′.

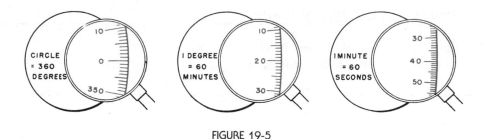

FIGURE 19-5

For greater accuracy, measurements are made in divisions finer than the minute. Each minute is divided into 60 equal parts, each of which is called a *second*. Seconds are written with two lines to the right and slightly above the number. An angle ending in 25 seconds is written 25″.

To summarize, angles are measured in degrees (°), minutes ('), and seconds (″). There are 360 degrees in a circle, 60 minutes in a degree, and 60 seconds in a minute (Figure 19-6).

Table of Angular Measure	
1 Circumference (or circle)	= 360 degrees
1 Degree	= 60 minutes
1 Minute	= 60 seconds

FIGURE 19-6

The circle is very often divided into four equal parts, sometimes called quadrants, each of which represents an angle of 90°. The term *right angle* is used to denote such an angle. When the circle is divided into two equal parts, the 180° for each part is called a *straight angle* (Figure 19-7).

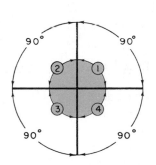

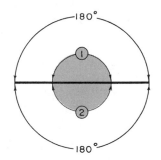

THE CIRCLE DIVIDED INTO 4 EQUAL PARTS OF 90° OR 4 RIGHT ANGLES OR QUADRANTS ①, ②, ③ AND ④

THE CIRCLE DIVIDED INTO 2 EQUAL PARTS OF 180° OR 2 STRAIGHT ANGLES ① AND ②

FIGURE 19-7

B. COMPUTING ANGULAR MEASURE

Problems in angular measure involve the four basic mathematical processes of addition, subtraction, multiplication, and division of angles and parts of angles. To do these operations, it is necessary to change degrees, minutes, and seconds to whichever one of the three units will expedite the mathematical processes involved. Angles given in fractional or decimal parts of a degree, minute, or second may be added, subtracted, multiplied, or divided the same as any fraction or decimal.

RULE FOR EXPRESSING DEGREES IN MINUTES

- Multiply the number of degrees by 60.
- Express the product in terms of minutes.

EXAMPLE: Express 15 degrees (15°) in minutes.
Step 1   Multiply 15 degrees (15°) by 60.   **15 × 60 = 900**
Step 2   Express answer in minutes (').   **15° = 900'   Ans**

RULE FOR EXPRESSING MINUTES IN SECONDS

- Multiply the number of minutes by 60.
- Express the product in terms of seconds.

EXAMPLE: Change 32 minutes (32') to seconds.
Step 1   Multiply 32 minutes (32') by 60.   **32 × 60 = 1920**
Step 2   Express answer in seconds (").   **32' = 1920"   Ans**

RULE FOR EXPRESSING AN ANGLE IN DEGREES AND MINUTES AS SECONDS

- Multiply the given number of degrees by 60.
  Note. The product denotes minutes.
- Multiply the product by 60.
  Note. The product denotes seconds in the given number of degrees.
- Multiply the given number of minutes by 60.
- Add the number of seconds in the given number of degrees to the number of seconds in the given number of minutes.
- Express the answer in terms of seconds.

EXAMPLE: *Case 1*. Change 3°10' to seconds.   **3 × 60 = 180'**
Step 1   Multiply 3° by 60.   **180 × 60 = 10,800"**
Step 2   Multiply 180' by 60.   **10' × 60 = 600"**
Step 3   Express ten minutes 10' as seconds.   **3° = 10,800"**
Step 4   Add values of 3° and 10' in seconds.   **+ 10' =   600"**

                                                      11,400"   Ans

EXAMPLE: *Case 2.* Express 2.75° as minutes.
- *Step 1*   Multiply 2.75° by 60.   **2.75 × 60 = 165.00**
- *Step 2*   Locate the decimal point.   **2.75° = 165′   Ans**

EXAMPLE: *Case 3.* Express 7.12° as minutes and seconds.   **7.12 × 60 = 427.20**
- *Step 1*   Multiply 7.12° by 60.   **.20 × 60 = 12.00**
- *Step 2*   Point off decimal places.   **7.12° = 427′12″   Ans**
  Note. The whole number of minutes is 427. The remainder (.20′) must be changed to seconds.
- *Step 3*   Multiply the decimal part of the minutes (.20) by 60 and locate the decimal point.
- *Step 4*   Combine the number of minutes and seconds.

RULE FOR EXPRESSING MINUTES AS DEGREES

- Divide the number of minutes by 60.
- Express thc quotient in terms of degrees.
  Note. Seconds are changed to minutes in the same way.

EXAMPLE: Express 45′ as degrees.
- *Step 1*   Divide the minutes (45′) by 60.
- *Step 2*   Express the quotient as degrees $\left(\frac{30}{4}\right)$. $\frac{45}{60} = \frac{30}{4}$ **Ans**
  Note. This quotient may also be written as (.75°).

C. COMPUTING CIRCULAR LENGTH

There are many cases where the length of an arc must be computed. This length is equal to the length of a portion of the circumference that is included in a given angle.

RULE FOR MEASURING THE LENGTH OF AN ARC (CIRCULAR LENGTH)

- Determine the number of degrees in the arc.
- Determine the diameter of the circle.
- Place the number of degrees in the angle over the number of degrees in the circle (360°) to determine what part the included angle is of the whole circle.
- Multiply this quantity by the circumference of the circle.
  Note. The circumference is equal to its diameter multiplied by 3.1416. The result is in the same dimension as the diameter.

EXAMPLE: What is the length of an arc included in an angle of 45° when the diameter of the circle is 2 inches?

Step 1   Determine size of angle (45°) and diameter of circle (2″).

Step 2   Place number of degrees in angle (45°) over the number of degrees in circle (360°).

$$\frac{45}{360} = \frac{1}{8}$$

Step 3   Compute circumference of (2″) circle.

Circumference =
2 × 3.1416 = 6.2832

Step 4   Multiply the circumference (6.2832) by the fractional part of circumference in the included angle $\left(\frac{1}{8}\right)$.

6.2832 × $\frac{1}{8}$ = .7854

Step 5   The product is the required length of the arc in inches.

.7854″   Ans

D. APPLICATION OF THE PROTRACTOR TO ANGULAR MEASUREMENTS

Three general types of instruments, called *protractors,* are used to measure and lay out angles. The simplest type of protractor used for comparatively rough work is flat, semicircular in shape, and graduated in degrees. Each tenth division is usually marked for ease in reading. The semicircular protractor finds wide application in laying out and measuring simple angles.

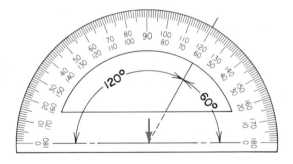

FIGURE 19-8

In the shop and on the job, the bevel protractor is the one most widely used. This protractor can be adjusted accurately to lay out or measure any angle from 0° to 180° by degrees. It is possible to estimate fairly accurately to a half degree or 30′. The number of degrees is read directly from the graduations on a movable turret. The swinging blade protractor is sometimes more convenient to use because of its simplicity of construction. Here again, the degrees from 0° to 180° are read at a center mark on one of the blades.

The third type of angle measuring tool is the universal vernier bevel protractor. This instrument is equipped with a vernier scale for very accurate measurement and layout of angles in terms of degrees and minutes.

ASSIGNMENT UNIT 19   REVIEW AND SELF TEST

PRETEST *The Review and Self-Test items that follow may be used as a pretest. Pretests are designed to measure a student's beginning level of mathematical skills competency and to determine the starting point of instruction.*

POSTTEST *The Review and Self-Test items may also be applied as a post test. Post tests are planned to establish the student's level of mathematical skills competency after instruction.*

A. Determining Equivalent Values of Angles (General Applications)

1. How many degrees are there in the following parts of a circle?
   a. $\frac{1}{6}$ b. $\frac{2}{9}$ c. $\frac{3}{8}$ d. $\frac{5}{36}$ e. $\frac{7}{24}$

2. What part of a circle is each of the following angles?
   a. 40°   b. 108°   c. 9°   d. 54°   e. 96°

3. Change each angular measurement (a) through (h) to its equivalent value in the unit of measure indicated.
   a. 30′ to degrees
   b. 45′ to degrees
   c. 75′ to degrees
   d. 5° to minutes
   e. $7\frac{1}{2}°$ to minutes
   f. 1.4° to minutes
   g. 5′ to seconds
   h. $5\frac{1}{2}'$ to seconds

B. Computation of Circular and Angular Measurements (General Applications)

1. Multiply and reduce each result to degrees, minutes, and seconds in whichever combination is needed.
   a. 5° × 6
   b. 17° × 15
   c. $12\frac{1}{2}°$ × 8
   d. 12.75′ × 6
   e. 9° 12′ × 9
   f. 23° 16′ 20″ × 4

2. Divide each angular measurement.
   a. 180° ÷ 9
   b. 135° ÷ 10
   c. 120° ÷ 9
   d. 90° 30′ ÷ 6
   e. 75.8° ÷ 10
   f. 144° 24′ 48″ ÷ 12

C. Measurement of Angles: Direct and Computed (Practical Problems)

1. Read angles Ⓐ through Ⓗ on the semicircular protractor.

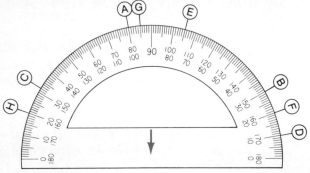

2. Measure angles Ⓐ through Ⓕ with a flat semicircular protractor.
   Note. It may be necessary to extend the sides of the angles to measure them.

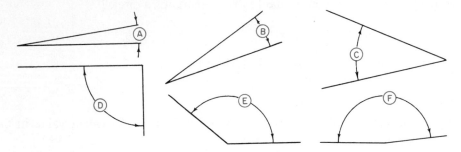

3. Lay out angles (a) through (d) with a semicircular protractor.
   a. 10°   b. 25°   c. 100°   d. 120°

4. The bevel protractor readings A, B, and C appear on one make of instrument; those at D, E, and F on another. Determine the reading for each setting.

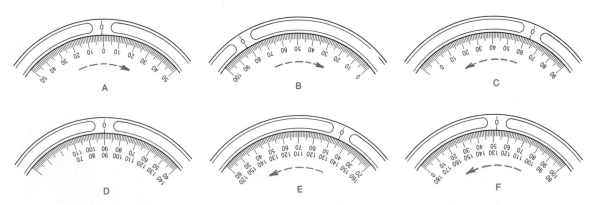

---

5. The table and the illustration relate to four engine strokes. The angular valve opening/closing or movement of the crankshaft are given. Compute the missing angular (X°) measurements for strokes A, B, C, and D.

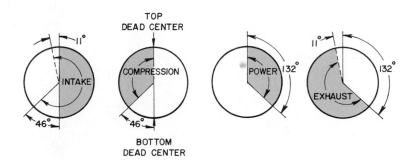

Stroke		Exhaust Valve		Angular Movement of Crankshaft
		Opening	Closing	
A	Intake	11° before top dead center	46° after top dead center	(X°)
B	Compression	top dead center	46° past bottom dead center	(X°)
C	Power	top dead center	(X°) before bottom dead center	132°
D	Exhaust	(X°) end of power stroke	11° before top dead center	132°

D. Computation of Circular Length (Practical Problems)

1. Compute the length of arcs A through D according to the given diameter for each part.

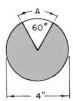 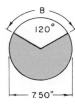 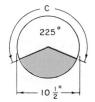 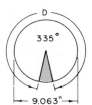

2. The diameters of three roof plates are (a) 7'-0" (b) 8'-6", and (c) 29'-1", respectively. Use $\pi = 3.1416$. Give the length (circumference) of each circular roof plate to the nearest inch.

3. Find the semicircular length of forms A, B, and C. State each length in feet and inches.

Form	Diameter	Semicircular Length
A	3'-0"	
B	$4\frac{1}{3}'$	
C	5.25'	

Unit 20 Principles of Surface Measure

OBJECTIVES OF THE UNIT

After satisfactorily studying this unit, the student/trainee will be able to
- *Understand the concept of square and surface measure.*
- *Change one unit value of surface measure to another unit.*
- *Express larger units of surface measure in terms of smaller units.*
- *Solve general and practical problems relating to finding the areas of rectangles, parallelograms, trapezoids, triangles, circles and sectors, and cylinders.*

PRETEST   *Use the Review and Self-Test items provided in the Unit Assignment to establish the level of mathematical skills competency and to determine the starting point of instruction.*

The term *surface measure* refers to the measurement of an object or part that has length and height. There are eight common objects or shapes for which area or surface measure must be computed. The list includes surfaces that are defined by pairs of lines like the square, rectangle, and parallelogram; by four lines like the trapezoid; by three lines like the triangle; then the circle and sector; and finally, the cylinder. The characteristics by which each of these shapes is recognized and defined are covered separately in this unit.

A. DEVELOPING A CONCEPT OF SQUARE OR SURFACE MEASURE

A surface is any figure that has length and height but no thickness. To measure a surface, its length and height must be in the same unit before they are multiplied. The result of this mathematical process is called the *area* of the surface. The area is expressed in square units of the same kind as the linear units. For example, if the length and height of an object are given in inches, the area will be in square inches. In this case, the surface contains a number of square inches. One square inch is the area a square figure measures that is one linear inch long and one linear inch high.

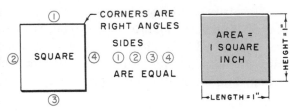

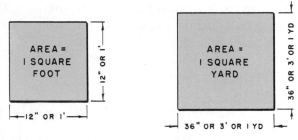

FIGURE 20-1

One square foot is the area a square figure measures that is 12 linear inches long and 12 linear inches high. One square yard is the area a square figure measures that is 36 linear inches long and 36 linear inches high.

In each case, the area of the square is found by multiplying the linear length by the linear height. Both dimensions must be in the same unit of measure.

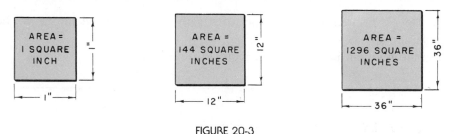

FIGURE 20-2

By comparison, the area of a square one inch on a side is 1 square inch. The area of a square 12 inches on a side is 144 square inches. The area of a square 36 inches on a side is 1,296 square inches.

FIGURE 20-3

The area of the second square illustrated in Figure 20-3 is 144 square inches. Since this value is obtained by multiplying the length and height (each of which is equal to one foot), the area may be given as 1 square foot. The square foot is, therefore, equivalent to 144 square inches.

By the same reasoning, in the third square in Figure 20-3 the 36 inches = 3 feet, which equals 1 yard. The area of the square, therefore, is equal to 1,296 square inches, or 9 square feet, or 1 square yard.

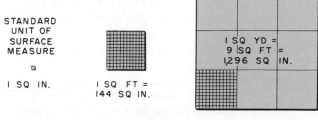

FIGURE 20-4

In place of continually writing the word "square" and the unit of measure, the symbol □ is used to indicate *square*. This symbol is followed by other symbols for inches (□″), feet (□′), or yards (□ yd). Another way of writing areas is to abbreviate the word square (sq) and the units of measure: (in.) for inches, (ft) for feet, and (yd) for yard. One square inch may be written 1 sq in. or 1 □″; one square foot, 1 sq ft or 1 □′; and one square yard, 1 sq yd or 1 □ yd.

The three common units of surface measure that find daily application are the square inch, the square foot, and the square yard. The value of each of these units is shown in Figure 20-5.

Table of Surface Measure
Unit of Surface Measure = 1 sq in.
144 sq in. = 1 sq ft
9 sq ft = 1 sq yd

FIGURE 20-5

B. CHANGING FROM ONE UNIT VALUE OF SURFACE MEASURE TO ANOTHER

A value expressed in square inches may be changed to square feet by dividing by 144 (144 sq in. = 1 sq ft). By the same process, an area given in square feet may be changed to square yards by dividing by 9 (9 sq ft = 1 sq yd). If an area is given in inches, the value may be changed to square yards or a fractional part by dividing by 1,296 (1,296 sq in. = 1 sq yd).

RULE FOR EXPRESSING A UNIT OF SURFACE MEASURE AS A LARGER UNIT

- Divide the given area by the number of square units contained in one of the required larger units.
- Express the quotient in terms of the required larger unit.

EXAMPLE: *Case 1*. Express 288 sq in. in sq ft.

Step 1   Divide given area by the number of sq in. in a sq ft (144).

$$\frac{288}{144} = 2$$

Step 2   Express quotient (2) in terms of required unit (sq ft).   **288 sq in. = 2 sq ft**   Ans

EXAMPLE: *Case 2*. Express 27 sq ft in sq yd.

Step 1   Divide given area by the number of sq ft in a sq yd (9).

$$\frac{27}{9} = 3$$

Step 2   Express quotient (3) in terms of required unit (sq yd).   **27 sq ft = 3 sq yd**   Ans

EXAMPLE: *Case 3*. Express 2,592 sq in. in sq yd.

Step 1   Divide the given area by the number of sq in. in a sq yd (1,296).

$$\frac{2592}{1296} = 2$$

Step 2   Express quotient (2) in terms of required unit.   **2,592 sq in. = 2 sq yd**   Ans

When the given area cannot be divided exactly, then the quotient may be expressed in more than one unit. For example, if 12 square feet is changed to square yards, the result may be given as $1\frac{1}{3}$ square yards or as 1 square yard and 3 square feet. In other words, the fraction $\left(\frac{1}{3}\right)$ is expressed in terms of a smaller unit.

RULE FOR EXPRESSING LARGER UNITS AS SMALLER UNITS OF SURFACE MEASURE

- Multiply the given unit by the number of smaller units contained in one of the required units.
- Express the product in terms of the required smaller unit.

EXAMPLE: *Case 1.* Express 5 sq ft in sq in.

Step 1   Multiply given unit (5) by the number of smaller $5 \times 144 = 720$
units contained in one of the required units (144).

Step 2   Express the product (720) in terms of the required   **5 sq ft = 720 sq in.**   Ans
unit (sq in.).

EXAMPLE: *Case 2.* Express 3 sq yd in sq in.

Step 1   Multiply given unit (3) by the number of smaller units con-
tained in one of the required units (1,296). $3 \times 1{,}296 = 3{,}888$

Step 2   Express the product (3,888) in terms of the required unit   **3 sq yds =**
(sq in.).   **3,888 sq in.**   Ans

An area given in terms of two or more units (2 square feet 9 square inches) may be changed in value to a smaller unit. Only that portion of the area that is not in terms of the required unit is multiplied by the number of smaller units contained in one of the given units. The remainder of the given area is added to this product.

EXAMPLE: Express 2 sq ft 9 sq in. in sq in.

Step 1   Multiply only that portion of the given area that is   **2 sq ft =**
not in terms of the required unit. Multiply by the   **2 × 144 (sq in. in**
number of smaller units contained in one of the   **1 sq ft) = 288**
given units.

Step 2   Express the product in terms of the required unit   **= 288 sq in.**
(sq in.).

Step 3   Add the remainder of the given unit (9 sq in.).   **288 + 9 = 297 sq in.**   Ans

C. APPLYING SURFACE MEASURE TO THE SQUARE AND RECTANGLE

The area of a square surface may be found by multiplying the length by the height. Since both of these dimensions are equal, the area is equal to the side multiplied by itself.

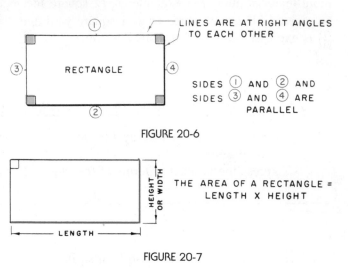

FIGURE 20-6

FIGURE 20-7

The term *rectangle* refers to a surface whose opposite sides are parallel. The adjacent sides are at right angles to each other. The area of this surface is the number of square units that it contains. In place of adding the number of square units contained in both the height and length, the process is simplified by multiplication.

RULE FOR FINDING THE AREA OF A RECTANGLE

- Express the dimensions for length and width (sometimes called *height*) in the same linear unit of measure.
- Multiply the length by the height.
- Express the product in units of surface measure.
- Express product in lowest terms, if required.

EXAMPLE: *Case 1.* Determine the area of a rectangle 9″ long by 3″ high.
Step 1 Multiply length by height. $9 \times 3 = 27$
Step 2 Express product (27) in terms of units of surface measure. **27 sq in.** Ans

EXAMPLE: *Case 2.* Determine the area of a rectangular surface 3′ long and 10″ wide.
Step 1 Express dimensions in same unit of linear measure. $3' = 36''$
Step 2 Multiply length (36″) by height (10″). $36 \times 10 = 360$ sq in.
Step 3 Express product (360) in units of surface measure.
Step 4 Express (360 sq in.) in lowest terms by dividing by (144). $\frac{360}{144} = 2$ sq ft and 72 sq in. Ans

D. APPLYING SURFACE MEASURE TO THE PARALLELOGRAM

A *parallelogram* has two pairs of sides that are parallel to each other. The sides are not necessarily at right angles as in the case of the rectangle. The parallelogram may be made into a rectangle, as illustrated by the shaded triangles in Figure 20-8, by cutting off the triangular surface (A) and placing it in position (B). It should be noted that the rectangle is a special kind of parallelogram in which the angles are right angles.

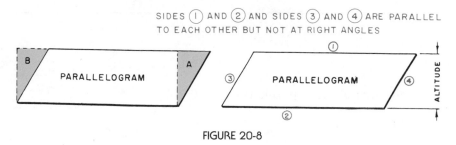

FIGURE 20-8

The rectangle thus formed has the same length base as the parallelogram. The height is the same as the altitude. Since the area of a rectangle is equal to the product of the length times the height, the area of a parallelogram is equal to the product of its base × altitude. The term *length* is often used instead of *base;* similarly, *height* is used for *altitude*. Regardless of which term is used, the process of finding the area is the same.

RULE FOR FINDING THE AREA OF A PARALLELOGRAM

- Express the dimensions for base and altitude in the same unit of linear measure, if needed.
- Multiply the base by the altitude.
- Express the product in units of surface measure.
- Express product, if needed, in lowest terms.

EXAMPLE: A parallelogram has a base 4 feet long and an altitude of 18 inches. Find its area.

Step 1   Express dimensions in same unit of measure.   **18 in. = $1\frac{1}{2}$ ft**
Step 2   Multiply the base and altitude.   **$4 \times 1\frac{1}{2} = 6$**
Step 3   Express product in units of surface measure.   **Area = 6 sq ft   Ans**

E. APPLYING SURFACE MEASURE TO A TRAPEZOID

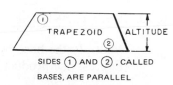

A *trapezoid* is a four-sided figure. Two of the trapezoid sides, called bases, are parallel (Figure 20-9).

FIGURE 20-9

RULE FOR FINDING THE AREA OF A TRAPEZOID

- Express the dimensions for the bases and altitude in the same unit of linear measure, if needed.
- Add the lengths of the two bases.
- Multiply the sum by one-half of the altitude.
- Express the product, if needed, in lowest terms.

EXAMPLE: Find the area of the trapezoid (Figure 20-10).

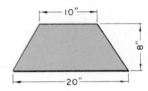

FIGURE 20-10

Step 1   Add the lengths of the two bases.   **10 + 20 = 30**
Step 2   Multiply the sum by $\frac{1}{2}$ of the altitude (8). $30 \times \frac{1}{2} \times 8 =$
                                                        Area = 120 sq in.   Ans

F. APPLYING SURFACE MEASURE TO A TRIANGLE

Two triangles of the same size and shape, when placed so that the longest side is their common side, form a parallelogram (Figure 20-11).

The opposite pairs of sides of this figure are parallel. The parallelogram thus formed has the same size base and altitude as the original triangle.

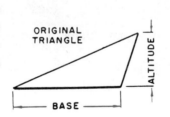
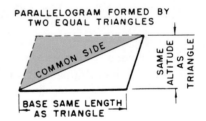

FIGURE 20-11

The area of the parallelogram is equal to the product of its base times altitude. Since the parallelogram is made up of two equal triangles, the area of each triangle is equal to one-half the area of the parallelogram. Thus, the area of a triangle may be computed directly by multiplying the base times $\frac{1}{2}$ the altitude.

RULE FOR FINDING THE AREA OF A TRIANGLE

- Multiply the base by $\frac{1}{2}$ the altitude.
- Express the product in units of surface measure.

EXAMPLE: Compute the area of a triangle having a 16-inch base
and an altitude of 6 inches.

Step 1   Multiply the base (16) by $\frac{1}{2}$ the altitude (6).   **$16 \times \frac{1}{2} \times 6 = 48$**

Step 2   Express the product in units of surface measure.   **Area $= 48$ sq in.   Ans**

G. APPLYING SURFACE MEASURE TO A CIRCLE AND A SECTOR

The Circle

If a circle is divided into an equal number of parts and the sectors thus formed are stretched out along two parallel lines and these two rows of sectors are brought together, a figure approaching a rectangle is formed. As the number of sectors into which the circle is divided is increased, the length of the rectangle approaches one-half the circumference of the circle.

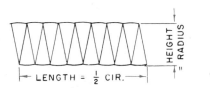
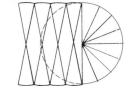

FIGURE 20-12

The area of the rectangle is equal to its *length* times *height*. The length is equal to $\frac{1}{2}$ the circumference of the circle. The *height* is equal to the radius of the circle. By substituting these values, the area of the rectangle formed is equal to its length $\left(\frac{1}{2}\right.$ the circumference$\left.\right)$ times height (radius of circle).

Area of Circle $= \dfrac{3.1416 \times \text{Diam of Circle}}{2} \times \text{Radius of Circle}$

Since the diameter of the circle is twice the radius, the area of a circle in terms of its diameter

$= \dfrac{3.1416 \times \text{Diam}}{2} \times \dfrac{\text{Diam}}{2}$ *or* $\dfrac{3.1416 \times \text{Diam} \times \text{Diam}}{4}$

$= .7854 \times \text{Diam} \times \text{Diam}$

The area of a circle may be expressed in simpler terms. However, the two forms as given will be used until these values are later expressed as formulas. Where the radius of a circle is given instead of the diameter

$$\text{Area} = \dfrac{3.1416 \times 2\,(\text{Radius})}{2} \times \text{Radius} = 3.1416 \times \text{Radius} \times \text{Radius}$$

In place of continually using 3.1416, the Greek letter pi (π) is used in writing a problem. The actual value is applied when working out a problem.

RULE FOR FINDING THE AREA WHEN THE DIAMETER IS GIVEN

- Multiply .7854 $\times$ diam $\times$ diam.
- Express the product (area of the circle) in units of surface measure.

EXAMPLE: Find the area of a circle whose diameter is 4 inches, correct to two decimal places.

Step 1	Multiply .7854 × Diam × Diam.	**.7854 × 4 × 4 = 12.5664**
Step 2	Round off product to two decimal places.	**= 12.57**
Step 3	Express the product in units of surface measure.	**Area = 12.57 sq in. Ans**

RULE FOR FINDING THE AREA WHEN THE RADIUS IS GIVEN

- Multiply 3.1416 × radius × radius.
- Express the product (area of the circle) in units of surface measure.

EXAMPLE: Find the area of a circle whose radius is 2.8″, correct to two decimal places.

Step 1	Multiply π × radius (2.8) × radius (2.8).	**3.1416 × 2.8 × 2.8 = 24.630144**
Step 2	Round off product to two decimal places.	**= 24.63**
Step 3	Express in units of surface measure.	**Area = 24.63 sq in. Ans**

The Sector

The *sector* of a circle is the surface or area between the center and circumference. The sector is included within a given angle.

The area of a sector is equal to the area of the circle divided by the fractional part of the whole circle occupied by the sector. The angle of the sector is usually expressed as an *included angle*.

The fractional part of a circle occupied by a sector equals the number of degrees in the included angle divided by the number of degrees in a circle (360°).

RULE FOR FINDING THE AREA OF A SECTOR

- Compute the area of the circle.
- Determine the fractional part of the circle that the sector occupies by dividing the angle of the sector by 360°.
- Multiply the area of the circle by this fraction.
- Express the result (area) in square measure.

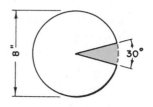

FIGURE 20-13

<u>EXAMPLE</u>:   Determine the area of the sector removed from the disc in Figure 20-13.

Step 1   Multiply (.7854 × Diam × Diam) to get the area of the circle.

$.7854 \times 8 \times 8 = 50.2656$

Step 2   Divide the number of degrees in the sector by 360°.

$\frac{30}{360} = \frac{1}{12}$

Step 3   Multiply the area of the circle (50.2656) by the fractional part occupied by the sector $\left(\frac{1}{12}\right)$.

$$\frac{\overset{4.1888}{\cancel{50.2656}}}{1} \times \frac{1}{\cancel{12}} = \mathbf{4.1888}$$

Step 4   Express the product in units of surface measure.

Area = 4.1888 sq in.   Ans

H. APPLYING SURFACE MEASURE TO THE SURFACE OF A CYLINDER

Occasionally, the area of a figure known as a *right cylinder* or a *cylinder* of *revolution* must be computed. These cylinders consist of two round bases of equal diameter and the vertical outside surface (called *lateral surface*) around the bases.

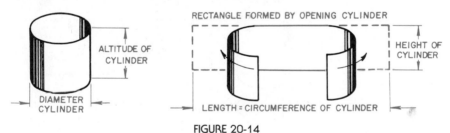

FIGURE 20-14

If the outside or lateral surface of the cylinder was unrolled, it would form a rectangle. The length of the rectangle is equal to the circumference of the cylinder. The rectangle height is the same as the altitude of the cylinder (Figure 20-14).

$$\text{The area of this lateral surface} = \text{Circumference} \times \text{Height}$$
$$= (\pi \times \text{Diameter}) \times \text{Altitude}$$

RULE FOR FINDING THE AREA OF THE SURFACE OF A CYLINDER

- Find the circumference of the base.
- Multiply the circumference by the altitude of the cylinder.
- Express the area of the lateral surface in units of surface measure.

EXAMPLE: Determine the area of the lateral surface of a cylinder 2 inches in diameter and 5 inches high.

Step 1	Circumference = 3.1416 × Diam.	**3.1416 × 2 = 6.2832**
Step 2	Multiply circumference (6.28232) by altitude (5).	**6.2832 × 5 = 31.4160**
Step 3	Express area (31.416) in units of linear measure.	**31.416 sq in.** Ans

The *total area of a cylinder* is equal to the area of the two bases plus the area of the lateral surface.

RULE FOR FINDING THE TOTAL AREA OF A CYLINDER

* Compute area of one base (.7854 × Diam × Diam).
* Multiply this area by 2 for two bases.
* Compute area of lateral surface (Circumference × Height).
* Add area of both bases to area of lateral surface.
* Express total area in units of surface measure.

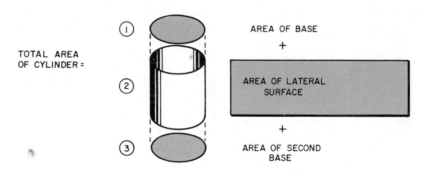

TOTAL AREA OF CYLINDER =

① AREA OF BASE

+

② AREA OF LATERAL SURFACE

+

③ AREA OF SECOND BASE

FIGURE 20-15

EXAMPLE: Determine the total area of a cylinder 4 inches in diameter and 10 inches in height, correct to two decimal places.

Step 1	Compute area of base (.7854 × Diam × Diam).	**.7854 × 4 × 4 = 12.5664**
Step 2	Multiply area of one base by 2.	**12.5664 × 2 = 25.1328**
Step 3	Compute area of lateral surface. Multiply the circumference of base times altitude.	**3.1416 × 4 × 10 = 125.6640**
Step 4	Add areas of two bases ⟶	**25.1328**
	to area of lateral surface ⟶	**125.6640**
	Total area =	**150.7968**
Step 5	Round off to two decimal places.	**Total area = 150.80 sq in.** Ans

ASSIGNMENT UNIT 20 REVIEW AND SELF TEST

PRETEST

The Review and Self-Test items that follow may be used as a pretest. Pretests are designed to measure a student's beginning level of mathematical skills competency and to determine the starting point of instruction.

POSTTEST

The Review and Self-Test items may also be applied as a post test. Post tests are planned to establish the student's level of mathematical skills competency after instruction.

A. Areas of Squares and Rectangles (General and Practical Problems)

1. Determine the areas of squares A, B, and C and rectangles D, E, and F.

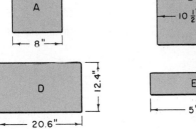

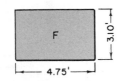
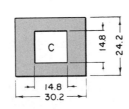

2. Find the cross-sectional areas of parts A and B and the area of the shaded portion of part C.

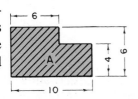
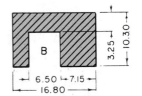

3. Find the total square foot area of the roof. The front portion is 46′ long × 18′ wide.

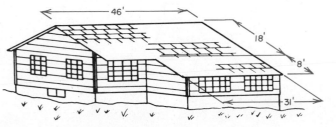

4. Assume the entire rectangular face of the carbon brush makes contact on a motor commutator. Calculate the contact surface areas of brushes A, B, and C. Round off answers to two decimal places.

Brush	Dimensions		Contact Area
	Length (L)	Width (W)	
A	$1\frac{1}{4}''$	$\frac{1}{2}''$	
B	$1\frac{1}{4}''$	$\frac{7}{8}''$	
C	$3\frac{3}{4}''$	$1\frac{1}{8}''$	

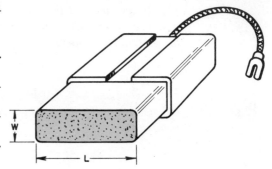

B. Areas of Parallelograms (Practical Problems)

1. Determine the areas of A, B, and C. All dimensions are in inches.

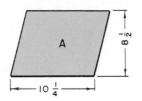
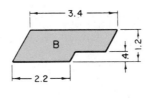
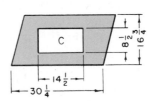

C. Areas of Trapezoids (Practical Problems)

1. Find the areas of stamped pieces A, B, and C that are shaped as trapezoids. Use the lengths of bases and altitudes given in each case. All dimensions are in inches.

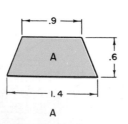

A

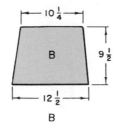

B

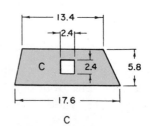

C

2. Determine the altitude of a trapezoid whose area is 122 square inches and whose bases are 12.2 inches and 18.3 inches.

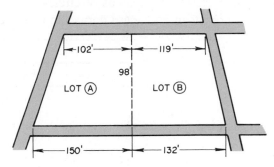

3. Compute the land area of each of the two trapezoidal-shaped lots Ⓐ and Ⓑ.

4. Compute the area in square feet in the end of the concrete retaining wall.

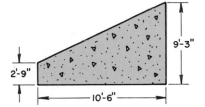

D. Areas of Triangles (Practical Problems)

1. Find the areas of triangles A, B, and C, and triangular stamping D. All dimensions are in inches.

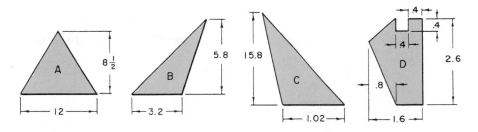

2. Compute the area of a triangu-lar louver that is 4'-6" wide × 2'-6" high.

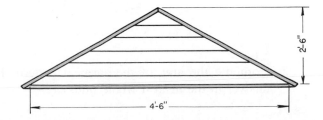

E. Areas of Circles (Practical Problems)

1. Determine the areas of circles A through F. The diameter or radius is given. Express the result in each case correct to two decimal places.

	A	B	C			D	E	F
Diameter	.8"	6"	5.25"	Radius		.5"	2"	3.8"

2. Find the area of stampings A and B. All dimensions are in inches.

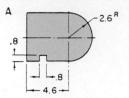

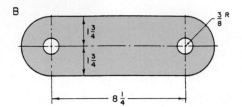

3. The brake lining of a magnetic disc motor brake has an outside diameter of 12". It is $2\frac{3}{4}''$ wide. Calculate the braking area of one disc.

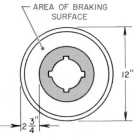

F. Areas of Sectors (Practical Problems)

1. Determine the area of the shaded areas A, B, and C, correct to three decimal places.

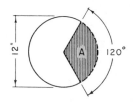
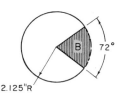
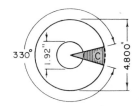

2. Compute the area of the garment pattern shown in the sketch. Give the answer in the nearest whole number.

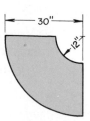

G. Areas of Lateral Surfaces and Cylinders (Practical Problems)

1. Determine the area of the lateral surface of cylinder A and the total area of cylinder B. The answers should be correct to two decimal places.

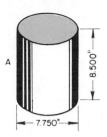

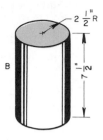

Unit 21 Principles of Volume Measure

OBJECTIVES OF THE UNIT

After satisfactorily studying this unit, the student/trainee will be able to
- *Understand the concept of volume measure.*
- *Express units of volume measure as larger, smaller, or combined units.*
- *Solve general and practical problems relating to the volumes of cubes, rectangular solids, cylinders, and irregular solid forms.*
- *Apply units of volume measure to solve problems involving liquid measure and interchange units of liquid and volume measure.*

 PRETEST   *Use the Review and Self-Test items provided in the Unit Assignment to establish the level of mathematical skills competency and to determine the starting point of instruction.*

Volume or cubic measure refers to the measurement of the space occupied by a body. Each body has three linear dimensions: length, height, and depth. The principles of volume measure are applied in this unit to three common shapes and the combinations of these three shapes: (1) the cube, (2) the rectangular solid, and (3) the cylinder.

A. DEVELOPING A CONCEPT OF VOLUME MEASURE

Volume measure is the product of three linear measurements. Each measurement must be in the same linear unit before being multiplied. The product is called the *volume* of the solid or body. Volume is expressed in cubic units of the same kind as the linear units. For example, if the length, height, and depth of a solid are given in inches, the volume will be in cubic inches.

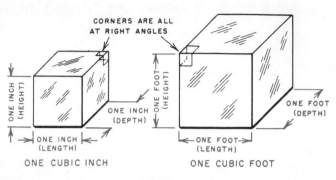

FIGURE 21-1

The standard unit of volume or cubic measure is the cubic inch. The *cubic inch* is the space occupied by a body (called a cube). The cube is one linear inch long, one inch high, and one inch deep.

One cubic foot is the space occupied by a cubical body that is one linear foot long, one foot high, and one foot deep. If 12 inches is used as the length of each side in place of the linear foot, the volume is equal to 1,728 cubic inches (12 × 12 × 12).

One cubic yard is the space occupied by a cube that is one linear yard long, one yard wide, and one yard deep. Using the equivalent value of the yard, namely, three feet, the volume is equal to 27 cubic feet (3 × 3 × 3).

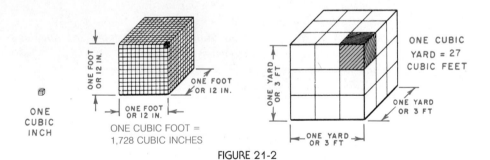

FIGURE 21-2

Cubic or volume measure is computed by multiplying three dimensions. By comparison, only two dimensions are multiplied in surface measure. The value of each of three standard units of volume measure may be compared. The number of cubic inches that the cubic foot and the cubic yard contain is illustrated (Figure 21-3).

The word cubic is abbreviated (cu) for ease in writing. It is followed by the symbol for the linear unit of measure. The three units of volume measure that are used most often are the cubic inch (cu in.), the cubic foot (cu ft), and the cubic yard (cu yd).

Table of Cubic or Volume Measure
Standard unit of measure = 1 cu in.
1,728 cu in. = 1 cu ft
27 cu ft = 1 cu yd

FIGURE 21-3

B. EXPRESSING UNITS OF VOLUME MEASURE

A volume in cubic inches may be expressed in cubic feet by dividing by 1,728 (1,728 cu in. = 1 cu ft). Volumes given in cubic feet may be expressed in cubic yards by dividing by 27 (27 cu ft = 1 cu yd).

RULE FOR EXPRESSING A UNIT OF VOLUME MEASURE AS A LARGER UNIT

- Divide the given volume by the number of cubic units contained in the required larger units.
- Express the quotient in terms of the required larger unit.

EXAMPLE: Express 5,184 cu in. as cu ft.

Step 1 Divide the given volume (5,184) by the number of cu in. in one cu ft. (1,728). $\frac{5184}{1728} = 3$

Step 2 Express the quotient (3) in terms of the required larger unit. **5,184 cu in. = 3 cu ft** Ans

Note. If the given volume cannot be divided exactly, the quotient may be expressed in more than one unit. A given number of cubic feet plus a stated number of cubic inches is an example.

RULE FOR EXPRESSING A LARGER UNIT OF VOLUME MEASURE AS A SMALLER UNIT

- Multiply the given unit by the number of smaller units contained in one of the required units.
- Express the product in terms of the required smaller unit.

E XAMPLE: Express 10 cu yd in cu ft.

Step 1	Multiply the given unit (10) by the number of cu ft in one cu yd (27 cu ft = 1 cu yd).	**10 × 27 = 270**
Step 2	Express the product (270) in terms of the required unit (cu ft).	**10 cu yd = 270 cu ft Ans**

RULE FOR EXPRESSING TWO OR MORE UNITS OF MEASURE

If a volume expressed in two or more units of measure is to be expressed in a smaller unit,

• Multiply those units of measure that are not in terms of the required unit by the number of smaller units equal to one given unit.

• Add the remaining units in the original given volume to this product.

E XAMPLE: Express 2 cu yd, 10 cu ft as cu ft.

Step 1	Multiply 2 cu yd by the number of cu ft in one cu yd.	**2 × 27 = 54 cu ft**
Step 2	Add to the product (54 cu ft) the remainder of the given volume (10 cu ft).	**10 cu ft**
		64 cu ft Ans

C. APPLYING VOLUME MEASURE TO THE CUBE AND RECTANGULAR SOLID

In volume measure the three linear dimensions that express length, height, and depth or their equivalents are multiplied to determine the cubical contents of a regular solid. The product is in cubic inches if the dimensions are in inches, cubic feet when the dimensions are given in feet, and cubic yards when the dimensions are given in yards.

When the area of one surface is extended in a third direction, a solid is formed. The space the solid occupies is measured in terms of the number of cubic units that it contains.

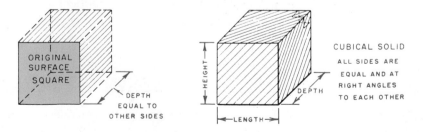

FIGURE 21-4

If the original surface is a square and its face is extended to add depth, the resulting figure is called a solid. When all corners of the solid are square and the linear length, height, and depth are equal, the object is called a *cube* or *cubical solid*. The volume of this cubical solid is equal to the product of its length times depth times height.

RULE FOR COMPUTING THE VOLUME OF A CUBE

- Express the dimensions for length, depth, and height in the same linear unit of measure, when needed.
- Multiply the length × depth × height.
- Express the product in terms of units of volume measure.
- Express the resulting product, if needed, in lowest terms.

EXAMPLE: *Case 1.* Find the volume of a cube, each side of which is 8 inches long.

Step 1 Multiply the length (8) × depth (8) × height (8).	$8 \times 8 \times 8 = 512$
Step 2 Express the product (512) in terms of volume measure.	**512 cu in.** Ans

EXAMPLE: *Case 2.* Determine the volume of a cube that measures 1′-9″ on a side.

Method I

Step 1 Express 1′-9″ as 21″.	$21 \times 21 \times 21 = 9261$
Step 2 Multiply length × depth × height.	**9,261 cu in.**
Step 3 Express product (9,261) in terms of volume measure.	$\frac{9,261}{1,728} = 5$ cu ft, 621 cu in.
Step 4 Express as cu ft and cu in.	**5 cu ft, 621 cu in.** Ans

Method II

Step 1 Express 1′-9″ as $1\frac{3}{4}'$.	$1\frac{3}{4} \times 1\frac{3}{4} \times 1\frac{3}{4} = 5\frac{23}{64}$ cu ft
Step 2 Multiply length × depth × height.	
Step 3 Express $\left(\frac{23}{64}\right)$ cu ft as cu in. by multiplying by (1,728).	$\frac{23}{64} \times \overset{27}{\cancel{1728}} = 621$ cu in.
Step 4 Add the number of cu in. to the cu ft to get volume of the cube.	**5 cu ft, 621 cu in.** Ans

A *rectangular solid* resembles a cube except that the faces or sides are rectangular in shape. The volume of a rectangular solid is equal to the product of the length × depth × height. Sometimes the volume is expressed as the product of the area of the base (length × depth) × the height.

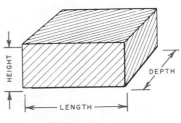

FIGURE 21-5

RULE FOR FINDING THE VOLUME OF A RECTANGULAR SOLID

- Express the dimensions of length, depth, and height in the same linear unit of measure if needed.
- Multiply the length × depth × height.
- Express the product in terms of units of volume measure and reduce to lowest terms, if needed.

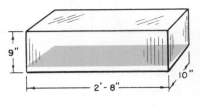

FIGURE 21-6

E XAMPLE: Find the volume of the block in Figure 21-6.

Step 1 Express all dimensions in the same linear unit.

Step 2 Multiply length × depth × height. $10 \times 32 \times 9 = 2,880$

Step 3 Express product as units of volume measure.   **2,880 cu in.**

Step 4 Express in lowest terms. $\frac{2,880}{1,728} = 1$ **cu ft, 1,152 cu in.**   Ans

The volume or weight of a hollow rectangular solid is computed by using the same rules and mathematical processes as for the rectangular solid. Such problems often require a double computation. One computation is required for the outer surface and one for a cored or cut-away section.

D. APPLICATION OF VOLUME MEASURE TO CYLINDERS

The volume of a cylinder is the number of cubic units of a given kind that it contains. This number is found by multiplying the area of the base by the length or height of the cylinder.

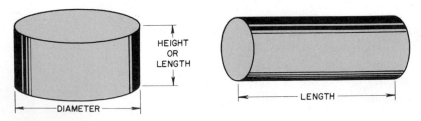

FIGURE 21-7

RULE FOR FINDING THE VOLUME OF A CYLINDER

- Compute the area of the base.
- Multiply this area by the height or length of the cylinder.
- Express the product (volume of cylinder) in units of volume measure.

EXAMPLE: Find the volume of a cylinder 3″ in diameter and 10″ long, correct to two decimal places.

Step 1 Compute the area of the base by multiplying **.7854 × 3 × 3 = 7.0686 sq in.** (.7854 × Diam × Diam).

Step 2 Multiply this area by the length of the cylinder (10″).   **7.0686 × 10 = 70.686**

Step 3 Express product (70.686), correct to two decimal places, in units of volume measure.   **70.69 cu in.**   Ans

E. APPLICATION OF VOLUME MEASURE TO IRREGULAR FORMS

In addition to regular solids like the cube, rectangle, and cylinder, many objects are a combination of these shapes in a modified form (Figure 21-8).

The volume of an irregular solid can be computed by dividing it into solids having regular shapes. The volume of each regular solid (or part of one) can be computed. The sum of the separate volumes equals the volume of the irregular solid.

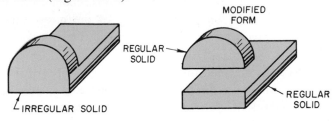

FIGURE 21-8

RULE FOR FINDING THE VOLUME OF AN IRREGULAR SOLID

- Divide the solid into regular forms.
- Compute the volume of each regular solid or part of one.
- Add the separate volumes.

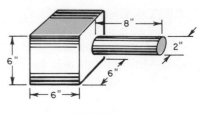

FIGURE 21-9

E XAMPLE: Determine the volume of the brass casting in Figure 21-9. Round the answer to two decimal places.

Step 1 Divide the irregular form into two regular solids (a cube and a cylinder).

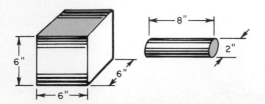

Step 2 Compute the volume of each regular solid.

  a. Volume of cube = length × depth × height

  b. Volume of the cylinder = area of base × height

$$6 \times 6 \times 6 = 216 \text{ cu in.}$$

$$(.7854 \times 2 \times 2) \times 8 = \underline{25.1328 \text{ cu in.}}$$

Step 3 Add the separate volumes.

Total volume of casting = 241.13 cu in. **Ans**

F. APPLICATION OF VOLUME MEASURE TO LIQUID MEASURE

Constant reference is made in the shop and laboratory to the measurement of liquids for cutting oils, oils for heat-treating metals, coolant solutions, marking fluids, cleaning agents, and lubricants. Also, an understanding of the measurement of liquids is essential in clinics and hospitals, business, merchandising, and the home.

Liquids are measured by cubical units of measure known as *liquid* measure. One common method of determining liquid capacity requires, first, computing the cubical contents of the object. Second, the resulting units of volume measure are changed to units of liquid measure.

Table of Liquid Measure
4 gills = 1 pint (pt)
2 pints (pt) = 1 quart (qt)
4 quarts (qt) = 1 gallon (gal) = 231 cu in.
$31\frac{1}{2}$ gallons = 1 standard barrel (bbl)

FIGURE 21-10

The standard units of liquid measure are the gill, pint, quart, gallon, and barrel. These units are sometimes abbreviated for ease in writing. The pint is written (pt), quart (qt), gallon (gal), and barrel (bbl). A comparison of values for each of these units is found in Figure 21-10.

The gallon as established by law contains 231 cubic inches of liquid (Figure 21-11). With this known value, it is possible to solve problems requiring the use of liquid measure by dividing the volume, expressed in cubic inches, by 231.

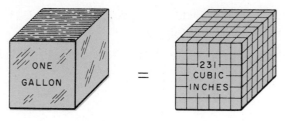

FIGURE 21-11

RULE FOR CHANGING UNITS OF VOLUME MEASURE TO UNITS OF LIQUID MEASURE

- Compute the volume of the object in terms of cubic inches.
- Divide this volume by 231 (231 cu in. = 1 gal).
- Express the quotient in terms of liquid measure (gal).

EXAMPLE: Determine the liquid capacity of a coolant tank whose volume is 1,155 cubic inches.

Step 1   Divide volume in cubic inches (1,155) by the number of cubic inches (231) in one gallon.

$$\frac{1,155}{231} = 5$$

Step 2   Express the quotient (5) in terms of liquid measure.

Capacity of tank = 5 gallons   Ans

RULE FOR EXPRESSING LARGER UNITS OF LIQUID MEASURE IN SMALLER UNITS (BARRELS TO GALLONS TO QUARTS TO PINTS TO GILLS)

- Determine the number of smaller units of liquid measure in one larger unit.
- Multiply the given units by this number.
- Express the product in terms of the required unit of measure.

EXAMPLE: *Case 1.* Express $4\frac{1}{2}$ gallons in quarts.

Step 1   Determine the number of smaller units (quarts) in one larger unit (gallons).

4 qt = 1 gal

Step 2   Multiply the given units ($4\frac{1}{2}$) by this number (4).

$$4\frac{1}{2} \times 4 = 18$$

Step 3   Express the product (18) in the required units of liquid measure (qt).

$$4\frac{1}{2} \text{ gals} = 18 \text{ qt}$$ Ans

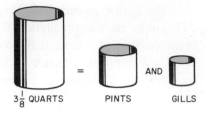

$3\frac{1}{8}$ QUARTS   PINTS   GILLS

EXAMPLE: *Case 2.* Express $3\frac{1}{8}$ quarts as pints and gills.

Step 1	Determine the number of smaller units (pints and gills) in one larger unit (quart).	**2 pt = 1 qt, 4 gills = 1 pt**
Step 2	Multiply the given unit ($3\frac{1}{8}$ qt) by the number of smaller units (2 pt) in one of the given units (1 qt). *Note.* The fractional part of a pint $\left(\frac{1}{4}\right)$ may be changed to gills by multiplying by 4 (the number of gills in one pint).	$3\frac{1}{8} \times 2 = 6\frac{1}{4}$ **pt** $\frac{1}{4} \times 4 = 1$ **gill**
Step 3	Combine both values (6 pints and 1 gill). The result is the equivalent of $3\frac{1}{8}$ quarts in terms of pints and gills.	$3\frac{1}{8}$ **qt = 6 pt, 1 gill** **Ans**

RULE FOR EXPRESSING SMALLER UNITS OF LIQUID MEASURE IN LARGER UNITS (GILLS TO PINTS TO QUARTS TO GALLONS TO BARRELS)

- Determine the number of smaller units of liquid measure in one of the required larger units.
- Divide the number of given units by this number.
  Note. Where the result is a mixed number, the fractional part is sometimes changed to the next smaller unit. For example $3\frac{1}{2}$ gallons may be written 3 gallons, 2 quarts.

EXAMPLE: *Case 1.* Express 24 pints in gallons.

Step 1	Determine the number of smaller units (pt) in one of the larger required units (gal).	**8 pt = 1 gal**
Step 2	Divide the number of given units (24) by this number (8) to get gallons.	$\frac{24}{8} = 3$ **24 pt = 3 gal** **Ans**

EXAMPLE: *Case 2.* Express 76 gills in gallons, quarts, and pints.

Step 1	Determine the number of smaller units of liquid measure (gills) in one of the required larger units (gallons and quarts).	**8 gills = 1 qt** **32 gills = 1 gal**

Step 2	Divide the 76 gills by 32 to determine the number of gallons.		$\frac{76}{32} = $ **2 gal, 12 gills**
Step 3	Express the 12 gills as quarts by dividing by 8.		$\frac{12}{8} = $ **1 qt, 4 gills**
Step 4	Express the remaining 4 gills as pints by dividing by 4.		$\frac{4}{4} = $ **1 pt**
Step 5	Combine all values. The result in gallons, quarts, and pints is the equivalent of 76 gills.		**76 gills = 2 gal, 1 qt, 1 pt** Ans

ASSIGNMENT UNIT 21   REVIEW AND SELF TEST

PRETEST

The Review and Self-Test items that follow may be used as a pretest. Pretests are designed to measure a student's beginning level of mathematical skills competency and to determine the starting point of instruction.

POSTTEST

The Review and Self-Test items may also be applied as a post test. Post tests are planned to establish the student's level of mathematical skills competency after instruction.

Note. Unless otherwise stated, all dimensions for all problems are expressed in inches.

A. Changing Values of Units of Volume Measure (General Applications)

1. Express each of the following volumes (a) through (l) in the unit of volume measure specified in each case.

   a. 2 cu ft in cu in.

   b. $1\frac{1}{2}$ cu ft in cu in.

   c. $3\frac{5}{8}$ cu ft in cu in.

   d. 10 cu ft, 19 cu in. in cu in.

   e. 3456 cu in. in cu ft

   f. 18.144 cu in. in cu ft

   g. 8640 cu in. in cu ft

   h. 1944 cu in. in cu ft and cu in.

   i. 3 cu yd in cu ft

   j. $4\frac{1}{3}$ cu yd in cu ft.

   k. 5 cu yd, 7 cu ft in cu ft

   l. 7 cu yd, 19 cu ft in cu ft

B. Applying Volume Measure to Cubes and Rectangular Solids (Practical Problems)

1. Determine the volume of cubes A, B, and C.

	A	B	C
Length	6	$8\frac{1}{2}$	$1'\text{-}6''$
Depth	6	$8\frac{1}{2}$	$1'\text{-}6''$
Height	6	$8\frac{1}{2}$	$1'\text{-}6''$

2. Compute the volume of rectangular solids A, B, and C, correct to two decimal places.

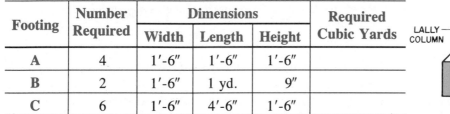

3. Determine the number of cubic yards of concrete mix that is needed to pour footings A, B, and C.

Footing	Number Required	Dimensions			Required Cubic Yards
		Width	**Length**	**Height**	
A	4	$1'\text{-}6''$	$1'\text{-}6''$	$1'\text{-}6''$	
B	2	$1'\text{-}6''$	1 yd.	$9''$	
C	6	$1'\text{-}6''$	$4'\text{-}6''$	$1'\text{-}6''$	

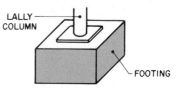

C. Applying Volume Measure to Rectangular Solids (Practical Problems)

1. Compute the number of cubic yards of earth that is to be removed for a basement. The basement dimensions are $8'$ deep × $36'$ wide × $48'$ long.

2. Determine the volume of concrete in the foundation wall. State the answer to the nearest cubic yard.

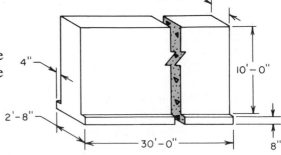

3. Compute the volume of hollow rectangular solids A and B. Round off the volume of B correct to one decimal place.

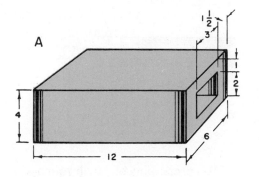
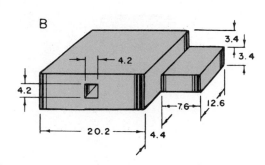

D. Applying Volume Measure to Cylinders (Practical Problems)

1. Determine the volume of cylinders A, B, and C correct to two decimal places.

	A	B	C
Diameter	4	12.5	
Radius			1.6
Length	10	24.5	6.4

2. Compute the liquid capacity of cisterns A, B, and C in gallons of water. Round off the value to one decimal place.

Cistern	Inside Diameter	Height
A	4'-0"	6'-0"
B	5'-6"	8'-0"
C	6'-6"	8'-6"

1 cu ft $= 7\frac{1}{2}$ gal

3. Find the volume of cored cylinders A and B, correct to one decimal place.

	A	B
Outside Diam	5	$4\frac{1}{4}$
Inside Diam	2	$1\frac{1}{2}$
Length	10	12

4. Find the weight of cored brass casting A and cored bronze casting B.

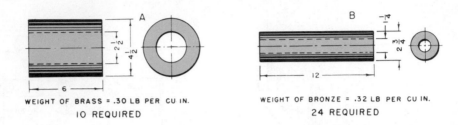

WEIGHT OF BRASS = .30 LB PER CU IN.
10 REQUIRED

WEIGHT OF BRONZE = .32 LB PER CU IN.
24 REQUIRED

5. Determine the cost of material needed to machine 250 special pins of $\frac{3}{4}$ inch round stock, $3\frac{1}{2}$ inches long. Allow $\frac{1}{8}$ inch for cutting off each pin and an additional 107 inches of the total length for waste and stock spoilage. The material weighs 0.28 pounds per cubic inch and costs $1.26 per pound. Find the cost to the nearest dollar.

6. Determine the weight of the aluminum parts shown at A and B, correct to two decimal places. The weight of aluminum is .09 pounds per cubic inch.

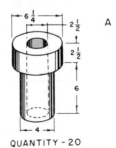

QUANTITY - 20

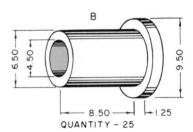

QUANTITY - 25

E. Applying Volume Measure to Irregular Shapes (Practical Problems)

1. Determine the weight of the rectangular cast iron blocks specified, correct to one decimal place.

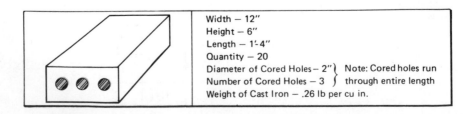

Width — 12"
Height — 6"
Length — 1'-4"
Quantity — 20
Diameter of Cored Holes— 2" } Note: Cored holes run
Number of Cored Holes — 3 } through entire length
Weight of Cast Iron — .26 lb per cu in.

QTY - 75

2. Determine the cost of 75 steel drop forgings conforming to the specifications given. The cost should be correct to two decimal places.

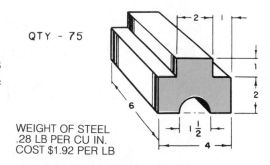

WEIGHT OF STEEL
.28 LB PER CU IN.
COST $1.92 PER LB

F. Expressing Units of Measure (General Applications and Practical Problems)

1. Express each of the following values in the unit or units of liquid or volume measure indicated in each case.

a.	4 gal	qt	i.	37 qt	gal and qt
b.	$6\frac{1}{2}$ gal	qt	j.	63 gal	bbl
c.	$3\frac{3}{4}$ gal	qt	k.	$96\frac{1}{2}$ gal	bbl and gal
d.	$6\frac{1}{2}$ qt	pt	l.	693 cu in.	gal and qt
e.	$5\frac{1}{4}$ qt	pt	m.	577.5 cu in.	gal and qt
f.	$8\frac{3}{4}$ qt	pt and gills	n.	5 gal	cu in.
g.	$3\frac{7}{8}$ qt	pt and gills	o.	4 gal, 3 qt	cu in.
h.	17 qt	gal and qt			

2. Determine the liquid capacity of the rectangular coolant tank A and the circular portable container B. Compute A to the nearest gallon and B to the nearest quart.

A

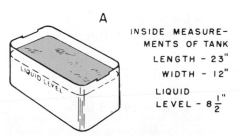

INSIDE MEASURE-
MENTS OF TANK
LENGTH – 23"
WIDTH – 12"
LIQUID
LEVEL – $8\frac{1}{2}$"

B

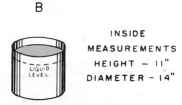

INSIDE
MEASUREMENTS
HEIGHT – 11"
DIAMETER – 14"

Section 4
Unit 22 Achievement Review on Measurement (Customary Units)

OBJECTIVES OF THE UNIT

This achievement review serves as an overall test for Section 4. The unit is designed to measure the student's/trainee's ability to

- *Express measurements in appropriate units of linear, circular, surface, and volume measure.*
- *Perform basic mathematical processes of addition, subtraction, multiplication, and division to solve problems involving linear, circular, surface, or volume measurements.*

SECTION PRETEST/ POSTTEST

The Review and Self-Test items that follow relate to each Unit within this section. The test items may be used as a Unit-by-Unit pretest and/or Section post test.

UNIT 18. APPLICATION OF LINEAR MEASURE (GENERAL APPLICATIONS AND PRACTICAL PROBLEMS)

1. Express each measurement in the unit indicated in each case.

   a. 15 ft to yd

   b. 8 ft to in.

   c. 3 yd 2 ft to ft

   d. 7 ft 5 in. to in.

   e. 176 in. to ft

   f. 12.5 ft to in.

   g. $6\frac{1}{2}$ yd to ft

   h. 9 ft 8 in. to in.

   i. 4 yd 6 ft 3 in. to in.

2. Add each series of measurement and reduce to lowest terms.

   a. $10' + 7' + 5'$

   b. 6 yd + 9 ft + 6 ft

   c. $.5'' + .375'' + .125''$

   d. $6.500'' + 1'\text{-}3\frac{1}{4}$ ft $+ 8$ in.

   e. $10\frac{1}{2}$ yd $+ 17\frac{1}{4}$ ft $+ 8$ in.

   f. $4\frac{3}{4}$ yd $+ 19$ ft 6 in.

3. Perform the arithmetical process required in each case. Give result in simplest form.

   a.   3 yd 6 ft 9 in.
       −1 yd 4 ft 5 in.

   c. 6 ft 2 in.
      × 28

   e.   280 in.
       ÷ 14 in.

   b. 9 yd 2 ft 4 in.
      −8 ft 6 in.

   d. 12′ - 3.5″
      ×   10

   f. 9′-10″
      ÷ 7″

4. The diesel engine plate gage illustrated is to be machined. The part is to be rough machined, finish machined, and ground to the finished sizes given on the drawing.

   a. Allow $\frac{1}{32}''$ on all faces and determine the rough machining dimensions for (A) through (H).

   b. Determine the size to which dimensions (A) through (H) are to be machined before grinding if .010″ is allowed on each dimension for the grinding operation.

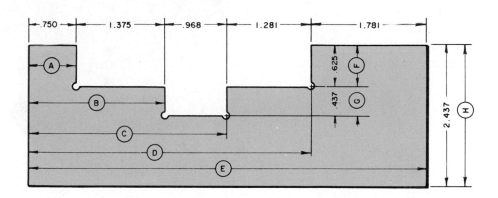

5. Determine the standard micrometer readings A, B, and C.

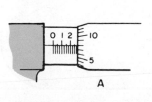

A

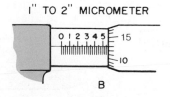

1″ TO 2″ MICROMETER

B

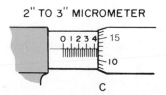

2″ TO 3″ MICROMETER

C

UNIT 19. APPLICATION OF ANGULAR AND CIRCULAR MEASURE (PRACTICAL PROBLEMS)

1. Give the bevel protractor readings A and B.

READING A _____   READING B _____

2. Determine angles Ⓐ through Ⓞ for the jig plate in order to machine the required slots and drill the holes.

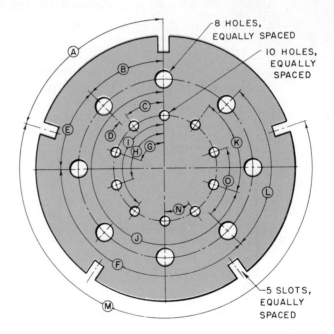

UNIT 20. PRINCIPLES OF SURFACE MEASURE

1. Find the area of the plastic part.

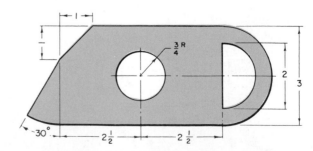

2. What is the cost of the 1-inch thick cored iron castings?

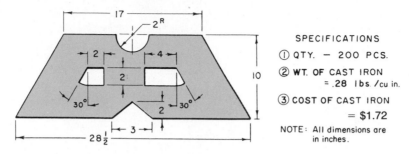

SPECIFICATIONS

① QTY. — 200 PCS.

② **WT. OF** CAST IRON
= .28 lbs./cu in.

③ COST OF CAST IRON
= $1.72

NOTE: All dimensions are
in inches.

UNIT 21. PRINCIPLES OF VOLUME MEASURE

1. Determine the amount of liquid held in a rectangular oil reservoir of a hydraulic machine for the liquid levels indicated in the table. Round off each answer to the nearest quart.

Gage Level	Height of Liquid
A	5"
B	$5\frac{1}{2}''$
C	6"
D	$6\frac{1}{2}''$
E	7"

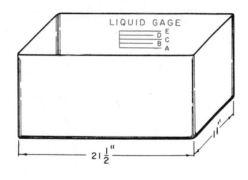

2. What is the weight of 2500 brass parts that are stamped from $\frac{1''}{16}$ sheet brass weighing .3 pound per cubic inch? Give the total weight to the nearest pound.

NOTE: All dimensions are
in inches.

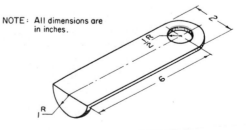

PERCENTAGE AND AVERAGES

Unit 23 The Concepts of Percent and Percentage

OBJECTIVES OF THE UNIT

After satisfactorily studying this unit, the student/trainee will be able to
- *Express numerical values in terms of percent.*
- *Determine any required percent of a measurement value.*
- *Change percent values to decimals and fractions.*

PRETEST   *Use the Review and Self-Test items provided in the Unit Assignment to establish the level of mathematical skills competency and to determine the starting point of instruction.*

Percents are given in catalogs, magazines, newspapers, handbooks for technicians, and other publications. Percents show how many parts of a total are taken out. Percents are used to make comparisons and compute wages, taxes, discounts, and increases or decreases in production. An ever-increasing number of applications of percents are found in health, business distribution and merchandising occupations, industry, and the home.

A. FORMS FOR EXPRESSING PERCENT

The word *percent* is a short way of saying ''by the hundred'' or ''hundredths part of the whole.'' A percent refers to a given number of parts of the whole which is equal to 100 percent. ''Fifteen percent'' is the same as writing 15%.

Percent may be shown graphically by two illustrations (Figure 23-1). The square is divided into 100 equal parts. Each of the small squares is one one-hundredth of the whole (100%) or $\frac{1}{100}$ of 100% = 1%.

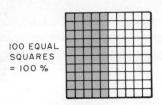

100 EQUAL SQUARES = 100 %

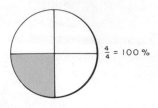

$\frac{4}{4}$ = 100 %

FIGURE 23-1

By the same reasoning, the 50 shaded squares are $\frac{50}{100}$ of the total = $\frac{50}{100}$ of 100% or 50%. In the circle, the shaded area is $\frac{1}{4}$ of the whole circle or $\frac{1}{4}$ of 100% or 25%.

Percent is another form of mathematical expression. A value like 50% may be represented as such. It also may be given in an equivalent form as the fraction $\left(\frac{1}{2}\right)$ or as the decimal (.5). The form in which a value is given depends on the requirements of the problem.

B. DETERMINING ONE HUNDRED PERCENT OF A NUMBER

One hundred percent of a number is the same as one hundred hundredths or 1. It represents the whole number or total. Therefore, 100% of a number is the number itself.

C. DETERMINING ONE PERCENT OF A NUMBER

RULE FOR FINDING ONE PERCENT OF A NUMBER

- Change one percent to an equivalent decimal and remove the percent sign.
  1% = one hundredth = .01.
- Multiply the given number by the decimal equivalent of 1% (.01).
- Label the answer in the required unit of measure.

EXAMPLE: Find 1% of 275 feet.

Step 1 Multiply the given number (275) by .01.	$\begin{array}{r} 275 \\ \times .01 \\ \hline 275 \end{array}$
Step 2 Point off, starting at the right, the same number of decimal places in the answer as there are in the multiplier and multiplicand.	**2.75**
Step 3 Label answer with correct unit of measure.	**2.75 feet** Ans

Since 1% is the same as .01, the process may be performed mentally by just placing a decimal in the original number. For instance, 1% of 120.33 is found by just moving the decimal point two more places to the left.

$$1\% \text{ of } 120.33 = .01 \times 120.33 = \mathbf{1.2033} \quad \textbf{Ans}$$

D. DETERMINING ANY PERCENT OF A NUMBER

A percent may be converted to any equivalent mathematical form. At times, the fractional equivalent is preferred. In other cases, the decimal form is best suited to a specific problem.

RULE FOR FINDING ANY PERCENT OF A NUMBER

- Convert the percent to either a fractional or decimal equivalent.
- Multiply the given number by this equivalent.
- Point off the same number of decimal places in the product as there are in the multiplier and multiplicand.
- Label answer with appropriate unit of measure.

EXAMPLE: Find 16% of 1,218 square yards.

Step 1 Change 16% to a decimal. $16\% = .16$

Step 2 Multiply the given number (1,218) by the decimal (.16).
$$\begin{array}{r} 1218 \\ \times\ .16 \\ \hline 19488 \end{array}$$

Step 3 Point off two decimal places in the product.   194.88

Step 4 Label answer.   **194.88 sq yd**   Ans

These same steps are followed when the percent is a mixed number.

EXAMPLE: Find $6\tfrac{1}{4}\%$ of 782 horsepower (hp).

Step 1 Change $6\tfrac{1}{4}\%$ to a decimal. $6\tfrac{1}{4}\% = .06\tfrac{1}{4} = .0625$

Step 2 Multiply the given number (782) by the decimal (.0625).
$$\begin{array}{r} 782 \\ \times\ .0625 \\ \hline 488750 \end{array}$$

Step 3 Point off four decimal places in the product.   48.8750

Step 4 Label answer.   **48.875 hp**   Ans

E. CHANGING PERCENTS TO DECIMALS AND FRACTIONS

A percent may be changed to its decimal equivalent. Simply move a decimal point two places to the left and take away the percent sign. The decimal value may then be changed to a fraction to get the fractional equivalent of the original percent. A few relationships between percents, decimals, and fractions are shown in Figure 23-2.

| Percent | Equivalent | | Parts of the Whole |
	Decimal	Fraction	
1%	.01	$\frac{1}{100}$	One hundredth
100%	1.00	$\frac{100}{100}$ or $\frac{1}{1}$	One hundred hundredths
325%	3.25	$3\frac{1}{4}$	Three hundred twenty-five hundredths
.1%	.001	$\frac{1}{1000}$	One tenth of one percent

FIGURE 23-2

ASSIGNMENT UNIT 23   REVIEW AND SELF TEST

PRETEST   *The Review and Self-Test items that follow may be used as a pretest. Pretests are designed to measure a student's beginning level of mathematical skills competency and to determine the starting point of instruction.*

POSTTEST   *The Review and Self-Test items may also be applied as a post test. Post tests are planned to establish the student's level of mathematical skills competency after instruction.*

A. The Concept of Percent (General Applications)

1. Give the percent of the whole that the shaded area of the rectangle (a), triangle (b), and circle (c) represents in each case.

a. 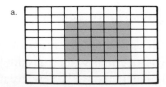   b. 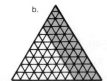   c.

2. Shade and indicate by percent the area of the whole square (a), rectangle (b), and circle (c) that is represented in each instance by 100%, 50%, 25%, 10%, and 1%.

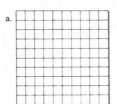 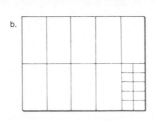 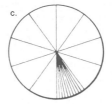

3. Write each percent (A through E) using the symbol (%).

A	B	C	D	E
65 percent	6.5 percent	.6 percent	$6\frac{2}{3}$ percent	12.25 percent

4. Draw a table similar to the one illustrated and insert the decimal equivalent of percents A through O.

	Percent	Decimal		Percent	Decimal		Percent	Decimal
A	50%		F	125%		K	.1%	
B	30%		G	205%		L	.25%	
C	25%		H	312%		M	.05%	
D	9%		I	101%		N	$3\frac{1}{2}$%	
E	1%		J	250%		O	$2\frac{1}{4}$%	

5. Prepare a table and insert the percent equivalent of each decimal A through F.

A	B	C	D	E	F
1.50	.75	.125	2.05	.025	.004

B. Determining 100% and 1% of a Number (General Applications)

1. Find 100% of values A through E.

2. Find 1% of the same values.

A	B	C	D	E
200	325	62.5	1.25 tons	.563 lb

C. Determining Any Percent of a Number (Practical Applications)

1. Find the percent indicated in each case for quantities A through F.

A	4% of $500		D	.2% of 325 cu yd
B	8% of 120 sheets		E	3.25% of 800 kW
C	$1\frac{1}{2}$% of 840 acres		F	$37\frac{1}{2}$% of 972 lb

2. The absenteeism in a large business is $1\frac{1}{2}$% of all employee-hours worked. If the work-week is 40 hours and there are 1,750 employees, determine the total number of workdays (of 8 hours each) that are lost during each 260 workday year.

3. Find the total number of pounds of tin and lead required to make three different solder compositions (A, B, and C).

Composition	A	50% tin 50% lead	B	90% tin 10% lead	C	60% tin 40% lead
Pounds Solder Required		212 lb		48 lb		125 lb

4. The area of a floor is 320 square feet. In laying this floor, there is a lumber allowance of 25% for waste and matching. Determine the amount of flooring required.

5. The power output of a motor is 58 horsepower. If 7.6% of the power is used to overcome friction and other losses, what horsepower is available under a full load?

6. It takes 3.35 yards of material to make a suit. An additional 12% is required for waste and matching. Determine the total yardage that is needed to make 12 suits.

Unit 24 Application of Percentage, Base, and Rate

OBJECTIVES OF THE UNIT

After satisfactorily studying this unit, the student/trainee will be able to
- *Apply base, rate, and percentage terms.*
- *Compute base, rate, or percentage values in practical occupational problems.*

PRETEST *Use the Review and Self-Test items provided in the Unit Assignment to establish the level of mathematical skills competency and to determine the starting point of instruction.*

A. DESCRIPTION OF PERCENTAGE TERMS AND RULES

Percentage refers to the value of any percent of a given number. In every percentage problem three numbers are involved. The first number is known as the *base* because a definite percent is to be taken of it. The second number, the *rate,* refers to the percent that is to be taken of the base. The third number is the *percentage.* The relationship of the base, rate, and percentage to one another may be stated in the rule: The product of the base times the rate equals the percentage.

In simplified form,

$$percentage = base \times rate.$$

Sometimes, the letter *B* is used for base, *R* for rate, and *P* for percentage. Using letters instead of words, the rule may be given as $P = B \times R$. Wherever this rule is used, the rate must always be in decimal form.

B. DETERMINING PERCENTAGE

RULE FOR FINDING PERCENTAGE (GIVEN: BASE AND RATE)

- Write the rule for percentage in simplified form *(P = B × R).*
- Change the given percent to its decimal equivalent.
- Substitute the given values for *B* and *R* in the rule.
- Multiply and label answer.
- Check by substituting the answer for P . Rework to see that the quantities on both side of the (=) sign are equal.

EXAMPLE: Find $12\frac{1}{2}\%$ of 360″.

Step 1	Write the rule.	$P = B \times R$
Step 2	Change $12\frac{1}{2}\%$ to its decimal equivalent.	$12\frac{1}{2}\% = .125$
Step 3	Substitute 360″ for the base and .125 as the rate.	$P = 360 \times .125$
Step 4	Multiply and label the answer.	$P = 45''$ **Ans**

C. DETERMINING BASE OR RATE

It is possible to use the percentage rule to determine the base or rate when the two other terms in the rule are known.

RULE FOR FINDING BASE OR RATE (GIVEN: PERCENTAGE AND BASE OR RATE)

- Write the rule in simplified form. $P = B \times R$
- Substitute the two known values in the rule.
  Note. If a percent is given, be certain that it is changed to its decimal equivalent.
- Divide the percentage by the base to get the rate as a decimal or by the rate to find the base.
- Convert the decimal or fractional value of the rate to a percent if this is required.
- Label the answer with the appropriate unit of measure.

EXAMPLE: *Case 1*. Two hundred (200) is what percent of 500?

Step 1	Write the rule.	$P = B \times R$
Step 2	Substitute known values: The base is 500 and the percentage 200.	$200 = 500 \times R$
Step 3	Divide P by B to get R.	$\dfrac{\overset{2}{\cancel{200}}}{\underset{5}{\cancel{500}}} = R$
Step 4	Change $\frac{2}{5}$ of the whole to a decimal.	$\frac{2}{5} = .4$
Step 5	Change the decimal .4 to its percent equivalent by moving the decimal two places to the right and adding the (%) symbol.	$.4 = 40\%$ **Ans**
Step 6	Check the values on both sides of the (=) sign.	$200 = 500 \times .4$
		$200 = 200$ **Proof**

EXAMPLE: *Case 2*. If 56 castings are 20% of the total, what is the total number?

Step 1	Write the rule.	$P = B \times R$
Step 2	Substitute the known values: The percentage is 56, the rate is 20%.	
	Note. Change the 20% rate to its decimal equivalent .20 before substituting.	$56 = B \times .20$
		$\frac{56}{20} = B$
Step 3	Divide P by R to get B.	$280 = B$
Step 4	Label answer in the appropriate unit of measure.	**280 castings** **Ans**

With the rule, *percentage = base × rate*, it is possible to solve any percentage problems when two of the three quantities are given and the rate is expressed in decimal form.

ASSIGNMENT UNIT 24   REVIEW AND SELF TEST

PRETEST
The Review and Self-Test items that follow may be used as a pretest. Pretests are designed to measure a student's beginning level of mathematical skills competency and to determine the starting point of instruction.

POSTTEST
The Review and Self-Test items may also be applied as a post test. Post tests are planned to establish the student's level of mathematical skills competency after instruction.

A. Determining Percentage

1. Find the percentage in each problem (A through E) for each value given for the base and rate.

	A	B	C	D	E
Base	2,400	1,875 gal	142.6 in.	3,268.5 sq ft	$296\frac{1}{2}$ sheets
Rate	80%	45%	3.8%	$4\frac{1}{2}\%$	$6\frac{1}{4}\%$

2. A piece of meat weighs 25.6 pounds before it is cooked and 23.2 pounds after cooking. Determine the percent of weight lost in cooking.

3. The total receipts of a merchandising store for one week total $3,800. The expenses include 40% wages, 12% rent, 9% for heating and cooling, and 27% for taxes and other overhead. Establish (a) the amount of profit and (b) the percent of profit.

4. A car speedometer registers 52 miles per hour. The actual car speed is 55 miles per hour. State the percent of error correct to two decimal places.

5. The original cost of 1,290 aluminum castings is 1.95 cents each. Five percent (5%) is scrapped as poor castings and another $7\frac{1}{2}\%$ is spoiled in machining. (a) How many *usable castings* are used and (b) what is the new unit cost (rounded to the nearest two-place value) of each usable casting?

B. Determining Base or Rate (Practical Problems)

1. Find the base in each problem (A through E) when the percentage and rate are given.

	A	B	C	D	E
Percentage	120 hp	2016 cables	137.8 lb	$78\frac{1}{2}$ bars	126.5 in.
Rate	90%	54%	7.2%	$6\frac{1}{2}\%$	$3\frac{1}{4}\%$

2. Find the rate in each problem (A through E) when the percentage and base are given.

	A	B	C	D	E
Percentage	30	360°	24" D	8.2	$92\frac{1}{2}''$
Base	30	30°	36" D	19.68	294"

3. What percent is wasted when 2.4 of every 120 sheets of plywood are spoiled?
4. What percent of metal is allowed for cut-off on each 2-inch length of stock?

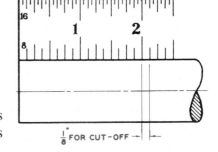

C. Determining Percentage or Base or Rate

1. One part of acid and four parts of water are mixed as an electrolyte for a storage battery. What percent is acid and what percent is water?
2. A generator rating is 42,500 kilowatts. If the output is 29,500 kilowatts, what percent of the rating is the generator delivering?
3. What is the operating spindle speed of a lathe spindle traveling at 346 RPM when 18% is lost through cutting force?
4. The cutting speed of a milling cutter is 85 feet per minute. Friction and other cutting losses amount to $12\frac{1}{2}\%$. Find the base cutting speed.
5. Two special bronze castings are composed of six different metals. The percent of each metal used in each casting and the casting weight are given. Determine the weight (in pounds and fractional parts) of each metal. Round off each decimal to one place.

Casting	Casting Weight	Composition (% by Weight)					
		Copper	Tin	Zinc	Phosphorus	Lead	Iron
A	400 lb	80	11	8.2	0.4	0.3	0.1
B	525 lb	83	8.7	7.38	0.34	0.52	0.06

Unit 25 Averages and Estimates

OBJECTIVES OF THE UNIT
After satisfactorily studying this unit, the student/trainee will be able to
- *Compute the average of several quantities.*
- *Find an unknown value when all quantities except the missing one and the average are given.*
- *Use estimating to check the mathematical accuracy of a computed value.*
- *Apply basic steps to estimating to compute quantities and values when the given data have a number of variable factors.*

PRETEST *Use the Review and Self-Test items provided in the Unit Assignment to establish the level of mathematical skills competency and to determine the starting point of instruction.*

Averages

Averages are used in linear, circular, angular, temperature, weight, and all other measurements. The *average* of given quantities and values is the starting point on which many other computations and factual data are based.

A. AVERAGE QUANTITIES

The average of two or more quantities that are in the same unit of measure is found by simple addition and division.

RULE FOR AVERAGING SEVERAL QUANTITIES

- Check the units of measure in each quantity to be sure they are the same.
- Arrange the quantities in a column and add.
- Divide by the number of quantities to get the average.

EXAMPLE: What is the average of the linear dimensions in Figure 25-1?

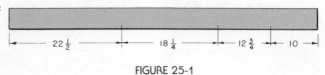

FIGURE 25-1

Step 1 Check each dimension and change, if necessary, to the same unit of measure.

Step 2 Arrange in a column and add.

Note. The mathematical processes may be simplified by expressing fractional values as decimals before averaging.

$$22\tfrac{1}{2} = 22.50$$
$$18\tfrac{1}{4} = 18.25$$
$$12\tfrac{3}{4} = 12.75$$
$$10 \phantom{\tfrac{3}{4}} = \underline{10.00}$$
$$63.50$$

Step 3 Divide the sum (63.50) by the number of quantities (4).

$$4\overline{)63.50}$$
$$15.875$$

Step 4 Express the decimal value (.875) as a fraction $\left(\tfrac{7}{8}\right)$, if necessary. The average of the four dimensions is $15\tfrac{7}{8}$.

$$15.875 = 15\tfrac{7}{8} \quad \text{Ans}$$

B. DETERMINING VALUES FROM AVERAGES

An unknown value may be computed when the average and all quantities but the unknown are given.

RULE FOR FINDING AN UNKNOWN VALUE

- Multiply the average by the number of quantities.
- Add the given quantities.
- Subtract this sum from the product. The difference is the missing value.
- Check by adding all quantities and dividing by the number of quantities. The numbers are correct when the given average equals the computed average in the check.

EXAMPLE: The average of four temperature readings is 1,672°. Three of the actual readings are 1,525°, 1,683°, and 1,726°. Give the fourth reading.

Step 1 Multiply the average (1,672) by the number of readings (4).

Step 2 Add the three given readings.

$$1,525 + 1,683 + 1,726 = 4,934$$

$$1,672$$
$$\times 4$$
$$\overline{6,688}$$

Step 3 Subtract this sum (4934) from the product of the average (6688). The difference of 1754° is the fourth temperature reading.

$$6,688$$
$$-4,934$$
$$\overline{1,754}$$

Step 4   Check the average of the four numbers with the given average.

$$\frac{1,525 + 1,683 + 1,726 + 1,754}{4} = \frac{6,888}{4} = 1,672 \qquad 1,754° \quad \textbf{Ans}$$

Estimating

Estimating has a twofold meaning. Estimating may refer to a shortcut mathematical process of determining a range against which an actual answer may be compared for accuracy. Estimating, in another instance, may require actual computations to determine cost, time, material, and other essential data. The estimate is as accurate as variations in materials and working conditions permit.

Estimating is valuable in checking computations and in arriving at quantities and values that vary as the basic conditions change.

C. ESTIMATING AS A MATHEMATICAL CHECK

Where estimating is used to check the accuracy of a solution to a problem, certain simple steps may be used.

RULE FOR ESTIMATING MATHEMATICAL ACCURACY

- Work the problem in a conventional way.
- Reread the original problem and determine what numbers may be rounded off.
- Use these numbers and perform the operations as required.
- Pay special attention to counting off decimal places in the answer, as most errors occur at this point.
- Check the computed value by comparing the original answer with the estimated value.
  Note. If both values are close, the answer is accurate in most instances.

E XAMPLE: What is $24\frac{1}{2}$% of 996?

Step 1   Compute answer in the conventional way.   996
Step 2   Estimate the answer.   ×.245
           Round off $24\frac{1}{2}$% to 25% or $\frac{1}{4}$ **244.02**
           Round off 996 to 1,000
           Take $\frac{1}{4}$ of 1,000 = 250 **Ans**
Step 3   Compare the estimated value (250) with the computed value (244.02).
        Note. Since the 25% and the 1,000 are larger than either of the original quantities, the computed value must be smaller than the estimate. If the answer were larger, the problem would require reworking.

D. ESTIMATING AS A BASE FOR ADDITIONAL COMPUTATION

Where there are a number of variable factors in given data, an estimate may provide the only sound basis for determining costs, appropriations, and employee-hours. On jobs where new materials are used, or all conditions are not known at any given time, the estimate furnishes about the best working information that is obtainable.

While estimates are determined in a number of ways, there are steps that are common to all estimating.

Step 1 Determine the data that must be computed.

Step 2 Analyze the available information to see what is given.

Step 3 Select averages (if available) or compute averages where needed.

Step 4 Perform the required mathematical operations and combine like quantities to get an accurate estimate.
Note. Take sufficient space for each part of an estimate. Label all answers for greater accuracy and speed.

Step 5 Total all computations. Check by reworking to see that all items are included.

Step 6 Estimate by rounding off quantities and amounts. Then perform the mathematical operations to get an estimated answer.

Step 7 Check the *accurate estimate* against the *rounded-off estimate* to see that the final result is within an acceptable range.

Step 8 Rework the original accurate estimate if the variation is too great.

Step 9 Label all answers with the appropriate units of measure.

ASSIGNMENT UNIT 25 REVIEW AND SELF TEST

PRETEST *The Review and Self-Test items that follow may be used as a pretest. Pretests are designed to measure a student's beginning level of mathematical skills competency and to determine the starting point of instruction.*

POSTTEST *The Review and Self-Test items may also be applied as a post test. Post tests are planned to establish the student's level of mathematical skills competency after instruction.*

A. Averaging Quantities and Estimating Accuracy (Practical Problems)

Solve each problem. Then check each solution by estimating.

1. Find the average length of the five rods.

2. What is the average weight of five castings that weigh $17\frac{1}{4}$ pounds, $12\frac{7}{8}$ pounds, $9\frac{1}{4}$ pounds, 4 pounds, and $8\frac{1}{2}$ pounds?

3. Micrometer measurements, taken at five places on a metal part, are recorded. Determine the average thickness of the part.

Measurements in Inches				
A	B	C	D	E
1.252	1.249	1.249	1.248	1.251

4. Five variations in temperature are recorded on a graph for a heat-treating operation. Find the average temperature.

Reading	A	B	C	D	E
Temperature	2,272°	2,346°	2,147°	2,286°	2,304°

B. Determining Missing Values with Known Averages (Practical Problems)

1. The weekly production of a mechanism averages 1,235 units. The number of units produced in each of four days is 212, 224, 232, and 275. How many units must be produced the fifth day to meet the required average?

2. The space available on three floors of a loft building is 900 sq ft, 1,475 sq ft, and 1,350 sq ft. If 140 production machines averaging 32 sq ft of space apiece are to be installed, how much additional space is required?

Section 5
Unit 26 Achievement Review on Percentage and Averages

SECTION PRETEST/ POSTTEST

The Review and Self-Test items that follow relate to each Unit within this Section. The test items may be used as a Unit-by-Unit pretest and/or Section post test.

UNIT 23. THE CONCEPTS OF PERCENT AND PERCENTAGE

1. Give the percent of the whole that the shaded area represents in a, b, and c.

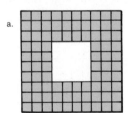
a.

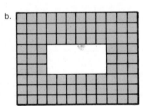
b.

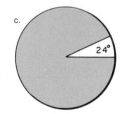
c.

2. Prepare a table similar to the one shown and insert the missing decimal, fractional, or percent equivalent.

	Percent	Decimal	Fraction
A	16.5%		
B	3.5%		
C		.5	
D			$\frac{1}{40}$
E		.0025	

UNIT 24. APPLICATION OF PERCENTAGE, BASE, AND RATE (PRACTICAL PROBLEMS)

1. A $191.10 charge for a set of forgings represents 22.5% of the total cost. Determine what the job costs.

2. Determine the percent by which the area of each square or round bar is increased or decreased for each size change given. (Use $\pi = 3.14$.)

	Square Stock				Round Stock	
	Original Dimension of Side	Increased or Decreased			Original Diameter	Increased or Decreased
A	2"	4"	E		2"	4"
B	2.5"	4"	F		2"	3"
C	.75"	.50"	G		1.500"	.750"
D	$1\frac{1}{4}''$	$\frac{3}{4}''$	H		$2\frac{1}{4}''$	$1\frac{1}{4}''$

3. The revolutions per minute of a milling machine spindle are reduced by friction and cutting pressure the percents indicated at A through E. In each instance, determine the rpm that the cutter actually turns.

	A	B	C	D	E
RPM	80	150	225	335	465
Speed Losses	10%	12.5%	$16\frac{1}{2}\%$	$12\frac{1}{4}\%$	$9\frac{1}{4}\%$

4. A restaurant buys three sides of beef. The original weight and trim waste are given. Compute (a) the percent of trim losses and (b) the average percent of losses for the three sides.

Sides of Beef	Original Weight (lb)	Trim Losses	
		Weight (lb)	Percent
A	106.54	38.35	
B	148.91	49.14	
C	173.44	60.7	

UNIT 25. AVERAGING QUANTITIES AND CHECKING BY ESTIMATING (PRACTICAL PROBLEMS)

Solve each problem and then check the answer by estimating.
1. Twenty-four castings weigh 272.5 pounds and cost $1.125 per pound. Determine the average weight of one casting and its cost.
2. Find the average weight of six sheets of metal that weigh $16\frac{1}{2}$ pounds, $12\frac{3}{4}$ pounds, 12 pounds, $12\frac{1}{4}$ pounds, $11\frac{1}{2}$ pounds, and $12\frac{1}{4}$ pounds.
3. Find the average length of rods measuring $2'-3''$, $3'-2''$, $2'-8''$, and $2'-6''$.

Determining Missing Values with Known Averages (Practical Problems)

1. At what temperature during the last hour must a piece of metal be held in a furnace to average $1,128°$? Hourly readings for the first four hours were $1,062°$, $1,110°$, $1,174°$, and $1,158°$.
2. During the first four days of a workweek, the total daily output reached 276, 320, 342, and 386 parts. The rejects each day of these totals were 5%, $4\frac{1}{2}$%, 6%, and 5%, respectively. The weekly quota to meet a contract is 325 perfect parts per day. How many parts must be produced the fifth day to meet the schedule? (Assume that the spoilage on the fifth day is the average percent of the four other days.)

GRAPHS AND STATISTICAL MEASUREMENTS

Unit 27 Development and Interpretation of Bar Graphs

OBJECTIVES OF THE UNIT

After satisfactorily studying this unit, the student/trainee will be able to
* *Translate technical information from tables for representation on bar graphs.*
* *Select appropriate scales to represent each value in practical problems in which X and Y axes are related to a common reference (origin) point.*
* *Prepare a bar graph from tabular data.*
* *Interpret bar graph values to meet specific conditions.*

PRETEST *Use the Review and Self-Test items provided in the Unit Assignment to establish the level of mathematical skills competency and to determine the starting point of instruction.*

Statistical data and other technical information are often represented in graphic form. In this way it is possible to compare one set of data with another. Equally important, values may be determined on the graph itself for various conditions. Graphs may be developed mechanically, by hand, or they may be generated by computer-aided design or other electronic equipment.

Types and Characteristics of Graphs

Many types of graphs are used in technical written materials, reports, and handbooks. The greatest number fall into one of four classifications: *picture graphs, bar graphs, circle graphs,* and *line graphs*.

Each type has certain advantages as well as disadvantages. The type of graph to use depends on the nature of the data to be presented and the skill of the person in portraying information graphically. The picture graph, sometimes called *pictogram*, is the easiest to read, but it is difficult to draw unless self-adhering, commercially available picture symbols are used. The steps required to produce and to read a bar graph are given in this unit. The development and interpretation of line graphs appear in Unit 28; the development and interpretation of circle graphs appear in Unit 29. Basic statistical concepts are covered in Unit 30.

Graph (coordinate) papers are obtainable with different spacings for varying conditions. These ruled sheets simplify the representation and interpretation of factual data. Graph papers often have scales printed on them. A 10×10 graph sheet indicates the number of equal spaces in a given area. In this instance, the 10×10 means 10 equal spaces vertically and 10 spaces horizontally.

Reference lines (sometimes called *base lines* or *axes*) of a graph intersect at a point called the *origin*. The horizontal lines on graph paper are generally associated with the *X axis;* the vertical lines, the *Y axis* values. Values on a graph are plotted according to a selected scale. The origin of the measurements may be zero (0) or any other appropriate starting value for those lines that are to be represented on the *X* and/or *Y* axes.

Wherever practical, graphs should be planned so that the units to be interpreted are read horizontally from left to right or vertically from the bottom up. Information presented horizontally is more easily readable than from a vertical position.

A. DEVELOPING BAR GRAPHS

The term bar graph merely signifies that solid lines or heavy bars of a definite length represent given quantities. Usually, graphs contain two *scales: a vertical (Y) scale* and a *horizontal (X) scale*. The scale indicates the value of each ruled line or lines. These specific values depend on the information to be presented.

RULE FOR DEVELOPING A BAR GRAPH

- Determine what information is to be presented and whether or not a bar graph is the best type to use.
- Range the data from smallest to largest or in some other logical sequence.
- Select a horizontal scale that makes it possible to represent the full range of data on the sheet.
- Select a vertical scale in the same manner.
- Determine the place on the sheet where both scales come together. Mark this point of origin as zero (0) or any other appropriate value.

- Write the vertical scale. Start at zero and add values at each major division on the graph paper.
  Note. The starting (baseline) point does not necessarily have to be zero because in many instances only higher values are needed.
- Repeat the same process for marking the horizontal scale on the graph paper.
  Note. Plan the spacing on both vertical and horizontal scales so the graph is balanced.
- Plot the values on the horizontal or vertical scale from an original table or compilation.
- Draw the bars as solid lines to furnish the required data.
- Label the scales and give the chart a descriptive title.

E XAMPLE: Develop a bar graph to show the variation in production for a one-year period from Figure 27-1.

Month	Jan.	Feb.	Mar.	Apr.	May	June	July	Aug.	Sept.	Oct.	Nov.	Dec.
Prod. Units	500	800	1,200	1,350	1,450	1,550	1,300	900	850	800	750	1,150

FIGURE 27-1

Step 1 Study the information in Figure 27-1 and determine the high and low ranges in production units. These are 1,550 and 500.
  Note. The months are arranged in a logical sequence as they will appear on the graph.

Step 2 Select a graph sheet with spaces ruled to meet the job requirements.
  Note. If no printed form is available, rule a sheet with light lines drawn to a predetermined scale.

Step 3 Determine whether to represent the months on the vertical or horizontal scale and the production units on the adjacent scale. In this case, represent the months on the horizontal scale (Figure 27-2).

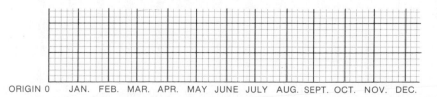

ORIGIN 0   JAN. FEB. MAR. APR. MAY JUNE JULY AUG. SEPT. OCT. NOV. DEC.

FIGURE 27-2

Step 4 Mark the major divisions or units on the vertical scale. Since the graph paper available is 10 × 10, each major vertical line is marked in intervals of 250 production units, starting at zero (Figure 27-3). Also, write what the scale represents.
Note. This scale is selected because, if a smaller one is used, the information will not fit the graph paper.

Step 5 Plot the height of the solid lines, starting with January. In this month, 500 units were produced. This value on the vertical scale is, as indicated, at the 500 point (Figure 27-4).

Step 6 Continue to plot the production units for the remaining months and recheck for accuracy.

Step 7 Label the graph so that the short descriptive title gives meaning to the facts.

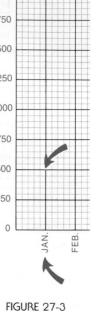

FIGURE 27-3

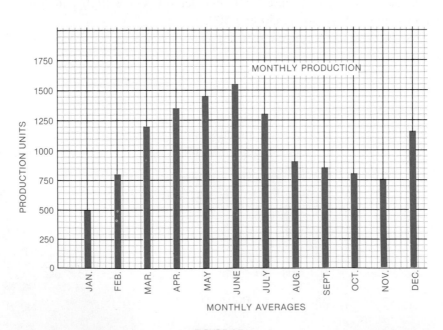

FIGURE 27-4

B. INTERPRETING BAR GRAPHS

The need for interpreting factual data from graphs already prepared makes the reading of graphs important. Regardless of the type of graph to be read, there are certain basic steps to be followed.

RULE FOR READING GRAPHS

- Study the problem to determine what information is given and what values must be determined from the graph.
- Read the title of the graph as a key to its organization and purpose.
- Determine the value which each major unit on the vertical scale represents.
- Read the horizontal scale for the value of each unit.
- Locate the given value or data on the horizontal scale.
- Follow an imaginary vertical line from this point to the end of the bar.
- Locate the length of the bar on the vertical scale.
- Continue to read other required values in the same manner.

Often the length of a bar does not fall on an even graduation. The extent to which a bar is above or below a graduated line may be estimated close enough for most practical purposes.

The same steps may be followed if the given values are on the vertical scale and the required value is represented somewhere on the horizontal scale.

EXAMPLE: Determine from the bar graph the number of pounds of brass plate used in manufacturing during the peak month and the lowest month.

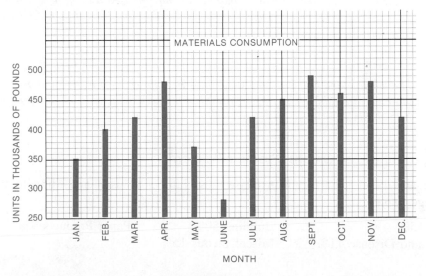

FIGURE 27-5

Step 1 Determine which bar and month represent the peak. **September**

Step 2 Read the value of each unit and subdivision on the vertical scale.
Note. The vertical scale reading = the unit reading (490 × 1,000). **490,000**

Step 3 Determine the shortest bar length. **June**

Step 4 Read the value of this bar on the vertical scale (280) and multiply by
1,000. **280,000**

Step 5 State results in specific units of measure. The greatest production month was
September when 490,000 pounds of brass plate were used. The lowest production
month was June with 280,000 pounds of brass plate.

Step 6 Continue to read the production for other months in the same manner (Figure 27-5).

ASSIGNMENT UNIT 27   REVIEW AND SELF TEST

PRETEST *The Review and Self-Test items that follow may be used as a pretest. Pretests are
designed to measure a student's beginning level of mathematical skills competency and
to determine the starting point of instruction.*

POSTTEST *The Review and Self-Test items may also be applied as a post test. Post tests are
planned to establish the student's level of mathematical skills competency after
instruction.*

A. Development of Bar Graphs (Practical Problems)

1. Make and label a bar graph to show how lumber is used according to the following
average percents.

Fuel 15%   Building 50%   Paper Products 9%
Fires and Disease 11%   Miscellaneous 15%

2. Make a bar graph to show the phenomenal motor car growth for periods A through J, according to the data given.

Period	Car Production	Period	Car Production
A	10,475,000	F	19,280,000
B	12,240,000	G	21,300,000
C	13,000,000	H	25,400,000
D	15,250,000	I	26,500,000
E	17,200,000	J	27,690,000

B. Interpretation of Bar Graphs (Practical Problems)

1. Study the horizontal bar graph.

   a. Select the greatest production year.

   b. Select the year with the smallest production.

   c. Determine the percent of increase in production between 1981 and 1984.

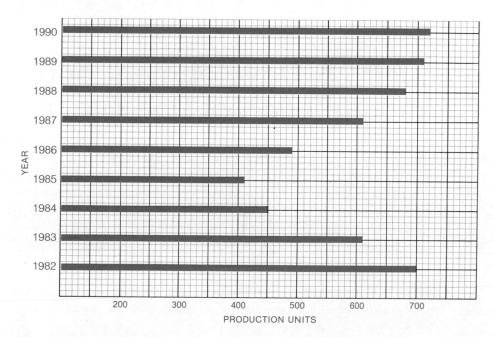

2. Study the bar graph of productivity.
   a. Select the three years of greatest productivity.
   b. Determine the average productivity for these three years, correct to two decimal places.
   c. Select the two years of lowest productivity.
   d. Determine the average productivity for these two years.
   e. Find the percent of increase in productivity between the average of the three highest years and the average of the three lowest years. (Round off answer to the nearest whole percent.)

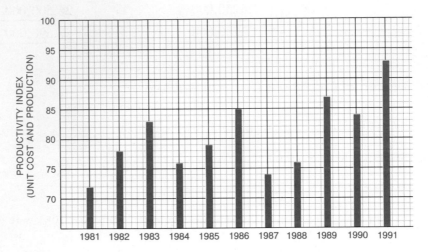

Unit 28 Development and Interpretation of Line Graphs

OBJECTIVES OF THE UNIT

After satisfactorily studying this unit, the student/trainee will be able to
- *Plan appropriate horizontal and vertical scales to represent different X- and Y-axis values on line graphs.*
- *Develop straight-line, curved-line, or broken-line graphs.*
- *Obtain technical information by interpreting X- or Y-axis quantities on line graphs.*

PRETEST   *Use the Review and Self-Test items provided in the Unit Assignment to establish the level of mathematical skills competency and to determine the starting point of instruction.*

The *line graph* is the most widely used graph. It is easy to make, presents facts clearly, and is simple to interpret. Line graphs are of three general types: straight line, curved line, and broken line.

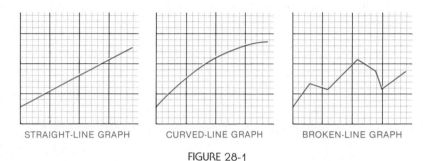

STRAIGHT-LINE GRAPH   CURVED-LINE GRAPH   BROKEN-LINE GRAPH

FIGURE 28-1

A. DEVELOPING LINE GRAPHS

The straight-line graph is used for related facts where there is some regularity in changes that take place. By contrast, the broken-line graph represents unrelated data where changes are irregular. The curved-line graph presents related information (Figure 28-1). In each instance, the graph is prepared from data that is either computed or available. One advantage of a graph is that once it is made, other values may be determined without additional computation.

RULE FOR DEVELOPING A LINE GRAPH

- Select appropriate horizontal and vertical scales so the facts may be presented clearly.
- Lay out the major graduations on graph paper for the vertical and horizontal scales and label each scale.
- Arrange the data in the same sequence as it will appear on the graph.
  Note. Sometimes these values must be computed from other given facts, using formulas.
- Plot the pairs of numbers on the horizontal and vertical scales. Place a dot or other identifying mark on the paper.
- Connect the points with either a straight edge or curve.
  Note. The line may be straight, broken, curved or a combination, depending on the relationship of data.

EXAMPLE: Make a line graph showing the variation in temperature from ground level to 8,000 feet.

Altitude in Feet	Ground Level	1,000	2,000	3,000	4,000	5,000	6,000	7,000	8,000
Air Temp. (°F)	85°	84°	79°	74°	64°	60°	47°	33°	10°

FIGURE 28-2

Step 1   Select suitable horizontal and vertical scales and label the graph.

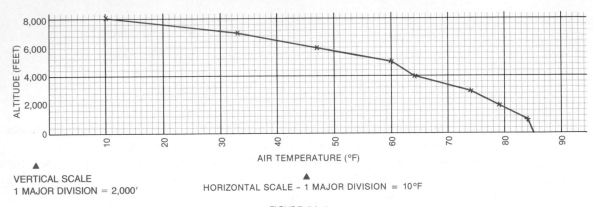

FIGURE 28-3

Step 2   Plot the points on the graph at which the pairs of numbers meet.

Step 3   Connect the points with a continuous line.

Step 4   Check the graph by taking altitudes of 2,500 feet and 6,500 feet, computing the air temperatures, and plotting these values on the graph.

Step 5   Check the original computations and plotting if the values being checked fall outside the graph line.

B. INTERPRETING LINE GRAPHS

The line graph furnishes information for comparisons of values without further computation. The results are, in most instances, accurate approximations, as it is impractical to represent data to too large a scale. The reading of a line graph is similar to that of a bar graph except that more data may be obtained.

RULE FOR READING A LINE GRAPH

- Determine what information is required.
- Locate the given value on either the horizontal or vertical scale.

- Visualize a horizontal or vertical line that passes through the given value and intersects the graph line.

- Determine the value of this point on the adjacent vertical or horizontal scale.

- Label the answer with an appropriate term.

EXAMPLE: Determine the air temperatures at altitudes of 7,500, 6,500, 5,500, 4,500, and 3,500 feet from the graph.

Step 1 Locate 7,500 feet on the vertical scale.

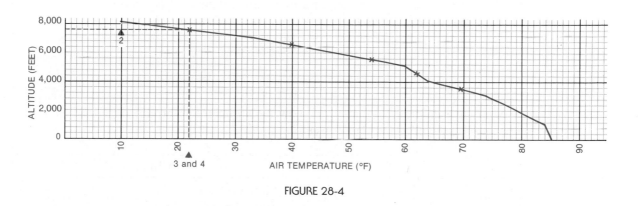

FIGURE 28-4

Step 2 Draw an imaginary horizontal line at 7,500 feet until it intersects the graph line.

Step 3 Drop another imaginary vertical line from the intersecting point to the horizontal scale.

Step 4 Read the air temperature on the scale (22°F).

Step 5 Repeat these steps for the remaining altitudes.

Step 6 Label all answers and check.

C. DEVELOPING LINE GRAPHS WITH TWO OR MORE SETS OF DATA

A graph with two or more line graphs that are prepared from different sets of technical information is widely used to make comparisons and obtain other data. While common axes are used to represent particular factors and conditions, each series of values is plotted to develop a separate line graph.

RULE FOR DEVELOPING A GRAPH THAT COMBINES TWO OR MORE LINE GRAPHS

- Determine the maximum range of the data that are to be represented on the graph.
- Mark off the major divisions of the X and Y axes on the graph paper for the items and quantities to be represented.
- Start with one set of data. Plot the value of each X and Y axis item.
- Draw the appropriate straight-, curved-, or broken-line graph to connect each point as plotted.
- Label the first line graph.
- Continue to plot the second set of values.
- Connect the points with short dashes or other distinguishing type of line.
- Label the second line graph.
  Note. Additional line graphs may be drawn using the same procedure.
- Interpret values from any one or combination of line graphs, as required.
  Note. Follow the same steps as those that are used to read a single-line graph.

E XAMPLE: Construct a multiple line graph to show the variations in productivity for machine steel parts according to the following data for cutting fluids A and B.

Cutting Fluid A					
Surface Feet per Minute	20	30	40	50	60
Productivity (Parts per Hour)	150	155	160	165	166

Cutting Fluid B					
Surface Feet per Minute	20	30	40	50	60
Productivity (Parts per Hour)	150	165	175	183	186

FIGURE 28-5

Step 1 Determine the maximum ranges of production (150 to 186) and surface feet per minute (sfpm of 20 to 60).

Step 2 Lay out the horizontal scale to represent production values; the vertical scale, surface feet per minute. Label each scale.

Step 3 Plot productivity and cutting speed values for machining with cutting fluid A.
Note. Use a solid line to represent this production.

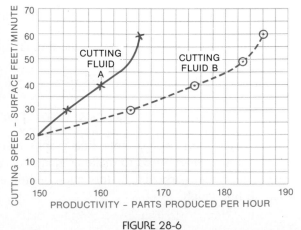

FIGURE 28-6

Step 4   Plot the values for machining with cutting fluid B.
           Note. Use a dash line for easy identification and contrast.
Step 5   Make whatever comparisons are required by interpreting the differences in performance values between each line graph.

ASSIGNMENT UNIT 28   REVIEW AND SELF TEST

PRETEST   *The Review and Self-Test items that follow may be used as a pretest. Pretests are designed to measure a student's beginning level of mathematical skills competency and to determine the starting point of instruction.*

POSTTEST *The Review and Self-Test items may also be applied as a post test. Post tests are planned to establish the student's level of mathematical skills competency after instruction.*

A. Development of Line Graphs (Practical Problems)

1. Plot a line graph that shows how the specific gravity of a battery changes as the voltage is decreased.

Voltage	2.00	1.98	1.96	1.92	1.88	1.84	1.80	1.72	1.50
Specific Gravity	1.300	1.295	1.285	1.280	1.275	1.270	1.265	1.253	1.250

2. Show by a curved line graph the number of British thermal units (heat units) required to produce a temperature range of 10°F to 140°F.

Temperature (°F)	10°	50°	80°	110°	130°	140°
Btu	500	1,000	1,500	2,000	3,500	5,000

B. Development and Interpretation of Line Graphs (Practical Problems)

1. Make a line graph of the surface speeds of grinding wheels. The diameters range from 8 inches to 16 inches inclusive. Increments in diameter are 1 inch. The constant speed is 2,500 revolutions per minute ($\pi = 3.14$)

   a. Use the formula, Surface Speed $= \dfrac{\pi \times D \times RPM}{12}$

   b. Round off surface speed values to the nearest 50 feet per minute.

2. Locate on the line graph the surface speeds of wheels worn to diameters of $8\frac{1}{2}$, $9\frac{1}{2}$, $10\frac{1}{2}$, and $11\frac{1}{2}$ inches.

   Note. Give answer to closest 50 feet per minute.

3. Make a line graph that shows the relationship of the cross-sectional area of a square pipe to its length. Use the dimensions in the table to establish the cross-sectional area and length of side.

Lengths of Side (in inches)	4	5	6	7	8	9	10

4. Locate on the graph in problem B. 3 the cross-sectional area of square pipes $4\frac{1}{2}$, $6\frac{1}{2}$, $8\frac{1}{2}$, and $9\frac{1}{4}$ inches on a side. Check the area of each measurement by computation.

5. The straight-line graph shows the relationship between the diameter of a driving pulley and the surface speed of a belt (in feet per minute ['/min]). The driving pulley revolves at 300 RPM.
   a. Determine the surface speed (in '/min) of the belt for the following drive pulley sizes (diameters): (1) $3\frac{1}{2}$, (2) 7, and (3) $10\frac{1}{2}$ inches.
   b. Locate the driver pulley diameter sizes required to produce surface speeds of (1) 475 '/min, (2) 750 '/min, and (3) 875 '/min (approximate to the nearest $\frac{1}{4}$ inch diameter).

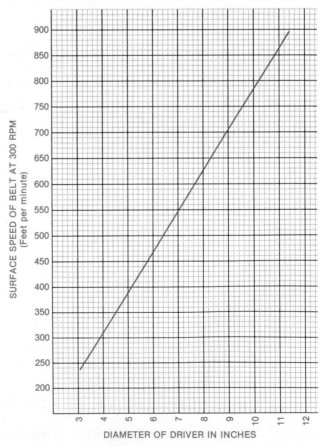

6. The curved (hyperbola) line graph represents the volume in cm³ of certain materials. The volume is related to the density in grams per cubic centimeter (g/cm³).

Estimate the density of materials A, B, C, and D.

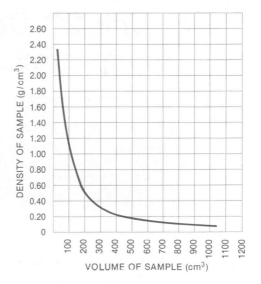

	Material	Volume (cm³)	Density (g/cm³)
A	balsa	850	
B	cork	400	
C	maple	175	
D	ice	110	

C. Interpreting Combination Multiple Line Graphs

1. The accompanying combination multiple line graph compares the production of electronic components from three different assembly systems. System A is semiautomatic. System B is numerically controlled using manual input. System C is computer-aided manufacturing.

   a. Identify the range of production hours involved in the study.
   b. Give the number of electronic components assembled in the first 42.5 hours in each system.
   c. State the productivity range of each system.
   d. Identify the system with the greatest variations in hourly production.
   e. Describe the effect on production among the three systems between 42.5 and 45.0 hours of operation and 45.0 to 47.5 hours.

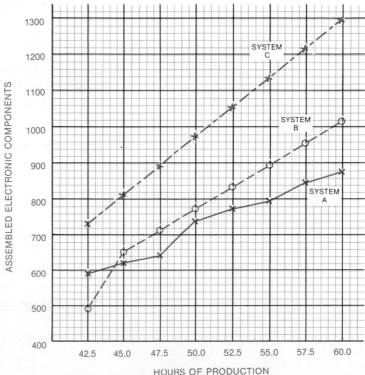

Unit 29 Development and Interpretation of Circle Graphs

OBJECTIVES OF THE UNIT

After satisfactorily studying this unit, the student/trainee will be able to
* Translate technical information found on circle graphs to equivalent sector angles for portions of circles represented.
* Interpret data in order to develop circle graphs.
* Obtain required data from the interpretation of circle graphs.

PRETEST Use the Review and Self-Test items provided in the Unit Assignment to establish the level of mathematical skills competency and to determine the starting point of instruction.

A circle graph, as the name implies, is a combination of a circle and the division of the circle into a given number of parts *(sectors)*. The circle graph (sometimes called a pie graph) is especially useful in showing how one part is related to another part and to the total.

A. DEVELOPING AND INTERPRETING CIRCLE GRAPHS

The circle graph differs from either the bar graph or the line graph in that it is used primarily for comparative purposes. It is impractical to determine other numerical values from the graph itself. One of the best uses of a circle graph is for items that equal 100%.

RULE FOR DEVELOPING CIRCLE GRAPHS

* Add together all of the items to be included on the graph. This sum is equal to 100%.
* Make a series of fractions of the data. The numerator represents one of the parts; the denominator, the total.
* Multiply the quotient of each fraction by 100 to get the percent equivalent each part is to the total.
* Multiply the quotient again, but this time by 360°, to get the equivalent angle of each part to the nearest degree.
* Draw a circle large enough to divide easily into the required number of parts so that the graph may be read easily.
* Divide the circle into the number of degrees in each part *(sector)*.

- Label each part and the chart itself with a descriptive caption and the percent which that part represents.
- Determine the relationship of each part of the total. Then compare one part with another.

As an example, one of the best ways to make quarterly comparisons graphically is to use a circle graph.

E XAMPLE: Develop a circle graph to show quarterly production according to the information given in the following table.

Quarter	First	Second	Third	Fourth
Production Units	12,500	25,000	31,250	18,750

Step 1 Total the production units.

$$12,500$$
$$25,000$$
$$31,250$$
$$\underline{18,750}$$
$$87,500$$

Step 2 Determine the fractional part of the total each quarter represents.

First	Second	Third	Fourth
$\frac{12,500}{87,500} = \frac{1}{7}$	$\frac{25,000}{87,500} = \frac{2}{7}$	$\frac{31,250}{87,500} = \frac{5}{14}$	$\frac{18,750}{87,500} = \frac{3}{14}$

Step 3 Determine the percent of the total that each quarterly production represents.

First	Second	Third	Fourth
$\frac{1}{7} = 14\frac{1}{3}\%$	$\frac{2}{7} = 28\frac{2}{3}\%$	$\frac{5}{14} = 35\frac{5}{6}\%$	$\frac{3}{14} = 21\frac{1}{6}\%$

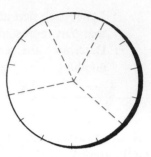

Step 4 Draw a circle and divide it into 14 parts. Then lay out each fractional part of the total.

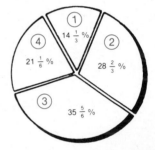

Step 5 Label the chart with a descriptive title and the percent production each quarter.

Step 6 Compare the data from the chart as may be required.

There are many ways of designating the parts of a circle graph. Another simple method is illustrated. Colors or screens may be used to emphasize and to add interest. The technique depends on the purpose and the persons who are to translate the information.

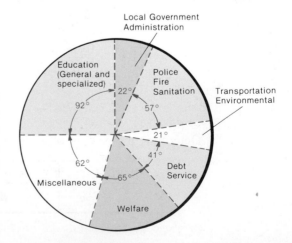

ASSIGNMENT UNIT 29   REVIEW AND SELF TEST

PRETEST *The Review and Self-Test items that follow may be used as a pretest. Pretests are designed to measure a student's beginning level of mathematical skills competency and to determine the starting point of instruction.*

POSTTEST *The Review and Self-Test items may also be applied as a post test. Post tests are planned to establish the student's level of mathematical skills competency after instruction.*

A. Development of Circle Graphs (Practical Problems)

1. Prepare a circle graph to show the relationship between the number of men and women employed in a plant where there are 1,300 men and 300 women.

2. Make a circle graph to illustrate how each dollar is spent in a particular industry.

Item	Expenditure
Admin./Eng.	$100,000
Development	250,000
Tooling	180,000
Production	160,000
Sales and Service	90,000
Taxes	220,000

B. Interpretation of Circle Graphs (Practical Problems)

1. Study the circle graph.
   a. Determine total labor force in one industry of an area labor market.
   b. What percent of the total number employed is men? Women?
   c. How many men are in the 18–35 and 36–55 year age groups? How many women?

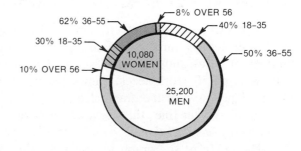

2. The following amounts were paid for materials: plastics, $7,500; steel castings, $17,500; brass castings, $14,000; paints, $2,500; and lumber, $8,500.
   a. Prepare a circle graph to show the relationship of costs.
   b. Determine the percent of the total that was spent on each material.

Unit 30 Statistical Measurements

OBJECTIVES OF THE UNIT

After satisfactorily studying this unit, the student/trainee will be able to
* *Understand the importance of statistical measurements in the manufacture and control of interchangeable parts.*
* *Translate the importance of normal frequency distribution graphs in relation to quality control of measurements of manufactured parts and products.*
* *Interpret the value of single sampling, double sampling, and sequential sampling plans in statistical measurement.*
* *Solve practical problems of central tendency measurement involving computations for range, average and mean, median, and mode.*

PRETEST *Use the Review and Self-Test items provided in the Unit Assignment to establish the level of mathematical skills competency and to determine the starting point of instruction.*

The term *statistical measurement,* as used in this unit, relates to applications of basic mathematical processes in working with recorded data and a large number of measurements. These are widely used in quality control in the manufacture of interchangeable parts and components and in other everyday applications in agriculture, business, health, and marketing. This means that statistical measurements are applied to establish standards of acceptance of mass-produced parts, assembled units, and other products.

A. NORMAL FREQUENCY DISTRIBUTION OF DIMENSIONAL MEASUREMENTS

Size variations of a given number of manufactured parts, fabrics, and other products may be established by directly measuring each unit. Graphs may then be prepared to project the number of cases and the actual sizes. For example, Figure 30-1, a graph of dimensional size variations of 50 workpieces, shows the number of pieces that range in size from 0.565″ to 0.559″. The curve of the graph also presents the greatest number *(concentration)* of workpieces to be at the center line. In this case, the greatest number of parts is machined to what is called the *nominal (design) dimension* of 0.562″.

The curve of the graph is known as a *normal frequency distribution curve.* Verified over the years by a number of studies, the curve may be plotted mathematically. The parts are distributed so that 34.13% of a total production is within the first allowable tolerance zone of say, + 0.001″. An equal number of pieces fall within the − 0.001″ allowable tolerance zone.

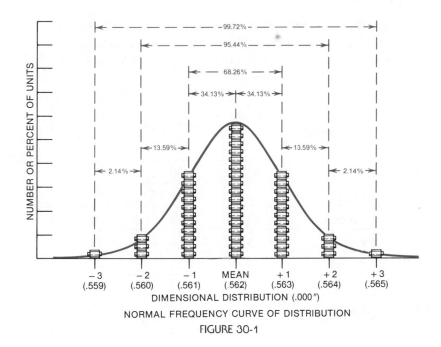

NORMAL FREQUENCY CURVE OF DISTRIBUTION

FIGURE 30-1

The total of 68.26% of the 50 parts on the graph is within ± 68.26%. Further, 95.44% of the parts are shown on the graph to be within the first and second allowable tolerance zones of ± 0.002″. In addition, 99.7% of the parts are within the maximum tolerance range of ± 0.003″. The cutoff point as to which parts meet the requirements of dimensional accuracy is determined by the parts or unit designer who specifies the maximum allowable tolerance that is acceptable.

B. SAMPLING PLANS AND STATISTICAL MEASUREMENT

Normal size variations are important to the designer, manufacturer, quality control inspector, and machine operator. Where parts are to be manufactured to extremely precise measurements, each part may be automatically inspected. Generally, a sampling plan, which is selected from technical tables, is used to establish the fixed number of parts to measure from the manufactured *lot (batch)*. From the lot size sample, it is possible to determine the *average outgoing quality limit* (AOQL). The AOQL represents the percent of defective parts. If this percent is equal to or less than the allowable number of defective parts in the sampling plan (acceptable quality level, AQL), the lot is accepted. An excessive spread of rejects based on dimensional accuracy measurements indicates the need for more precise tooling, adjustments of machine conditions, or changes in other production practices.

**Inspection Sampling Plans for Quantity Manufacturing
in Lot Sizes between 750 and 1,250 Units (Parts)
(Based on an Acceptable Quality Level of .85 to 1.75)**

Sampling Plans	Sample and Sequence	Sample Size	Combined Samples		
			Cumulative Size	Lot Acceptance Quantity	Lot Rejection Quantity
Single	Single	50	50	3	4
Double	First	34	34	1	5
	Second	68	102	4	5
Sequential	First	14	14	0*	2
	Second	14	28	0	3
	Third	14	42	1	4
	Fourth	14	56	3	5
	Fifth	14	70	3	5
	Sixth	14	84	3	5
	Seventh	14	98	4	5

*Two sample batches are required for acceptance.

FIGURE 30-2

Single Sampling Plan

There are three basic sampling plans: *single, double,* and *sequential.* The sampling plan table of lot sizes from 750 to 1,250 parts (Figure 30-2) shows that a random sample of 50 parts is required for a single sampling. If there are three or less parts in the 50 parts sampled that are defective, the whole lot (750 to 1,200 parts) is accepted. If there are four or more rejects, the lot is rejected.

Double Sampling for Dimensional Accuracy

When the lot is rejected, each item in the sample is inspected. In the *double sampling plan,* a first sample of 34 parts is used. If one part in 34 is defective, the lot is accepted. But, if 2, 3, or 4 parts are rejected, a second sample of 68 additional parts is screened. The lot is accepted if the actual measurements of parts in the second sample produce up to four rejects; otherwise, the lot is rejected. The total number of parts *(cumulative size)* inspected in this double sampling example is 102.

Sequential Sampling of Measurements

In sequential sampling in this example (Figure 30-2), there are seven different samples of 14 parts each for a total of 98 parts. If there is zero or one part defective in the first sample, a second sample must be used. Two or more rejects in the first sample calls for the lot to be rejected. Sampling continues as long as the parts fall within the lot acceptance and lot rejection numbers for each sample.

The product designer, engineer, or production department decides which sampling plan to use. The sampling plan depends upon the complexity of the part and the required degree of precision. Single sampling requires the inspection of the greatest number of units. This plan is used when manufacturing extremely accurate parts with high precision surface finish or other exacting specifications. Double and sequential sampling plans are practical with large lots where the part design has a comparatively wide range of acceptable tolerances.

C. CENTRAL TENDENCY TERMS AND APPLICATIONS

There are four common statistical terms used to describe representative variations in scores or measurements. The terms include *range, mean, mode,* and *median.*

Measurement Range

The simplest and most frequent way to report variations is to give the *range* betwcen the highest and lowest measurement or score.

RULE FOR DETERMINING A MEASUREMENT RANGE

- Arrange the values in sequence from the lowest or smallest measurement or score to the highest or largest.
- Express the *range* (difference) from the highest to the lowest values.

EXAMPLE: Determine the range of sizes of the eight die cast part batches from the dimensions given in Figure 30-3.

Batch Number	A	B	C	D	E	F	G	H
Number of Parts	30	26	24	18	16	2	2	2
Dimensional Measurements (inches)	1.125	1.126	1.124	1.127	1.123	1.128	1.122	1.120

FIGURE 30-3

Step 1 Arrange the measurements in sequence.	**1.128″ 1.124″** **1.127″ 1.123″** **1.126″ 1.122″** **1.125″ 1.120″**
Step 2 Determine the largest measurement.	**(1.128″)**
Step 3 Determine the smallest measurement.	**(1.120″)**
Step 4 State the range of the measurements from the largest to the smallest.	**1.128″ to 1.120″ Ans**

Determining the Average or Mean

The average is widely used in daily computations. The average is more technically known in statistical measurement as the *mean* (M).

RULE FOR DETERMINING THE AVERAGE OR MEAN

- Add all the scores or measurements.
- Divide the sum by the number of cases *(N)*.
- Label the measurement average in terms of the measurement unit.

E XAMPLE: Determine the average (mean) fuel consumption per hour from the data given in Figure 30-4 for six diesel engines.

Diesel Engine	A	B	C	D	E	F
Fuel Consumption (gal/hr)	75.4	82.9	76.9	85.8	96.2	75.4

FIGURE 30-4

Step 1   Add the separate fuel consumption values.

$$\begin{array}{ll} 75.4 & 85.8 \\ 82.9 & 96.2 \\ 76.9 & 75.4 \\ & \overline{492.6} \end{array}$$

Step 2   Divide the sum (492.6) by the number of diesel engines (6).

$$6\overline{)492.6}$$

Step 3   Express the mean or average in terms of the measurement unit.

82.1

82.1 gal/min   Ans

Determining the Mode of a Measurement Series

The *mode* is another measurement of central tendency. Mode means the most frequently used score in a measurement series. Mode is the midpoint of the measurements. Mode is also referred to in statistics as the *class interval*. In the previous example, 75.4 gallons per hour represents the most frequent value even though there are only two cases. If there are a great number of values between a range, like 42 and 45, the mode is the value that is midway between the two values. In this instance, the mode is 43.5.

Determining the Median

The *median* is another measure of central tendency. The median is the midpoint in a distribution of measurements or other values that are arranged in the order of size. This means the distribution is bisected: 50% fall above the median; 50% fall below.

RULE FOR DETERMINING THE MEDIAN

- Arrange the scores or measurements (Figure 30-4) in sequence according to size.

$$50\% \text{ of Scores} \begin{cases} 75.4 \\ 75.4 \\ 76.9 \end{cases}$$

- Determine the midpoint in the series of scores.

Midpoint - - - - - - - - -

$$50\% \text{ of Scores} \begin{cases} 82.9 \\ 85.8 \\ 96.2 \end{cases}$$

- Add the value of the score immediately above and below the midpoint and divide the sum by 2.

$$2)\overline{76.9 + 82.9}$$
$$79.9$$

- Label the median answer in the required measurement units.

79.9 gal/hour   Ans

ASSIGNMENT UNIT 30   REVIEW AND SELF TEST

PRETEST *The Review and Self-Test items that follow may be used as a pretest. Pretests are designed to measure a student's beginning level of mathematical skills competency and to determine the starting point of instruction.*

POSTTEST *The Review and Self-Test items may also be applied as a post test. Post tests are planned to establish the student's level of mathematical skills competency after instruction.*

A. Normal Frequency Distribution of Dimensional Measurements (Practical Problems)

1. Refer to the graph, which shows the distribution of measurements for the production of 10,000 precision-ground ball bearings. The measurements and numbers are plotted and form a normal frequency distribution curve. First-quality bearings fall within the +1 to

−1 distribution range. Second-quality bearings, having a higher dimensional tolerance, are identified between +1 and +2 and −1 and −2 ranges. Bearings with higher dimensional variations are rejected. The normal frequency distribution curve follows.

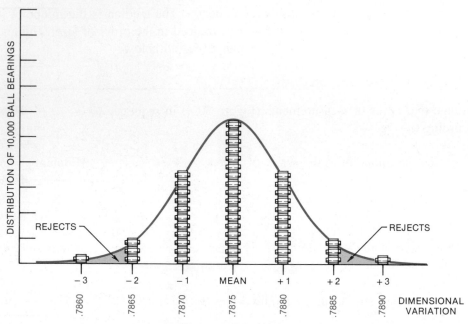

NORMAL FREQUENCY CURVE ACCORDING TO DIMENSIONAL VARIATIONS

a. (1) State the percent of first-quality bearings that fall into the + and − ranges. (2) Calculate the number of first-quality bearings in the + range of the distribution and the − range.
b. Calculate the total number of second-quality bearings that is acceptable.
c. Determine the total number of rejects.

B. Sampling Plans for Statistical Measurements (Practical Problems)

1. Use the data from the accompanying table of three sampling plans.
   a. Identify the size of single sample required for a 1,500-parts lot within the acceptable quality level of the table.
   b. State the conditions under which the quantity is accepted or rejected.
   c. Identify the number of parts to be inspected in the first sample if a double sampling plan is used.
   d. Explain when a second sampling is required. Give the total number to inspect in the double sampling plan.
   e. Assume a sequential sampling plan is used. (1) Indicate the number of sequences and (2) the total number of the 1,500 parts that are to be inspected.

Inspection Sampling Plans for Lot Sizes between 1,300 and 2,000
Parts with an Acceptable Quality Level of 1.0 to 2.2

Sampling Plans	Sample and Sequence	Sample Size	Cumulative Size	Lot Acceptance Quantity	Lot Rejection Quantity
Single	Single	75	75	4	5
Double	First	42	42	2	5
	Second	84	126	5	6
Sequential	First	16	16	0*	2
	Second	16	32	0	3
	Third	16	48	1	4
	Fourth	16	64	3	5
	Fifth	16	80	3	5
	Sixth	16	96	3	5
	Seventh	16	112	4	5
	Eighth	16	128	4	5

*Acceptance is based on the inspection of two samples.

C. Determining Central Tendency Measurements (Practical Problems)

1. Refer to the table of results of physical laboratory tests of tensile strength of seven brass/
   bronze alloy specimens. The alloys are listed according to SAE specification numbers.
   a. Determine the range of tensile strength among the specimens.
   b. Compute (1) the average (mean) tensile strength, (2) the median tensile strength, and
   (3) the mode.

Brass/Bronze Alloys	A	B	C	D	E	F	G
SAE Specification Number	40	41	43	430	62	64	65
Tensile Strength (lb/sq in.)	26,000	20,000	65,000	90,000	30,000	25,000	20,000

2. Use the data given in the following table for the weekly production of electronic devices
   for a six-week period.
   a. Give the (1) range of units produced each week and (2) the defects.
   b. Determine the average number of units produced each week.

c. Identify the production mode.

d. Calculate the median of the production number.

e. Find the number of defective parts for each weekly production run.

f. Determine the (1) mean, (2) median, and (3) mode percents for the defective electronic units produced over the six-week production period.

Weekly Production Units	1	2	3	4	5	6
	1,000	3,000	1,400	1,800	1,000	2,000
Number of Defects	2	9	5	6	3	6

Section 6
Unit 31 Achievement Review on Graphs and Statistical Measurements

OBJECTIVES OF THE UNIT

This achievement review serves as an overall test for Section 6. The unit is designed to measure the student's/trainee's ability to

- *Understand the function of graphs and practical applications in industry.*
- *Prepare and read bar graphs from written and tabular data.*
- *Develop and interpret line graphs.*
- *Apply word statements and quantities in developing and reading circle graphs.*
- *Solve problems relating to central tendency and statistical quality control measurements.*

SECTION PRETEST/ POSTTEST

The Review and Self-Test items that follow relate to each Unit within this Section. The test items may be used as a Unit-by-Unit pretest and/or Section post test.

UNIT 27. DEVELOPMENT AND INTERPRETATION OF BAR GRAPHS

1. Compute the Celsius melting points of metals A through E.

   Note. Find each Celsius reading by multiplying the Fahrenheit reading minus 32, by $\frac{5}{9}$; $\left(C = (F - 32) \times \frac{5}{9}\right)$.

	Metals	Fahrenheit	Celsius
A	Chromium	2,740°	
B	Cast Iron	2,300°	
C	Copper	1,940°	
D	Aluminum	1,200°	
E	Lead	620°	

2. Prepare a bar graph to show the melting point temperatures on both the Fahrenheit and Celsius scales for all metals in the table.
   Note. Use a solid bar for the Celsius readings and a dotted outline for Fahrenheit.
3. Determine from the graph what the differences are in Celsius readings between chromium and lead, cast iron and aluminum, and copper and lead.

UNIT 28. DEVELOPMENT AND INTERPRETATION OF LINE GRAPHS

1. Statistics compiled on the exhaust temperatures of low-, medium-, and high-compression ratio engines, running at different RPM, are given in table form. Translate these facts into a line graph.
   Note. Use a solid line for medium compression, dotted line (– – –) for low compression, and an alternate solid-dash line(—–) for high compression.

Compression Ratio	Exhaust Temperature (°F)					
Low	780	870	950	1,030	1,100	1,160
Medium	825	920	1,000	1,080	1,150	1,220
High	880	970	1,060	1,130	1,210	1,275
RPM	1,000	1,500	2,000	2,500	3,000	3,500

2. What is the average difference in exhaust temperature between a medium- and a low-compression ratio engine as shown on the graph?
3. Locate on the graph and mark the exhaust temperature of a small gasoline engine whose RPM is 1,750, 2,250, 2,750, 3,250, and 3,750.
4. Determine from the graph what the RPM of a low-compression engine is for these exhaust temperatures: a) 950°F; b) 1,050°F; c) 1,150°F; d) 1,250°F.
5. What is the difference in exhaust temperatures between a high- and a low-compression engine turning at 3,250 RPM?

UNIT 29. DEVELOPMENT AND INTERPRETATION OF CIRCLE GRAPHS

1. Draw a circle graph to compare the productive capacity of each country to manufacture a specific product.

Country	Productive Capacity
United States	175,000
France	90,000
Great Britain	125,000
Canada	100,000
All Others	60,000

2. Give the percent that each country produces of the total.
3. Determine the percent of the total that is produced by the two largest producers (to the nearest two decimal places).
4. Show by a circle graph the percent of the total produced by the two largest, two smallest, and all other producers.

UNIT 30. STATISTICAL MEASUREMENTS (PRACTICAL PROBLEMS)

1. a. State the purpose of using sampling plans in the production of automotive parts in quantities of 5,000 or more.
   b. Identify two differences between a single sampling plan and a sequential sampling plan in the manufacture of interchangeable parts.
2. a. Prepare a broken-line graph that shows the °F (and corresponding °C) tempering temperature range for the different plain carbon steel tools as recorded in the table.

Plain Carbon Steel Tools		Tempering Temperatures	
A	Roughing mills	430	221
B	Counterbores	460	238
C	Knurls	485	251
D	Tube cutters	485	251
E	Taps	500	260
F	Threading dies	530	277
G	Pneumatic tools	580	302
H	Noncutting tools	640	338

   b. Compute the average (mean) tempering temperature in °F and °C.
   c. Determine (1) the median temperature and (2) the mode temperatures in °F and °C.

PART TWO

Fundamentals of SI Metric Measurement

Section 7 METRICATION: SYSTEMS, INSTRUMENTS, AND MEASUREMENT CONVERSIONS ■ 221

METRICATION: SYSTEMS, INSTRUMENTS, AND MEASUREMENT CONVERSIONS

Unit 32 SI Metric Units of Measurement

OBJECTIVES OF THE UNIT

After satisfactorily studying this unit, the student/trainee will be able to

- *Understand why SI metrics is an evolving international system of measurement standards that is affected by high-technology developments.*
- *Translate the advantages of SI metrics over customary units of measure in mathematical computations.*
- *Apply the Scientific Notation System using prefixes, symbols, and power of ten values.*
- *Use established guidelines of writing problems involving SI metric units of measure.*
- *Solve practical problems in SI metrics relating to direct and computed linear measurements and computations of areas, volumes (including liquid quantities), mass (weight), and temperatures in Fahrenheit and Celsius degrees.*

 PRETEST *Use the Review and Self-Test items provided in the Unit Assignment to establish the level of mathematical skills competency and to determine the starting point of instruction.*

This section has five units, including an end-of-section achievement review. SI metrics is introduced in the first of the units. Some advantages are given in contrast with customary inch and other British and American units of measurement. Applications are made of the Scientific Notation System, writing styles, and basic mathematics. These are related to problems involving linear, circular, area, volume and liquid, and temperature measurements.

Direct precision instruments and gages are applied in Unit 33 to linear measurements. Principles of metric measurement with inside, outside, and depth micrometers are covered.

Experience with vernier micrometers, height gages, and gage blocks is extended to cover the use of metric measuring instruments.

Unit 34 deals with the seven base units, common derived units, and supplementary units in SI metrics. Attention is directed to soft and hard conversion of measurements, tables of conversion values, and converting dimensions and other quantities between different measuring systems.

A. ADVANTAGES OF SI METRICS

SI metrics is recognized as an international system of accurately measuring and quantitatively defining all measurable objects. "SI" was universally adopted as the abbreviation of the International System of Units *(Système International d'Unités)* at the 1960 CGPM. SI is also referred to as "SI metrics," the term used in this book. Terms and certain units are spelled according to accepted American standards. The following are some of the most important advantages of SI metrics.

- Seven base units are used in the *measurement of every known physical quantity*. These seven include the *meter* (length), *kilogram* (mass), *second* (time), *kelvin* (temperature), *ampere* (electric current), *candela* (light intensity), and the *mole* (substance of a system). In addition, there are two supplementary units, *radian* and *steradian,* and a number of derived units.
- A well-defined set of symbols and abbreviations is established, which is adequate to define all conditions and phenomena.
- Mathematically, a set of prefixes, multiples, and submultiples is used to simplify computations involving large values or number of digits.
- SI is a "coherent system." The product or quotient of any two unit quantities in the system is a unit of the resulting quantity. For example, in a coherent system, multiplying unit length by unit length produces unit area.
- SI base units are precisely defined and are reproducible in laboratories in each country. The one exception is the kilogram (mass) standard. This is still preserved in the International Bureau of Weights and Measures.
- The SI system may be related by powers of ten to other units that are not a part of the system.
- SI provides for a single international standards system that affects the interchangeability of parts, processes, components, and systems.

B. SCIENTIFIC NOTATION SYSTEM: SI PREFIXES, SYMBOLS, AND POWER OF TEN VALUES

The term *SI* or *SI metric* is used in this textbook as an abbreviation of *le Système International d'Unités*. The evolving system was named by the *Conference Général des Poids et Mesures* (CGPM) in 1960. Quantities are measured and computed in SI metrics according to the *scientific notation system*. This system has many advantages. Mathematical processes involving many multiple-digit values are simplified. Errors in calculations are reduced. Quantities are easier to read and write.

The quantity 10 is used as the foundation of the scientific notation system. The writing of mathematical quantities and mathematical computations is simplified by using what are called *power of ten multiple* (greater than 1) *and submultiple* (smaller than 1) *values*. A number of the powers are further identified by a *prefix* or a *symbol* for the prefix. A prefix identifies a specific value or quantity associated with a particular unit of measurement. For instance, *centi* in centimeter means one-hundredth part of a meter. *Centi* is called the prefix.

Selected prefixes, symbols, and power of ten values are given in Figure 32-1. Section (A) shows decimal *multiple values* of SI units. Note the simplified way of writing one million (1,000,000) as (10^6). The 10^6 is the same as the value obtained by multiplying 1 by 10 for 6 times. The designation for 1,000,000 is *mega*. *Mega* (M) is called a prefix.

Prefix	Symbol	Value as Power of Ten	Multiplication Factor
(A) Power of Ten Multiple Values			
deka-	da	10	10
hecto-	h	10^2	100
kilo-	k	10^3	1 000
mega-	M	10^6	1 000 000
giga-	G	10^9	1 000 000 000
tera-	T	10^{12}	1 000 000 000 000
(B) Power of Ten Submultiple Values			
deci-	d	10^{-1}	0.1
centi-	c	10^{-2}	0.01
milli-	m	10^{-3}	0.001
micro-	μ	10^{-6}	0.000 001
nano-	n	10^{-9}	0.000 000 001
pico-	p	10^{-12}	0.000 000 000 001
femto-	f	10^{-15}	0.000 000 000 000 001
atto-	a	10^{-18}	0.000 000 000 000 000 001

FIGURE 32-1 Selected SI prefixes, symbols, and power-of-ten multiple and submultiple values

Every measurement is stated in terms of (1) a *quantity* and (2) a *unit of measure*. An answer of 10.2×10^6 states a quantity. It must be more descriptive, however, and also relate to a specific unit of measure. If the problem deals with waveforms that are measured by cycles (or frequency), the unit of measure is *hertz* (Hz). Thus, the answer is stated as 10.2 MHz (megahertz). This quantity quickly defines 10,200,000 hertz.

Section (B) in Figure 32-1 shows decimal *submultiple* values of SI units. Note that there is a different set of prefixes and symbols required. In either instance, a prefix and a symbol are followed by a particular unit of measure. The many mathematical processes and practical applications that are based on the scientific notation system follow.

C. GUIDELINES FOR COMMUNICATING IN SI METRICS

As a coherent system, SI metrics requires a uniform set of rules for communication. *Numerical* and *literal* (letter) *quantities* are identified with specific SI units of measure. The use of SI metric units follows prescribed guidelines.

- Specific measurements (quantities) are given in terms of base units, derived units, supplementary units, and combinations of these.
- Digits and decimals, in excess of a required degree of precise measurement, are eliminated.
- Multiple and submultiple power of ten quantities are used in computations.
- Prefixes express the "order of magnitude" of a quantity. For example, 200 kg and 24.26 MV define precisely a 200 kilogram mass and a 24.26 megavolt circuit, respectively.
- A prefix is considered to be combined to the symbols to which it is attached.
  Example: $1 \text{ cm}^3 = (10^{-2} \text{ m})^3 = 10^{-6} \text{ m}^3$
- Prefix values of 1 000 are preferred over lesser or greater values. For example, milli- (m, 10^{-3}), micro- (μ, 10^{-6}), and pico- (p, 10^{-12}) are used in expressing submultiple values. The most common and preferred power of ten multiples are the kilo- (k, 10^3), mega- (M, 10^6), and giga- (G, 10^9).
- Units of measure that are derived from proper names are capitalized; for example, volts (V), siemens (S), and henry (H).
- Numerical prefixes of mega-, giga-, and tera- are capitalized: (M), (G), and (T), respectively.
- SI metric symbols of units of measure are written in singular form. For example, 125 megavolts is written as 125 MV; 12.7 kilometers as 12.7 km.
- Numerical quantities are written in groups of three digits. Commas are omitted. A quantity like 96243.8275 pascals of pressure is written 96 243.827 5 Pa.

D. APPLYING METRIC UNITS OF LINEAR MEASURE

The standard units of linear measure in the metric system are given in Figure 32-2. Unit values appear with appropriate symbols that simplify the writing of each unit.

Metric System			
	Common Linear Units	Symbol	Value of Unit in Terms of the Meter
	1 meter	m	Standard Unit of Length
Submultiple Values	1 decimeter $= \frac{1}{10}$	dm	0.1 meter
	1 centimeter $= \frac{1}{100}$	cm	0.01 meter
	1 millimeter $= \frac{1}{1000}$	mm	0.001 meter
Multiple Values	1 dekameter $= 10$	dam	10. meters
	1 hectometer $= 100$	hm	100. meters
	1 kilometer $= 1\ 000$	km	1 000. meters

FIGURE 32-2 Values of metric units of linear measure

Where dimensions in the metric system are expressed in more than one kind of unit in a system of measure, the different kinds may be combined to simplify the mathematical processes. For example, a dimension 3 decimeters 4 centimeters 6 millimeters long (3 dm 4 cm 6 mm) may be expressed as 3.46 decimeters. This value is based on the fact that 10 cm = 1 dm and 100 mm = 1 dm. The 4 cm is equivalent to $\frac{4}{10}$ and the 6 mm to $\frac{6}{100}$ of a decimeter. The dimension then is equal to $3 + \frac{4}{10} + \frac{6}{100}$ decimeters. Expressed as a decimal, this value is 3.46 dm.

Dimensions in the metric system that are given as units of linear, surface, or volume measure may be simplified in the same way.

Direct Linear Measurements in the Metric System

Metric measurements may be computed or taken directly. Measuring tools and precision instruments are similar to those used for customary units. The measurements are different, however. Whole and fractional part values in the customary system are related to the inch as the base unit. Metric measurements are in terms of the meter as the base unit.

Comparatively rough measurements may be taken directly with meter sticks and rulers graduated in centimeters and fractional parts. More accurate measurements are made with other line-graduated measuring tools. Metric triangular engineering and flat rules, drafting machine rules, and metal rules are examples. Some of these tools are graduated for measurements in one millimeter to $\frac{1}{2}$ millimeter (mm) range (Figure 32-3). Measurements to accuracies of 0.02 mm, 0.002 mm, and finer are covered in Unit 33 in relation to vernier measuring tools and gage blocks.

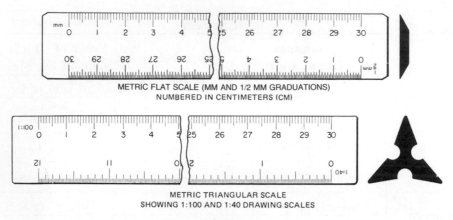

METRIC FLAT SCALE (MM AND 1/2 MM GRADUATIONS)
NUMBERED IN CENTIMETERS (CM)

METRIC TRIANGULAR SCALE
SHOWING 1:100 AND 1:40 DRAWING SCALES

FIGURE 32-3

E. COMPUTING SQUARE MEASURE (AREAS) IN SI METRICS

The principles of surface measure that are used with the British customary units are applied in the same manner with the metric units of measure. The names of the units and the values of each differ in both systems, however.

The area of a surface in the metric system is the number of square metric units that it contains. The three metric units most commonly used for small areas are the square meter, square centimeter, and square decimeter (Figure 32-4).

One square meter represents the area of a square figure that is one meter long and one meter high. The linear meter, in turn, is equal to 10 decimeters or 100 centimeters. By substituting these values for the meter, the area of a square 10 decimeters on a side is 100 square decimeters. The area of a square 100 centimeters on a side is 10,000 square centimeters. Thus, the value of the square meter, which is the standard unit in terms of decimeters and centimeters, is 100 square decimeters or the equivalent, 10,000 square centimeters.

The three common metric units of square measure and the value of each unit in the metric system are given in Figure 32-5.

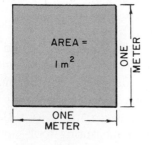

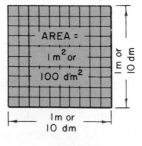

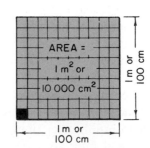

FIGURE 32-4

Metric Unit of Surface Measure	Value of Unit in Metric System
1 sq meter (m^2)	Standard Unit of Measure
1 sq decimeter (dm^2)	0.01 square meter (sq m)
1 sq centimeter (cm^2)	0.000 1 sq m
1 sq millimeter (mm^2)	0.000 001 sq m

FIGURE 32-5 Values of metric units of square measure

The principles of surface measure that apply to the area measurement of squares, rectangles, parallelograms, trapezoids, triangles, circles, sectors of a circle, and cylinders in British units are the same for the metric system. The only difference is in the name of the unit and its size. The same rules for determining the area of a surface apply for both the British and metric systems (Figure 32-5).

F. DETERMINING VOLUME MEASURE IN SI METRICS

The volume of a body is the measurement of its cubical contents. These are expressed in cubic units of the same kind as the linear units. When the linear dimensions are given as metric units, the volume is the number of cubic metric units that the body contains. The common metric units are the meter, decimeter, centimeter, and millimeter (Figure 32-6). The volume is computed as so many cubic meters (m^3), cubic decimeters (dm^3), cubic centimeters (cm^3), or cubic millimeters (mm^3).

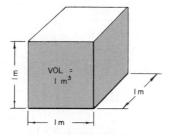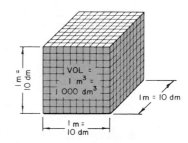

FIGURE 32-6

Since 1 meter = 10 decimeters, the volume of a cube 10 decimeters on a side is 1 000 cubic decimeters (dm^3), which is equivalent to 1 cubic meter (m^3).

One cubic decimeter is the volume of a cube one decimeter long, one decimeter high, and one decimeter deep. One cubic decimeter is equivalent also to the volume of a cube that measures 10 centimeters on a side (10 cm = 1 dm) or 1 000 cubic centimeters.

One cubic centimeter is the volume of a cube one centimeter long, one centimeter high, and one centimeter deep. Since the centimeter is equal to 10 millimeters, one cubic centimeter is equal to the volume of a cube 10 millimeters on a side, or 1 000 cubic millimeters. Values in the metric system are given in Figure 32-7 for the cubic meter, cubic decimeter, and cubic centimeter.

Metric Unit of Volume Measure	Value of Unit in Metric System
1 cu meter (m^3)	1 000 cu decimeters (dm^3)
1 cu decimeter (dm^3)	1 000 cu centimeters (cm^3)
1 cu centimeter (cm^3)	1 000 cu millimeters (mm^3)

FIGURE 32-7 Values of common metric units of volume measure

The measurement of the cubical contents of squares, rectangular solids, and cylinders is the same for both the British and metric systems of volume measure. The volume of a solid is always computed in the same manner regardless of the system in which the linear dimensions are expressed.

G. PRINCIPLES OF VOLUME MEASURE APPLIED TO LIQUID MEASURE (METRIC SYSTEM)

The standard unit of liquid measure in the metric system is the liter (ℓ). This unit, as defined by law, is equivalent to a volume of 1 000 cubic centimeters or 1 cubic decimeter.

Metric Unit of Liquid Measure	Value of the Liter — Value of Unit in Metric System
1 liter (ℓ)	Standard Unit of Liquid Measure
	1 000 cm^3 1 dm^3

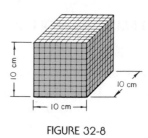

FIGURE 32-8

The same methods of computing the liquid capacity of a container are used regardless of whether the volume is expressed in British or metric units. Known values in one system may be changed readily to a desired unit in the other system.

H. DETERMINING METRIC UNITS OF MASS (WEIGHT) MEASUREMENT

Mass in physical science terms means the measurement of the earth's force in attracting a material. Mass and weight are often used interchangeably. The *gram* (g) is the base unit of mass measurement in SI metrics. Technically, one gram represents the mass (weight) of one cubic centimeter of water at 4° Celsius temperature.

Prefixes are used to identify larger and smaller quantities than one gram. The prefixes *deka, hecto,* and *kilo,* and the *metric ton* are used for larger units. Smaller weight units are identified by the prefixes *deci, centi,* and *milli.* The value of each unit of mass measurement in relation to the gram is recorded in the table.

1 000 kilograms	= 1 metric ton	
1 000 grams	= 1 kilogram (kg)	
100 grams	= 1 hectogram (hg)	
10 grams	= 1 dekagram (dag)	
1 gram (g)	= Base Unit of Measure	
0.1 gram	= 1 decigram (dg)	
0.01 gram	= 1 centigram (cg)	
0.001 gram	= 1 milligram (mg)	

FIGURE 32-9 Common units of metric mass (weight) measure

RULE FOR CHANGING METRIC UNITS OF MASS (WEIGHT) MEASURE

- Determine whether the given unit of measure is to be changed to a larger or a smaller unit.
- Multiply by 10 (or add a zero digit while moving the decimal point one place to the right) to change from one SI metric unit of mass measure to the next smaller unit.
  Note. When changing to successively smaller units of measure, add additional zeros to the right according to the quantity relationship.
- Divide by 10 to change from one unit of mass measure to the next larger unit or by a multiple of ten depending on the required larger unit.
- Express the changed value in terms of the required unit of measure.

EXAMPLE: *Case 1*. Change 1.24 grams to equivalent centigram units.

Step 1	Determine the relationship between a gram and a centigram.	**1 centigram (cg) = 0.01 gram (g)**
Step 2	Change the larger gram unit to the smaller centigram unit by multiplying by 100.	**100 cg = 1 g** **1.24 × 100 = 124**
Step 3	Express the answer in the required measurement unit.	**124 grams (g) Ans**

EXAMPLE: *Case 2*. Change 4 746 kilograms to the equivalent metric tons.

Step 1	Determine the number of kilograms in a metric ton.	**1 000 kg = 1 metric ton**
Step 2	Divide by 1 000 by moving the decimal point three places to the left	**4 746**
Step 3	Label the answer.	**4.746 metric tons Ans**

I. DETERMINING CELSIUS AND FAHRENHEIT TEMPERATURE MEASUREMENTS

The metric Celsius and the Fahrenheit temperature systems of measuring temperature changes are the two most commonly used systems. The unit of temperature measurement in both systems is called *degree* and is denoted by the symbol (°). However, the value of 1° on the Fahrenheit scale differs from 1° on the Celsius scale.

Temperature scales are calibrated in relation to the boiling and freezing points of water at a standard pressure. On the Fahrenheit scale, 32° is used to indicate the temperature at which water freezes; 212° indicates the boiling point. The corresponding points on a Celsius scale are 0° for freezing and 100° for boiling. By comparison, the freezing points are 0°C and 32°F and the boiling points are 100°C and 212°F. This means there are 100 one-degree graduations on the Celsius scale and 180 one-degree graduations on the Fahrenheit scale, beginning at 32°F. Celsius degree temperatures that are below the equivalent 32°F are always negative. For example, a temperature like 27°F is equal to −5°C. Formulas requiring simple mathematical processes are used to convert temperature values from one scale to the other.

RULE FOR CONVERTING TEMPERATURE MEASUREMENTS BETWEEN SYSTEMS

- Use the formula $°C = (°F - 32) \times \frac{5}{9}$ to change °F to °C.
- Use the formula $°F = \left(°C \times \frac{9}{5}\right) + 32$ to change °C to °F.
- Substitute given temperature values in the appropriate formula.
- Perform the mathematical processes. Label the answer in the required temperature system.

EXAMPLE: *Case 1*. Determine the equivalent temperature in degrees Celsius for subzero heat treating steel gages at −103°F.

Step 1	Write the formula for changing °F to °C.	$°C = (°F - 32) \times \frac{5}{9}$
Step 2	Substitute °F values.	$°C = (-103 - 32) \times \frac{5}{9}$
Step 3	Perform the required mathematical processes.	$= (-135) \times \frac{5}{9}$
Step 4	Express the answer in the required temperature measurement.	$= 75°C$ Ans

EXAMPLE: *Case 2*. Compute the equivalent °F temperature required to anneal an SAE 1040 steel that requires heating to 665°C.

Step 1	Use the formula.	$°F = \left(°C \times \frac{9}{5}\right) + 32$
Step 2	Substitute the given °C temperature value of 665.	$°F = \left(665 \times \frac{9}{5}\right) + 32$
Step 3	Perform the mathematical processes.	$= (1197) + 32$
Step 4	Label the answer in °F.	$= 1229°F$ Ans

ASSIGNMENT UNIT 32   REVIEW AND SELF TEST

PRETEST
The Review and Self-Test items that follow may be used as a pretest. Pretests are designed to measure a student's beginning level of mathematical skills competency and to determine the starting point of instruction.

POSTTEST
The Review and Self-Test items may also be applied as a post test. Post tests are planned to establish the student's level of mathematical skills competency after instruction.

A. Applying the Scientific Notation System and SI Metric Units of Measurement (General Applications)

1. Match each term in column I with its corresponding value in column II.

Column I	Column II
Term	*Value*

   Column I — *Term*
   1. Multiple power of ten
   2. Submultiple power of ten
   3. M (mega)
   4. k (kilo)
   5. deci (d)
   6. micro (μ)

   Column II — *Value*
   a. 10 (power of -1 or less)   f. 10 (power of -1 or more)
   b. 0.000 1   g. 100
   c. 0.000 000 1   h. 10^3
   d. 10^{-1} i. 10^6
   e. 10^{-6} j. 1 billion

2. Take each of the four computed quantities in the table.
   a. Write each one as an SI metric quantity.
   b. Round off each value to the indicated degree of accuracy. State this new quantity.

	Computed Quantity	Written as an SI Metric Quantity	Required Accuracy (Decimals)	Rounded-off SI Metric Quantity
A	7624.2796		2	
B	1729685.499		0	
C	300002.9768392		4	
D	1753.0501		1	

B. Direct Linear Measurements in the Metric System

1. Read and record each metric measurement Ⓐ through Ⓙ as indicated. The first rule has graduations in millimeters. On the second rule the numbered graduations are in centimeters. Each division represents 1 millimeter. State dimensions Ⓕ through Ⓙ in terms of centimeter values.

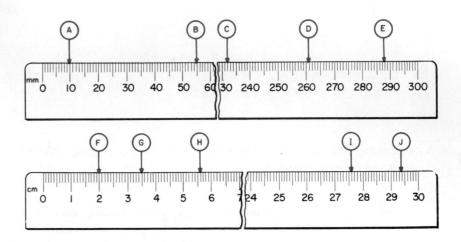

2. Measure the length of lines (a) through (i) to the degree of accuracy indicated. Measurements are to be metric.

a. ⊢————————————————————————————┤ ⎫
b. ⊢——————————————————┤   ⎬ Nearest cm
                                      ⎭
c. ⊢——————————————————————————┤   ⎫
d. ⊢————————————————————————————————┤ ⎬ Nearest .5 cm
                                      ⎭
e. ⊢————————————————————————————┤   ⎫
f. ⊢————————————————————————┤   ⎬ Nearest .2 cm
                                      ⎭
g. ⊢——————————————————————————————————┤ ⎫
h. ⊢——————————————┤   ⎬ Nearest mm
i. ⊢————————————————————————————————┤ ⎭

3. Measure and record lengths Ⓐ, Ⓑ, Ⓒ, and Ⓓ. Dimensions are in mm. Add the distance to the first center line to the sum of Ⓐ, Ⓑ, Ⓒ, and Ⓓ. Check this overall measurement with dimension Ⓔ.

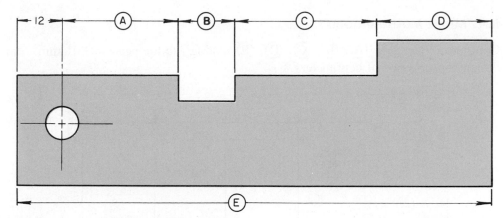

4. Measure the outside diameters of bars A, B, C, D, and E with a caliper. Transfer this measurement to a rule. Record the diameter to the nearest 0.1 centimeter.

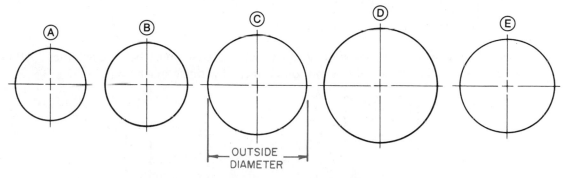

5. Use a metric architect's scale of 1:100.
   a. Measure the lengths of sections Ⓐ, Ⓑ, Ⓒ, and Ⓓ. Record each measurement to the nearest centimeter (cm).
   b. Give the overall length Ⓔ of the structure. Then, add each dimension and check the sum against the overall length.

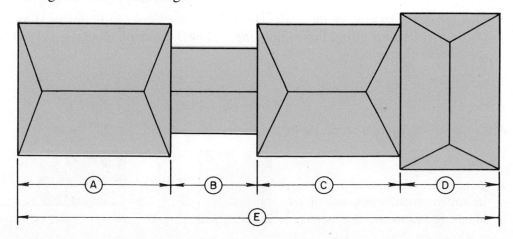

C. Indirect Metric Measurements

1. Compute dimensions Ⓐ, Ⓑ, Ⓒ, Ⓓ, Ⓔ, and Ⓕ to the nearest 0.01 mm. State each dimension in terms of centimeters.

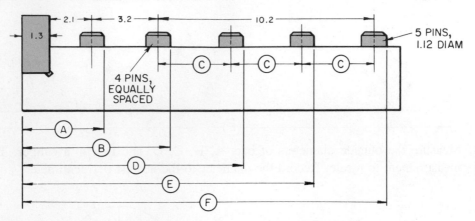

2. Calculate dimensions Ⓐ through Ⓙ to which the shaft must be rough turned. Allow 1.6 mm on all diameters and 0.8 mm on all faces for finish machining.

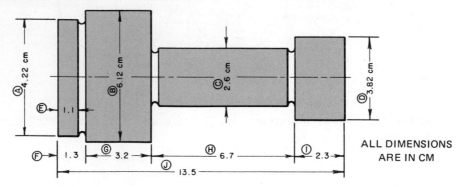

3. Use the same drawing of the shaft. Determine dimensions Ⓐ through Ⓙ to which the part must be finish turned before grinding. Allow 0.2 mm on all diameters and 0.16 mm on all faces for grinding.

D. Application of Square Measure in the Metric System

1. Determine the area of square Ⓐ in square centimeters and of rectangle Ⓑ in square decimeters.

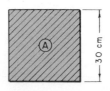

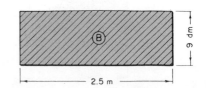

2. Determine the area of circles A and B from the given diameters and the area of C and D from the given radii.

	A	B	C	D
Diameter	4 dm	7.5 cm	—	—
Radius	—	—	2 m	22.4 cm

3. Determine the area of the sector of circle Ⓐ and the shaded area of Ⓑ.

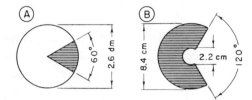

4. Compute the cross-sectional area of the link. Express the result in square inches correct to two decimal places.

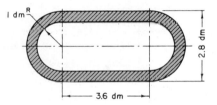

5. A form is to be built for a circular pool. The pool is 4.34 meters in diameter by 1.5 meters high. Find (a) the lateral surface area and (b) the number of boards required to construct the form. The boards are 7.62 cm wide. Round off the surface area to two decimal places and the boards to the next board.

E. Application of Volume Measure in SI Metrics

1. Determine the volume of cube A in cubic decimeters and rectangular solid B in cubic meters.

	A	B
Length	6.2 dm	1.2 m
Depth	6.2 dm	18 dm
Height	6.2 dm	9.4 dm

2. Find the volume of the rectangular parts Ⓐ and Ⓑ.

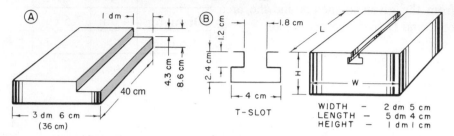

3. Determine the volume of cylinders A, B, and C. Express the volume correct to two decimal places in the unit of measure indicated in each case.

			Volume in
A	Diam	8.2 cm	cm³
	Length	24.6 cm	
B	Radius	2.4 dm	dm³
	Length	3.2 dm	
C	Diam	6 cm 4 mm	cm³
	Length	4 dm 4 cm 8 mm	

4. Find the volume of the bronze bushing in cubic decimeters.

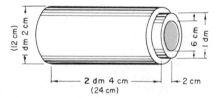

F. Application of Liquid Measure in the Metric System

1. Find the liquid capacity of the rectangular coolant tank to the nearest half liter.

INSIDE DIMENSIONS
LENGTH – 7.4 dm
WIDTH – 3.2 dm
LIQUID HEIGHT–2.6 dm

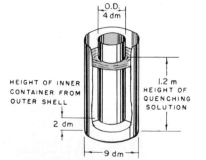

2. Determine the liquid capacity of the outer shell of the quenching tank to the nearest gallon.

3. Compute the liquid capacity of containers A, B, C, and D. State each capacity in the unit of measure indicated in the table, rounded to two decimal places.

Container	Diameter (*d*) or Radius (*r*)	Height (*h*)	Required Unit of Measure
A	(*d*) 25.4 cm	38.1 cm	cm^3
B	(*d*) 15.24 cm	11.43 cm	liters
C	(*r*) 64.77 cm	76.84 cm	qt
D	(*d*) 1.3 m	2.36 m	gal

G. SI Metric Units of Mass (Weight) Measurement (General Applications)

1. a. Identify two metric mass (weight) measurement units that are (a) larger than a gram and (b) two smaller units than a gram.

	Larger Units		Smaller Units	
a. Designation				
b. Unit Symbol				
c. Value in Relation to 1 Gram				

   b. Give the measurement symbol for cach larger and smaller unit.
   c. State the value of each larger and each smaller unit in relation to one gram.

2. Change each of the following weight measurements (a through f) to the unit of measurement specified in each case.

   a. 4 978 milligrams to grams   d. 22.5 kilograms to grams
   b. 396.2 centigrams to grams   e. 6.8 metric tons to kilograms
   c. 52.5 grams to decigrams   f. 96 hectograms to kilograms

H. Converting Celsius and Fahrenheit Temperature Measurements (Practical Problems)

1. Convert the °F at which the normalizing and hardening heat treating processes are carried on to equivalent degree Celsius temperatures.

SAE Steel Number 1080	Heat Treating Processes in °F			
	Normalizing	Annealing	Hardening	Tempering
	1,550		1,450	
	Equivalent Heat Treating Temperatures in °C			
		760		232.2

2. Compute the equivalent °F temperatures to those Celsius degree heat treating temperatures given in the table for annealing and tempering the SAE 1080 steel part.

Unit 33 Precision Measurement: Metric Measuring Instruments

OBJECTIVES OF THE UNIT

After satisfactorily studying this unit, the student/trainee will be able to

- *Understand how the linear movement of a spindle or vernier beam (in relation to a fixed anvil or base) is transferred to become a precision measurement.*
- *Read and set metric inside, outside, and depth micrometers to an accuracy of ±0.01 mm.*
- *Read and set metric vernier micrometers to accuracies of 0.002 mm and 0.001 mm.*
- *Read and set metric vernier calipers and height gages to within 0.01 mm.*
- *Use the vernier bevel protractor to read angular dimensions within tolerances of ±5 minutes for applications in any system of measurement.*
- *Establish metric gage block combinations for setting instruments and other gages within accuracies of ±0.000 02 mm.*

PRETEST   *Use the Review and Self-Test items provided in the Unit Assignment to establish the level of mathematical skills competency and to determine the starting point of instruction.*

Design features, principles used in graduating instruments, and applications of metric micrometers, verniers, and gage blocks are similar to those employed in the customary inch system. The main difference is that metric instruments are graduated in relation to the millimeter as the standard unit of measure.

The most commonly used linear metric measuring instruments, besides the steel rule and metal measuring tape, include the standard (0.01 mm) and vernier (0.002 mm) micrometers. The graduated standard micrometer head (consisting of a barrel and thimble) is adapted to inside, outside, depth, and other special micrometers. The metric calibrated vernier caliper beam and scales are also used on height and depth gages, gear-tooth calipers, and other precision calipers.

This unit deals with metric graduations on measuring instruments and practical applications to linear measurements in the metric system. The vernier principle is applied to the bevel protractor for precision angle measurements. Previous experiences with inch-standard gage blocks are extended to include metric gage block combinations.

A. GRADUATIONS: READING AND SETTING THE STANDARD (0.01 mm) METRIC MICROMETER

The standard metric micrometer is graduated to read to an accuracy of 0.01 mm (equivalent to 0.0004″). The addition of a vernier scale on the barrel makes it possible to improve the accuracy to within 0.002 mm (0.000 08″). The principles of reading measuring rules, micrometers, vernier instruments, and gage blocks are the same, regardless of the system. The standard metric micrometer has line graduations along the barrel at intervals of 0.5 millimeter. Enlarged graduations are shown on the line drawing (Figure 33-1). The graduations indicate the distance the thimble and spindle move each complete revolution $\left(\frac{1}{2} \text{ mm or } 0.5 \text{ mm}\right)$. Each graduation above the index line represents a measurement of 1 mm. Each graduation below the index line represents 0.5 mm.

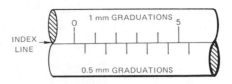

FIGURE 33-1 Graduations on barrel (enlarged)

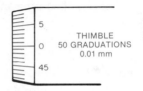

FIGURE 33-2 Graduations on micrometer thimble (enlarged)

The thimble (Figure 33-2) has 50 graduations cut into and around the beveled edge. Each graduation represents $\frac{1}{50}$ of 0.5 mm or 0.01 mm. The linear measurement is read directly by combining the barrel and sleeve readings. The line drawing (Figure 33-3) shows a reading of 5.5 mm on the barrel and another 41 (mm) graduations of the sleeve. The micrometer reading is 5.5 mm plus 41 mm, totaling 5.91 mm.

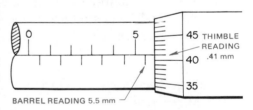

FIGURE 33-3 Combined barrel and thimble reading (enlarged)

RULE FOR READING AND SETTING A STANDARD (0.01 mm) METRIC MICROMETER

- Read the last visible 1 mm graduation above the index line on the barrel.
- Add 0.5 mm for any visible graduation below the index line.
- Read and add the 0.01 mm graduation line on the thimble at the index line of the barrel.

  Note. The sum of the readings represents the linear measurement correct to 0.01 mm.

EXAMPLE: *Case 1.* Read the standard metric micrometer head setting as illustrated.

    Step 1 Record the number of whole millimeters on the top barrel scale.

    Step 2 Add the reading of the division on the thimble as the 0.01 mm value.

          Note. Since no half-mm value shows on the scale below the index line, the top reading (mm) and thimble division represent the actual measurement.

    Step 3 Express the sum of the readings as the required 0.01 mm metric micrometer reading.

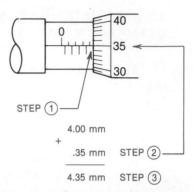

EXAMPLE: *Case 2.* Set a standard metric micrometer head to a measurement of 7.92 mm.

    Step 1 Turn the thimble to the 7 mm graduation (above the index line) and slightly beyond the next 0.5 mm graduation (below the index line).

    Step 2 Continue to turn the thimble until the 42 graduation (0.42 mm) cuts the index line.

    Step 3 Read the barrel and thimble settings for the 7.92 mm measurement.

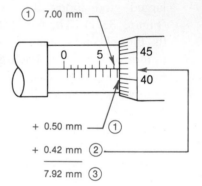

Estimating Readings of 0.005 mm on a Standard Metric Micrometer

It is possible to estimate to a fractional part of one division on the sleeve. A judgment is made about the additional distance between two graduations on the thimble at the index line. The illustration (Figure 33-4) shows a reading of 5.915 mm. The additional precision of 0.005 mm is the estimated quantity.

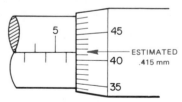

FIGURE 33-4 Reading of 5.915 (enlarged)

B. READING AND SETTING METRIC VERNIER (0.002 mm and 0.001 mm) MICROMETERS

As stated before, the vernier principle of measurement applies equally to instruments graduated in customary inch (0.0001″) and SI metric millimeter units of measure. A metric vernier micrometer is read in the same manner as the standard micrometer. The three-place decimal millimeter reading is read on the vernier scale. The reading is taken at the graduation on the vernier scale on the barrel that coincides with one of the graduations on the thimble.

The representative line drawing (Figure 33-5) shows a 0.001 mm metric vernier micrometer reading of 3.563 mm. The barrel reading is 3.5 mm (3.0 mm + 0.5 mm). The sleeve reading is 0.06 mm. The vernier reading is 3 (0.003 mm) because the 3 line coincides with a graduation on the thimble. The final reading of 3.563 mm consists of the 3.0 mm + 0.5 mm + 0.06 mm + 0.003 mm.

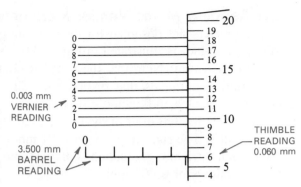

FIGURE 33-5 Reading of 3.563 mm on a 0.001 mm metric vernier micrometer

E XAMPLE: Read the metric vernier micrometer setting as illustrated.

Step 1   Read the last visible whole millimeter graduation on the barrel.

Step 2   Determine whether a half millimeter graduation appears on the lower scale on the barrel. Add 0.50 mm if the graduation is visible.

Step 3   Read the dimension on the thimble at the index line.

Step 4   Determine which vernier line coincides with a graduation on the thimble. Read the vernier graduation value.

Step 5   Combine the separate readings of 6.000 mm + 0.500 mm + 0.330 mm + 0.008 mm = **6.838 mm**   Ans

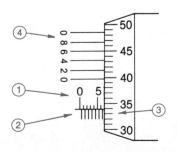

Direct Reading Metric Vernier Micrometers

Micrometers are designed for direct reading. Such a micrometer, showing sections of the barrel and thimble, is illustrated in Figure 33-6. Each graduation shown on the barrel represents readings of 1.0 mm. Note that the barrel reading shows 5+ millimeters. A mechanism in the thimble moves numerals that display the additional fractional parts of a millimeter in the measurement. The numeral in the left digit of the thimble represents 0.10 mm units. The right digit shows 0.01 mm units. Further, each 0.01 mm graduation is divided into five parts for readings of 0.002 mm.

The direct reading of the last exposed numeral on the barrel and the numerals and fractional part read on the thimble is the linear measurement of the part. In this example, the total reading is 5.374 mm. This three-place millimeter value carries the same degree of precision measurement as a vernier micrometer graduated in customary units of 0.000 1″.

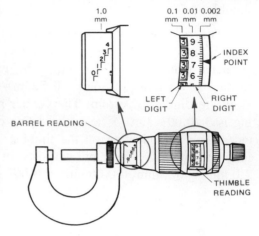

FIGURE 33-6 Direct reading micrometer

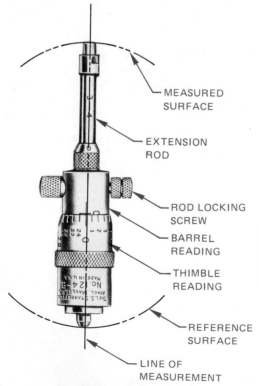

FIGURE 33-7 Setup for taking an internal measurement (*Courtesy of the L. S. Starrett Company*)

C. READING AND SETTING INSIDE METRIC MICROMETERS

The inside micrometer consists of a head (barrel and thimble), precision extension rods, and a rod locking screw. Inside micrometers are used to take internal measurements between a reference surface and a measured surface as shown in Figure 33-7.

RULE FOR MEASURING WITH AND READING CUSTOMARY-INCH AND METRIC-GRADUATED INSIDE MICROMETERS

- Select the appropriate extension rod. Test the inside micrometer for accuracy.
- Place one leg of the micrometer against the reference point.
- Adjust the thimble. Turn it slowly while swinging a slight arc to bring the leg of the rod to the measured point. Continue until a slight contact is made.
- Remove the micrometer and read the measurement.
  Note. The inside measurement includes the graduation readings on the barrel + thimble + the length of the extension rod.

EXAMPLE: Determine the inside diameter reading of the measurement shown in Figure 33-7.

Step 1   Take the reading on the barrel.

Step 2   Add the reading on the thimble.

Step 3   Add the size of the extension rod. The sum of these readings represents the inside micrometer measurement.

Note. The same procedure is followed for inside measurements using metric graduated instruments.

① - - - - - - - - - - - **0.125**
② - - - - - - - - - - **0.020**
③ - - - - - - - - - - **3.000**
④ - - - - - - - - - - **3.145″** Ans

D. READING MEASUREMENTS ON A METRIC-DEPTH MICROMETER

The graduations on the micrometer head portion of a depth micrometer are in a *reverse order* from the conventional micrometer. In other words, the zero reading is at the topmost position of the barrel. The usual one inch of 25.0 mm depth measurement range may be increased by changing to an appropriate size interchangeable measuring (extension) rod.

RULE FOR MEASURING WITH A DEPTH MICROMETER AND READING MEASUREMENTS

- Select the length measuring rod to use according to the depth of the measurement.
- Position the depth micrometer. Slowly turn the thimble to lower the measuring rod until it almost touches the surface to be measured. Then use the ratchet mechanism to bottom the rod on the measured surface.
- Read the numbered graduation and any other exposed graduation value on the barrel.
- Add the reading on the thimble and the length of the measuring rod.
- State the depth as the total of the readings and the rod length.

<u>E XAMPLE</u>: Determine the depth as represented by the readings shown in Figure 33-8.

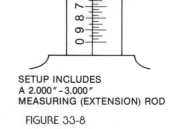

0.022″ ②

0.525″ ①

    Step 1 Read the value of the last graduation shown on the barrel.

    Step 2 Read the value of the thimble graduation at the index line.

    Step 3 Add the length of the measuring (extension) rod.

2.000″ ③

SETUP INCLUDES
A 2.000″–3.000″
MEASURING (EXTENSION) ROD

FIGURE 33-8

    Step 4 State the measurement in the required unit of measure.

.525 + .022 + 2.000 =
2.547″ Depth
Measurement Ans

E. MEASURING WITH METRIC VERNIER CALIPERS

Vernier calipers with either metric or customary inch units of measure are read in the same manner. In the case of metric vernier calipers, each graduation represents 1 mm. Each graduation on the vernier scale equals 0.02 mm. One of the vernier scales is positioned for outside (external) measurements; the other scale is positioned for inside (internal) measurements. Regardless of the system, dimensional readings are taken by combining the beam and vernier reading on either the inside or outside scale, as required.

<u>R</u>ULE FOR READING VERNIER CALIPER MEASUREMENTS

- Note which graduation on the appropriate vernier scale coincides with a graduation on the beam.
- Read the beam graduation at the zero index point.
- Add the vernier scale reading. The sum equals the required vernier caliper measurement.

E̲X̲A̲M̲P̲L̲E̲:   Read both the external and internal metric measurements as represented by the line drawing.

> *Step 1*   Read the external millimeter measurement (80 mm) at the zero (0) vernier scale graduation.

> *Step 2*   Determine which graduation on the beam scale coincides with a graduation on the vernier scale.

> *Step 3*   Add the vernier scale reading (0.08 mm). The sum equals the required outside measurement (80.08 mm).

> *Step 4*   Repeat the steps 1 through 3 to obtain an internal metric measurement. In this instance, if the measurement were an internal one, the reading would be 88.08 mm.

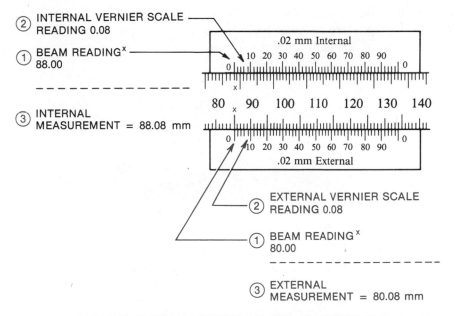

EXTERNAL/INTERNAL VERNIER CALIPER MEASUREMENTS

Metric and Customary Inch Combination Vernier Calipers

Another functionally designed vernier caliper includes graduations in millimeters on one edge of the beam. The other edge is graduated in customary inch units of measurement of 0.050″. The vernier plate attachments permit measurements to be taken on the metric scale to an accuracy of 0.02 mm and to 0.001″ on the inch standard vernier.

F. READING ANGLES WITH A VERNIER BEVEL PROTRACTOR

The vernier principle is also applied to measuring angles with a vernier bevel protractor (Figure 33-9). This instrument permits the measurement of angles to a precision of 5 minutes or $\frac{1}{12}$ of one degree.

The vernier bevel protractor consists of a body, a turret, an adjustable blade, and a clamp. Degree graduations appear near the outer rim of the body. The vernier scale is attached to a movable, rotating turret. The measuring blade is positioned for measuring an angle and is locked to the turret. The angle formed is measured by the combination of the outer scale and vernier scale readings. There are 24 divisions on the vernier scale shown in Figure 33-10. The divisions are numbered 60 to 0 to 60 in increments of 5 minutes.

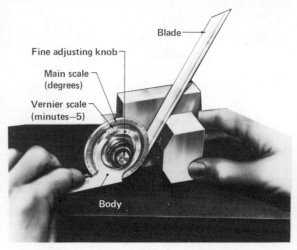

FIGURE 33-9 Application of vernier bevel protractor in measuring angles to 5' (*Courtesy of the L. S. Starrett Company*)

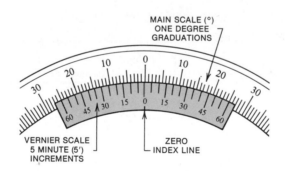

FIGURE 33-10 Graduations on the vernier bevel protractor

RULE FOR READING VERNIER BEVEL PROTRACTOR MEASUREMENTS

- Read the number of degrees on the main scale at the zero index line.
- Follow in the same direction and read the minute graduation on the vernier scale which coincides with a graduation on the main scale.
- Combine the main scale degree reading and the vertical scale (5') reading. This value represents the angle being measured in degrees and minutes.

EXAMPLE: Read the angle for the vernier bevel protractor setting as illustrated.

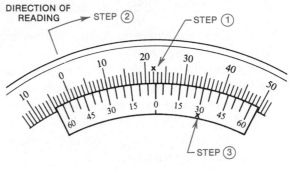

Step 1   Read the whole number of degrees on the main scale at the 0 index line.

Step 2   Note the direction from which the reading is made.

Step 3   Continue in the same direction and determine which graduation on the vernier scale coincides with a graduation on the main scale.

Step 4   Combine the degree and minute values to establish the measured angle.

Reading = 22° 30″   Ans

G. DETERMINING METRIC GAGE BLOCK COMBINATIONS

Metric gage blocks are commercially available in sets that cover the same range of measurements as the customary inch gage blocks. Metric gage blocks are produced to dimensional accuracies of ± 0.000 05 mm (0.05 microns, 0.05 μm) for size, flatness, and parallelism. Generally, the gage blocks in a set permit measurements in increments of 0.001 mm to 25.0 mm. Figure 33-11 gives the size increment for each series of blocks in a set and the range of sizes in each series.

Increment for Each Series in a Set	Range of Lengths for Each Increment Series
0.001 mm	1.001 mm to 1.009 mm
0.01 mm	1.01 mm to 1.49 mm
0.5 mm	1.0 mm to 24.5 mm
1.0 mm	1.0 mm to 9.0 mm
10.0 mm	10.0 mm to 100.0 mm
25.0 mm	25.0 mm to 100.0 mm

FIGURE 33-11

Metric gage blocks are selected, combined, and wrung together following the same steps as for inch standard precision gage blocks. In general practice, *wear blocks* are placed on each end of a gage block combination. Also, the largest possible sizes and the least number of blocks are combined.

RULE FOR ESTABLISHING A PRECISION MEASUREMENT WITH METRIC GAGE BLOCKS

- Follow the same procedure in selecting the gage block combination as for inch standard gage blocks.
- Select the first gage block from the series that has the same number of decimal places as the last right decimal digit.
- Select one or more gage blocks from the series having the same number of decimal places as the next decimal digit.
- Proceed to select gage blocks from blocks in the 0.5, 1.0, 10.0, or 25.0 mm series, as may be required.
- Add the combination of blocks. Check this measurement against the required measurement.
- Clean and ''wring'' the gage blocks together.

E XAMPLE: Select the metric gage block combination for a measurement of 120.887 mm. *Note*. Use 2 mm wear blocks as the end blocks. Refer to a table of metric gage block thicknesses in a set.

Step 1 Subtract the two wear blocks from the required measurement.

Step 2 Select a gage block in the 0.001 series to eliminate the last right decimal digit.

Step 3 Proceed to eliminate each remaining decimal digit.

Step 4 Add the blocks in the combination. Then, subtract the sum from 116.887 mm to find the remaining blocks.

Step 5 Eliminate the two right digits (__ 1 3) by selecting two blocks in the 1 mm series that total 13 mm.

Step 6 Eliminate the last digit (1 __ __) by selecting the 100 mm block in the 25.0 mm series.

Step 7 Recheck the combination of blocks against the required measurement.

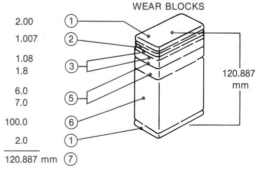

2.00	①
1.007	②
1.08	
1.8	③
6.0	
7.0	⑤
100.0	⑥
2.0	①
120.887 mm	⑦

WEAR BLOCKS

120.887 mm

ASSIGNMENT UNIT 33   REVIEW AND SELF TEST

PRETEST
The Review and Self-Test items that follow may be used as a pretest. Pretests are designed to measure a student's beginning level of mathematical skills competency and to determine the starting point of instruction.

POSTTEST
The Review and Self-Test items may also be applied as a post test. Post tests are planned to establish the student's level of mathematical skills competency after instruction.

A. Reading Standard Metric Micrometer Measurements

1. Read the linear measurements displayed on the standards metric micrometer settings at Ⓐ, Ⓑ, and Ⓒ. Estimate the third decimal-place measurement for settings Ⓓ and Ⓔ. (*Note.* The barrel graduations in the illustrations are enlarged for easier reading.)

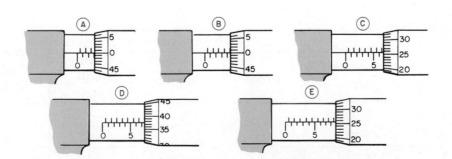

B. Reading and Setting Metric Vernier Micrometers

1. Read the metric vernier micrometer settings as represented by line drawings Ⓐ, Ⓑ, and Ⓒ to 0.01 mm; Ⓓ and Ⓔ to 0.002 mm; and Ⓕ to 0.001 mm.

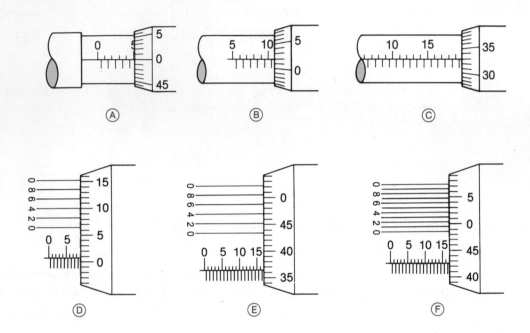

2. Indicate the required barrel, thimble, and vernier scale settings of a 0.002 mm outside micrometer for measurements A, B, and C.

	Required Measurement (mm)	Barrel Graduations		Thimble Setting (0.01 mm)	Vernier Scale Setting (0.002 mm)
		1 mm	0.5 mm		
A	7.50				
B	12.77				
C	125.926				

C. Reading Depth Micrometer Measurements

1. Read the 0.001″ depth micrometer measurements as illustrated at Ⓐ and Ⓑ.

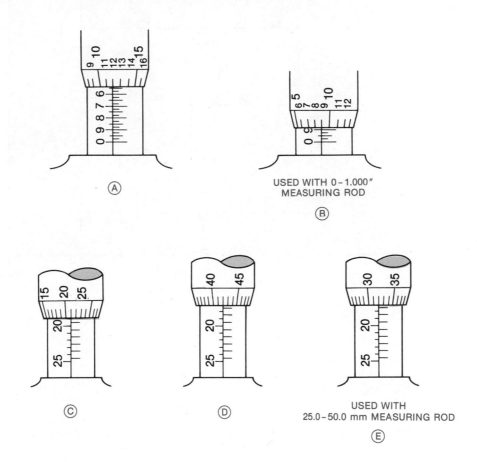

USED WITH 0–1.000″
MEASURING ROD

Ⓑ

USED WITH
25.0–50.0 mm MEASURING ROD

Ⓔ

2. Read the SI metric depth micrometer measurements at Ⓒ, Ⓓ, and Ⓔ.

D. Reading Metric Vernier Caliper Measurements

1. Compute the overall linear measurements Ⓐ, Ⓑ, and Ⓒ from the information provided on the drawing and in the table of hole sizes.

Hole Diameters	
Hole #	Nominal Hole Diameter (mm)
A	25.4
B	37.12
C	28.58

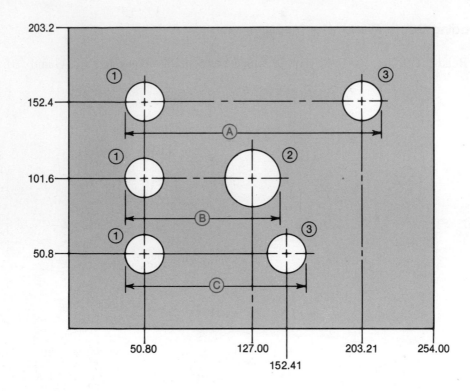

2. Give the beam (main scale) and inside vernier scale settings for each overall measurement.

Hole #	Overall Measurement (mm)	Vernier Caliper Scale Settings	
		Main Scale (mm)	Vernier Scale (0.02 mm)
A			
B			
C			

E. Reading Metric Vernier Height Gage Measurements

1. Read the 0.02 mm vernier height gage settings Ⓐ, Ⓑ, and Ⓒ. Each vernier scale graduation is equal to 0.02 mm.

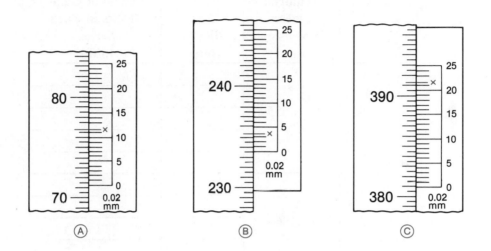

F. Reading Angular Measurements with the Vernier Bevel Protractor

1. Read the angular settings as displayed on the main beam and vernier scales for bevel protractor measurements Ⓐ, Ⓑ, and Ⓒ.

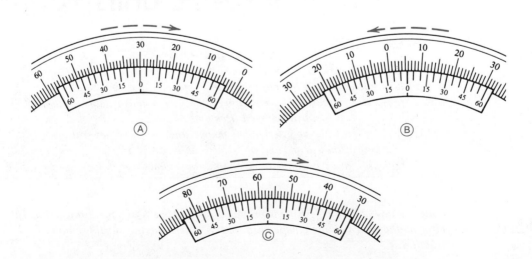

G. Determining Metric Gage Block Combinations

1. Establish the series and determine the sizes of metric gage blocks to combine to produce measurements A and B.

	Required Linear Dimension (mm)	Series of Gage Blocks (mm)	Size(s) of Gage Blocks in Each Series (mm)
A	58.555		
B	155.863		

Unit 34 Base, Supplementary, and Derived SI Metric Units

OBJECTIVES OF THE UNIT

After satisfactorily studying this unit, the student/trainee will be able to
- *Understand how each of the seven base units in SI metrics is defined and used for length, mass, time, electric current, temperature, luminous intensity, and amount of substance measurements.*
- *Apply supplementary and derived SI metric units to solve general, practical industrial, and physical science problems.*
- *Work with nonsignificant zeros and dual dimensioning.*

PRETEST   *Use the Review and Self-Test items provided in the Unit Assignment to establish the level of mathematical skills competency and to determine the starting point of instruction.*

Each SI metric unit may be converted to a customary unit or other conventional metric unit of measure. Tables are readily available that give

- The *quantity to be measured* (for example, density, *D*)
- The *unit of measure* (kilograms per cubic meter, kg/m^3)
- The *formula* ($D = $ kg/m^3)
- The *conversion units* that may be in SI metrics and conventional metric or customary units (for example, g/cm^3 or lb (mass)/ft^3)
- The *conversion factor and mathematical processes,* for example,

$$kg/m^3 = lb \ (mass)/in.^3 \cdot (2.768 \times 10^4)$$

In Unit 34, attention is directed to seven base units, two supplementary units, and a series of derived units in the SI metric system. The combination of base, supplementary, and derived units in the full system is adequate to measure all known physical quantities.

A. SEVEN BASE UNITS OF MEASURE IN SI METRICS

The base units of measure cover (1) length, (2) mass, (3) time, (4) electric current, (5) temperature, (6) luminous intensity, and (7) the amount of a substance. Each of these base units is identified and, in some cases, defined technically.

1. Unit of Length

Figure 34-1 shows how the meter, as a measure of length, is standardized. In October 1983 the *Conference Général des Poids et Measures* (GCPM) adopted a new definition for the meter as the unit of length in the International System of Measurement (ISO). The *new meter* is defined as the distance traveled by light in a vacuum in 1/299,792,458 of a second. This combination of scientific activity, transferred to the measuring bar, represents the precise measurement of the meter.

The *meter* (m) is the standard unit of length in the SI metric and all other conventional metric measurement systems. One major difference among the standards for the early meter standard, the SI metric standard until late 1983, and the present one is in the laboratory method of producing a standard meter measuring bar by using time as the most accurate measurement to define length. The last meter standard was defined in terms of a wavelength of orange-red light emitted from a krypton$_{86}$ lamp.

The early meter represented a fixed value in relation to the earth's circumference. Initially, standard meter measur-

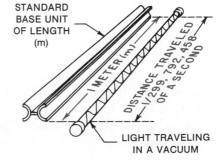

FIGURE 34-1

ing bars were compared for precision and accuracy against a standard bar held by the International Bureau and the predecessor academy. The duplicate measuring bars were called *prototypes*. The prototypes became the standards for each participating nation. The advantage of the new meter is that it may be reproduced anywhere in the world in a laboratory.

Measurements must be taken that range all the way from minute particles finer than a billionth part of a meter to earth-outer space dimensions that represent multidigit quantities. A series of more practical units of measure, derived from the meter as the base unit, are described later.

2. Unit of Mass

The *kilogram* still represents a base *unit of mass*. The kilogram (kg) is approximately the equivalent of 2.2 pounds. Other recommended multiples of the SI kilogram unit are the *megagram (Mg), gram (g), milligram (mg),* and *microgram (μg)*. The *metric ton (t)* is another unit that is often used (1 t = 10^3kg). Additional units that relate to mass are discussed later.

3. Unit of Time

The *second* remains as the unit of time. This unit is defined scientifically in terms of periods of radiation of a cesium atom and two levels of activity. The activity is overly simplified in Figure 34-2.

The duration or time interval of a second is still 1/86 400 part of the mean solar day. The designation for the second is (s). Recommended decimal multiples and submultiples of the SI unit include the *kilosecond* (ks), *millisecond* (ms), *microsecond* (μs), and the *nanosecond* (ns). Other units that may be used are the customary ones of *minute, hour,* and *day*.

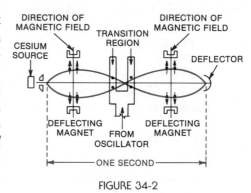

FIGURE 34-2

4. Unit of Electric Current

The *ampere* is the unit of intensity of electric current. As a common name, the ampere is designated by the capital (A). Other recommended values of this SI unit include the *kiloampere* (kA), *milliampere* (mA), *microampere* (μA), *nanoampere* (nA) and *picoampere* (pA). Note that most of these examples express submultiple decimal values of an ampere.

The illustration (Figure 34-3) shows two straight parallel conductors of negligible cross section and infinite length. These are placed one meter apart in a vacuum. The *ampere* is that constant, which, if maintained in the two conductors, produces a force equal to 2×10^{-7} newton/meter of length.

5. Unit of Temperature

The *kelvin* (K) is the SI unit of thermo-dynamic temperature. A kelvin temperature is expressed without the degree (°) symbol. The boiling point of water on the kelvin scale is 373.15 K degrees (373.15 K). Water freezes at 273.15 K.

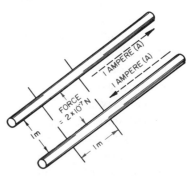

FIGURE 34-3

The more commonly used temperature measurement unit is the derived *Celsius* degree. Temperatures are readily convertible between the kelvin and Celsius and the kelvin and Fahrenheit readings by transposing values in these two formulas.

$$t_K = t_C + 273.15 \qquad t_K = \frac{(t_F + 459.67)}{1.8}$$

6. Unit of Luminous Intensity

The *candela* is the SI unit of luminous intensity. The candela was adopted at the 13th CGPM in 1967. By scientific definition the candela represents

- the luminous intensity
- of a 1/600 000 square meter surface of a black body
- at the temperature of freezing platinum under a pressure of
- 101 325 newtons per square meter.

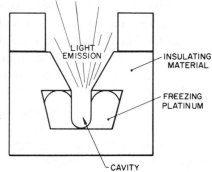

FIGURE 34-4

7. Units for the Amount of Substance of a System

The *mole* (mol) is the most recent SI base unit. It was introduced to physical chemistry and molecular physics in relation to mass, internal energy, and heat capacity. The mole represents a unit of measure of the quantity or the *amount of substance* in such an elementary unit as a molecule, atom, ion, or electron.

B. SUPPLEMENTARY UNITS OF ANGULAR MEASURE

There are two types of angles in SI metrics for which supplementary units of measure have been developed. The *radian* unit (rad) is used for measurements of *plane angles* (Figure 34-5). The *steradian* (sr) unit provides for the measurement of *solid angles*.

In simple mathematical terms, 1 rad $= 180°/\pi$, or 57.295 78°. The SI recommended units for radian are *milliradian* (mrad) and *microradian* (μrad). Other customary units that may be used include *degree* (°), *minute* ('), *second* ("), and *grad* (^{g}).

By contrast, the steradian encloses an area of the spherical surface as shown in Figure 34-6. The enclosed area is equal to that of a square. The sides of the square are equal in length to the radius.

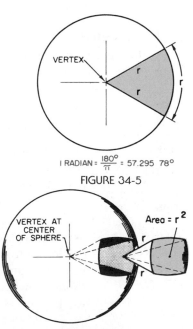

I RADIAN = $\frac{180°}{\pi}$ = 57.295 78°

FIGURE 34-5

FIGURE 34-6

C. DERIVED UNITS OF MEASURE

There are many instances when base or supplementary units of measure are not suitable and *derived* units are required. These are derived from base or supplementary units. In general, the preferred units of measure in SI metrics are *multiples of three* (^{3}): *kilo* (k, 10^3), *mega* (M, 10^6), *giga* (G, 10^9) or *milli* (m, 10^{-3}), *micro* (μ, 10^{-6}), and *pico* (p, 10^{-12}).

1. Derived Length, Area, and Volume Units

The decimal multiple and submultiples of SI units that are used for linear measurements include the *kilometer* (km), *millimeter* (mm), *micrometer* (μm) and *nanometer* (nm). The *decimeter* (dm) and *centimeter* (cm) are also accepted.

The derived units of area and volume are also related to the meter as the base unit. In area measure, the *square kilometer* (km^2) and *square millimeter* (mm^2) are recommended multiples. The dm^2 and cm^2 are accepted. For larger measurements the *are* and *hectare* are other units. One are (a) = $10^2 m^2$ (100 square meters); 1 hectare (ha) = $10^4 m^2$ (10 000 square meters).

In volume measure, the *cubic millimeter* (mm^3) is a recommended multiple of the base unit of m^3. The *cubic decimeter* (dm^3) and (cm^3) are accepted units of volume measure. The *liter* (ℓ) has been declared an acceptable SI unit. One liter equals one cubic decimeter (1 ℓ = 1 dm^3). In terms of the meter, 1 ℓ = 10^{-3} m^3 or 0.001 m^3. The *centiliter, milliliter,* and *hectoliter* are other accepted units of volume measure.

$$1 \text{ centiliter } (c\ell) = 10^{-5} m^3 \text{ (or 0.000 01 } m^{3)}$$
$$1 \text{ milliliter } (m\ell) = 10^{-6} m^3 \text{ (or 0.000 001 } m^{3)}$$

2. Units Derived from the Unit of Mass

Units derived from the unit of mass deal with force, work, power, pressure and other physical science measurements. The units are derived from the *kilogram* (kg) as the base unit of mass. Recommended multiples of the kg are the *megagram* (Mg), *gram* (g), *milligram* (mg), and *microgram* (μg). Other widely used common units, which are derived from the kilogram as the unit of mass, are discussed.

FORCE Force is measured in *newtons* (N). The SI unit of force is N $\cdot$ m. A force of one newton applied to a mass of one kilogram produces an acceleration of one meter per second. Expressed as a formula

$$N = kg \cdot m/s^2$$

WORK, ENERGY, QUANTITY OF HEAT Work is measured in *joules* (J). One joule of work is produced when a force of one newton is applied through a distance of one meter (J = N $\cdot$ m). Work is usually expressed in *gigajoules* (GJ), *megajoules* (MJ), *kilojoules* (kJ), and *millijoules* (mJ). Another practical unit of measure is the *kilowatt-hour* (kW·h) of work.

$$1 \text{ kW·h} = 3.6 \times 10^6 \text{ J or 3.6 MJ}$$

POWER   The *watt* (W) is the SI unit of power. It represents one joule of work completed in one second.

$$W = J/s \text{ or } N \cdot m/s$$

PRESSURE   The *pascal* (Pa) is the SI unit for measuring pressure. A pascal is equal to one newton per meter squared.

$$Pa = N/m^2$$

3. Units Derived from the Unit of Time

Three commonly needed units of measure that are derived from the second relate to frequency (f), velocity (v), and acceleration (a).

FREQUENCY   Frequency denotes the number of cycles per second. The *hertz* (hz) is the SI unit of measure of frequency.

$$1 \text{ Hz} = 1 \text{ s}^{-1}$$

VELOCITY   Velocity relates to distance and time. It is measured in terms of *meters per second* (v = m/s). Velocity is also expressed in the derived unit of *kilometer per hour*.

$$1 \text{ km/h} = \frac{1}{3.6} \text{ m/s}$$

ACCELERATION   Acceleration (a) denotes a rate of change in velocity. Acceleration is equal to distance (meters) divided by time in seconds squared.

$$a = m/s^2$$

4. Units Derived from the Base Electrical Unit

Six common electrical measurements relate to *potential, resistance, capacitance, quantity, inductance,* and *magnetic flux*. These characteristics are measured by units that are derived from the *ampere* (A) as the base unit. Figure 34-7 indicates the derived unit in each case and the formula.

Base Electrical Unit: Ampere (A)		
Electrical Measurement	Derived Unit of Measurement	Formula
Potential	volt (V)	V = W/A
Resistance	ohm (Ω)	Ω = V/A
Capacitance	farad (F)	F = A·(s/V)
Quantity	coulomb (C)	C = A·s
Inductance	henry (H)	H = Wb/A
Flux	weber (Wb)	Wb = V·s

FIGURE 34-7

5. Units Derived from the Base Unit of Luminous Intensity

The *candela* (cd) is the base unit of luminous intensity. There are two major derived units from the candela: *illumination* lux (lx) and the lumen (lm, luminous flux). The *watt* (W), as a common unit of electrical power, equals 17 lumens (1 W = 17 lm).

6. Units Derived from the Base Unit of Temperature

The *degree Celsius* is the derived unit of temperature. It is used in general everyday applications. Unlike the kelvin (K) SI unit, a temperature on the Celsius scale is followed by the °C symbol. The degree Celsius replaced the centigrade designation. The freezing and boiling points of water on the Celsius scale are 0°C and 100°C, respectively. The mathematical relationships between computing Celsius and customary Fahrenheit degrees remain the same.

Many problems still require the conversion of temperature units between customary and other metric systems. The same simple formulas that were used between the former Centigrade and Fahrenheit scales apply to Celsius/Fahrenheit conversions.

$$t_C = (t_F - 32) \cdot \tfrac{5}{9} \qquad\qquad t_F = \left(t_C \times \tfrac{9}{5} + 32\right)$$
$$\text{or} \qquad\qquad\qquad \text{or}$$
$$t_C = (t_F - 32) \div 1.8 \qquad t_F = 1.8 t_C + 32$$

7. Units Derived from the Base Unit of Time for Ordinary Time Measurement

The *day* (d), *hour* (h), *minute* (min) and *week, month* and *year* are commonly used units of time. Each is derived from the *second* as the base unit. The value of each of these derived units and its use in mathematical computations follows the customary or the conventional metric system measurement in the problem.

8. Units Derived from the Basic Molar Unit

The *mole* (mol) is the most recent base unit in SI metrics. Other derived mole units that measure quantity are related to the nature of each quantity. For example, the kg/mol is a derived unit for measuring *molar mass;* j/mol for *molar internal energy;* and J/(mol K) or J/(mol °C) for *molar heat capacity.*

D. NONSIGNIFICANT ZEROS IN MEASUREMENT

The term *nonsignificant zeros* refers to the practice of adding zeros to decimal digits. Nonsignificant zeros appear primarily on drawings where a tolerance is specified. One or more nonsignificant zeros may be added to the decimal value of a dimension or a tolerance. This means that the dimension and the tolerance have the same number of digits. For instance, in a dimension like 5.3125" ± 0.0010" the last right decimal digit is a nonsignificant zero.

It has no value since the tolerance is not changed by its addition. The last decimal digit is important in stating precisely what the tolerance is.

Nonsignificant zeros are not added in SI metrics. It is common practice to see values or quantities in tables or on drawings with a different number of decimal digits in a dimension and the tolerance. For instance, an object may be dimensioned 272.38 mm ± 0.025 mm. The three-place decimal tolerance of 0.025 mm tells that the part can be accepted if machined between 272.38 + 0.025 mm or 272.405 mm as the upper limit and 272.38 − 0.025 mm or 272.355 mm as the lower limit.

E. SI METRIC AND CUSTOMARY UNIT DUAL DIMENSIONING

Dual dimensioning implies that a product or part is dimensioned in equivalent SI metric and customary units of measure. This practice applies particularly to industry drawings where the worker must interpret and use the dimensions to produce a part precisely using either system of measurement.

Special dimensioning guidelines are followed. Two sets of dimensions, SI metric and customary units, are used. Generally, a dimension that appears first and that is separated from the dimension that follows by a slash or bracket or a dimension that appears on a drawing above another dimension is considered the *controlling dimension*. This dimension represents the system of measurement in which the product is designed. Dimensions that are given after a slash symbol or within brackets are usually noncontrolling dimensions. A note is often used on a drawing to indicate the controlling dimensioning system or the position of the dimensions.

Each technique of positioning a controlling dimension or indicating the system of dimensioning is illustrated in Figure 34-8. Placement of dimensions one above another is shown at (A); the use of slash and bracket symbols at (B); and the identification of measurement systems as a note at (C).

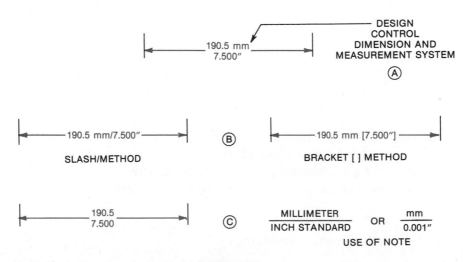

FIGURE 34-8

ASSIGNMENT UNIT 34 REVIEW AND SELF TEST

PRETEST | *The Review and Self-Test items that follow may be used as a pretest. Pretests are designed to measure a student's beginning level of mathematical skills competency and to determine the starting point of instruction.*

POSTTEST | *The Review and Self-Test items may also be applied as a post test. Post tests are planned to establish the student's level of mathematical skills competency after instruction.*

A. Base, Supplementary, and Derived Units of Measure (Practical Problems)

1. Identify four different kinds of information that are generally found in SI metric tables.
2. a. Give two applications of derived units of measure for each base unit.
   b. Identify the measurement symbol that is used with each application of a derived unit.

	Base Units			Derived Unit Applications			
	Measure	**Unit**	**Symbol**	**Application 1**	**Measurement Symbol**	**Application 2**	**Measurement Symbol**
1	Length	meter	(m)				
2	Mass	kilogram	(kg)				
3	Electric Current	ampere	(A)				
4	Luminous Intensity	candela	(cd)				
5	Time	second	(s)				
6	Temperature	degree kelvin	(K)				
7	Substance	mole	(mol)				

3. a. Name the two SI supplementary units of angular measure.
   b. Identify the symbol for each unit of measure.
   c. Explain briefly what each supplementary unit measures.
   d. Give the mathematical value of either supplementary unit.

B. Nonsignificant Zeros (Practical Problems)

1. Identify one purpose for using nonsignificant zeros in customary unit dimensioning.

2. State one difference in dimensioning practices followed in SI metrics as contrasted with customary unit dimensioning in relation to the use of nonsignificant zeros.
3. Give an example of dimensioning and tolerancing in SI metric and another example for customary unit practices in using nonsignificant zeros.

C. SI Metric and Customary Unit Dual Dimensioning (Practical Problems)

1. State what function is served by the controlling dimension on a drawing.
2. Give the controlling dimension and identify the measurement system in examples Ⓐ, Ⓑ, and Ⓒ.

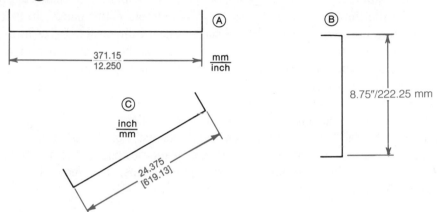

Unit 35 Conversion: Factors and Processes (SI and Customary Units)

OBJECTIVES OF THE UNIT

After satisfactorily studying this unit, the student/trainee will be able to
* *Interpret measurements obtained by soft or hard conversion.*
* *Use conversion tables and reciprocals to obtain measurements and values in either SI metrics or customary units of measure or in both systems.*
* *Solve practical problems requiring the conversion of linear, square, mass, and other volume measurements and quantities.*

PRETEST   *Use the Review and Self-Test items provided in the Unit Assignment to establish the level of mathematical skills competency and to determine the starting point of instruction.*

Today, it is common practice to use dimensions and other measurements in either the SI metric or customary system or in both systems at one time. The measurement values are usually computed, although, in some applications, it is possible to use direct-reading conversion scales and rules.

A. SOFT AND HARD CONVERSION OF SI METRIC AND CUSTOMARY UNITS OF MEASURE

Equivalent measurements, which are computed directly without any changes, are known as *soft conversion*. While the conversion process involves simple mathematics, in most manufacturing processes there are few instances where each standard dimensional size in one system is commercially available in the other system. Further, there is the question of dimensional accuracy. For example, the exact equivalent metric linear measurement to a dimensional reading of 3.125″ on a standard inch micrometer is 79.375 mm. The decimal 0.375 mm identifies a higher degree of accuracy.

In this case, the three decimal-place reading is equivalent to an accuracy of 0.000 04″. This precision may be far beyond production requirements. Equal precision in both systems requires that one less decimal digit be used with an SI metric measurement. Thus, the 79.375 mm becomes 79.38 mm. For everyday conversion applications, it is understood that soft conversion is required unless stated otherwise.

Hard Conversion

Hard conversion means that a product is designed to conform to measurements in one system, regardless of comparable sizes in the other system. A quart in British measure is equal to 0.9463 liters in metric measure. One millimeter in linear measure equals 0.03937″. A table of drill sizes shows that the closest metric size drill to $\frac{15}{32}″$ (0.4688″) is 12 mm, which is the equivalent of 0.4724″. These values provide examples of problems in production resulting from soft conversion.

Hard conversion requires complete design changes. As a result, products designed to conform to standards in one system are not interchangeable in the second system. Preferred sizes are used in hard conversion, such as 25 mm in place of 25.4 mm; 4 liters, not 3.785, and 1 kg, not 453.6 g. It is apparent that ISO standards and those adopted by the American National Standards Institute (ANSI), the Canadian Standards Association (CSA), and special bodies like the American Gear Manufacturer's Association (AGMA) and others are not identical.

B. CONVERSION FACTORS AND PROCESSES IN METRICATION

It is possible to remember some of the commonly used values and formulas for computing specific quantities. However, it is a more accurate practice to use tables containing both formulas and numerical values. These tables are identified as ''tables of conversion factors.''

Different tables are available depending on the degree of precision required. In engineering and scientific problems, tables are used that have a great number of least significant digits (decimal places). Tables ranging from six- to eight-place decimal values are adequate for general applications. The final result is then rounded off, depending on the required accuracy.

A smaller unit of measure is changed to a larger unit in the metric system by moving the decimal point to the left. For example, in changing 1 700 millimeters to meters, the value is stated as 1.7 m. In reverse, zeros are added and the decimal point is moved to the right (12 MHz = 12 000 000 Hz).

When a conversion table is used, a conversion factor like 1 mm = 0.039 37″ may be changed to 1 cm = 0.393 7″, or 1 km = 39 370″. The prefix indicates the relationship of the numerical value to the unit of measure.

Reciprocal Processes in Conversion

While a table may give one set of conversion factors, other units of measure may be computed. The reverse process is performed by using reciprocals.

Mathematical processes involving reciprocals are covered in later units. For the present, "using the reciprocal" means dividing (1) by a numerical value. For example, to convert feet to meters the conversion factor is 0.304 8. Thus, 10 ft = 10 · (0.304 8) or 3.048 m. The answer in meters may be converted to its feet equivalent by using the reciprocal of the multiplier (0.304 8). In this instance, the reciprocal is 1/0.304 8. Thus, 3.048/0.304 8 = 10 ft.

Tables are available for conversion of common quantities from SI or conventional metrics to customary units and the reverse. Other tables are detailed for specialized application. Part of Conversion Table 6 that appears in the Appendix is reported in Figure 35-1. Selected units of measure are grouped according to use by categories (for example, acceleration, area, and density). Conversion factors are indicated for units of measure in the three common measurement systems.

	Conversion of Customary or Metric Units to SI Metrics			Conversion of SI Metrics to Customary and Conventional Metric Units		
Category	From Customary or Conventional Metric Unit	To SI Metric	Factor (A) (Multiply by)	From SI Metric	To Customary or Conventional Metric Unit	
Acceleration	ft./s^2	m/s^2	0.304 8	m/s^2	ft./s^2	Factor
	in./s^2		$2.540\ 0 \times 10^{-2}$*		in./s^2	Multiply by the reciprocal of
Area	ft.^{2}	m^2	$9.290\ 3 \times 10^{-2}$	m^2	ft.^{2}	the multiplier (Factor A)
	in.^{2}		$6.451\ 6 \times 10^{-4}$*		in.^{2}	which is used in conversion
Density	g/cm^3	kg/m^3	$1.000\ 0 \times 10^{3}$*	kg/m^3	g/cm^3	to the SI Metric unit
	lb. (mass)/ft.^{3}		16.018 5		lb. (mass)/ft.^{3}	
	lb. (mass)/in.^{3}		$2.768\ 0 \times 10^{4}$		lb. (mass)/in.^{3}	

*Exact Values

FIGURE 35-1 Partial table of conversion factors for SI and conventional metric and customary units of measurement

The abstracted contents of Table 6 from the Appendix summarize the information presented to this point. Symbols, powers of ten decimal multiple and submultiple quantities, positive and negative values, and simple mathematical processes are illustrated in Figure 35-1. The balance of this unit now deals with soft conversions of commonly used measurement units in practical problems involving linear, surface, volume, and liquid measures.

C. CHANGING UNITS OF MEASURE FROM ONE SYSTEM TO ANOTHER

No uniform relationship exists between the units of measure in the British (customary) and metric systems. In changing values from one system to the other it is necessary to know what the equivalent of one unit is in terms of the other system. The desired unit may then be computed by either multiplying or dividing.

Units that are commonly used to convert a measurement from the British (customary) to the metric system are given in Figure 35-2.

British Units	Equivalent in Metric Units (Approximate)
One inch (in.)	2.54 centimeters (cm) or 25.4 millimeters (mm)
One foot (ft)	0.304 8 meter (m)
One yard (yd)	0.914 4 meter (m)

FIGURE 35-2 Comparison of British and metric units of linear measure

RULE FOR CHANGING CUSTOMARY UNITS OF MEASURE TO METRIC UNITS (INCHES, FEET, OR YARDS TO METERS, CENTIMETERS, DECIMETERS, OR MILLIMETERS)

• Determine the equivalent value of one metric unit in the customary system.

• Divide the given customary unit by the metric equivalent of one unit.

• Express the result in the required metric unit.

Note. The same result is obtained if the number of metric units in one customary unit is determined and the given number is multiplied by the metric units. This second method is also illustrated.

E XAMPLE: Change 9″ to centimeters (cm).

Method 1

Step 1 Determine the value of one centimeter in terms of inches. **1 cm = 0.393 7″**

Step 2 Divide the given unit (9) by this number (0.3937). $\frac{9}{0.393\,7} = 22.86$

Step 3 Express the result (22.86) in terms of the required unit of metric measure (cm). **9″ = 22.86 cm** Ans

Method 2

Step 1 Determine the number of metric units (cm) in one given unit (1″). **1″ = 2.54 cm**

Step 2 Multiply the given unit (9) by this number (2.54). **9 × 2.54 = 22.86**

Step 3 Express the product 22.86 in terms of the required unit (cm). **9″ = 22.86 cm** Ans

The mathematical processes required to change units of measure from one system to the other are the same. This condition applies regardless of whether the unit is expressed in terms of square or volume measure.

RULE FOR CHANGING METRIC UNITS OF MEASURE TO CUSTOMARY UNITS (MILLIMETERS, DECIMETERS, CENTIMETERS, OR METERS TO INCHES, FEET, OR YARDS)

• Determine the number of equivalent metric units in one of the required units.
• Divide the given metric unit by this number.
• Express the result in terms of the required customary unit.

E XAMPLE: Change 228.6 millimeters (mm) to inches.

Step 1 Determine the number of metric units (mm) in one required unit (1″). **1″ = 25.4 mm**

Step 2 Divide the given metric unit (228.6) by (25.4) mm. $\frac{228.6}{25.4} = 9$

Step 3 Express the result (9) in terms of the required customary unit (inches). **228.6 mm = 9″** Ans

D. CONVERTING SQUARE (SURFACE) MEASURE VALUES

The same principles of surface measure for determining customary units are applied in the same manner to SI and conventional metric units of measure. The names and value of the units differ in each system, however. The area of a surface in the metric system is the number of square metric units that it contains. The three metric units most commonly used for small areas are the square meter, square centimeter, and square decimeter. Examples are given in Figure 35-3 of comparative values for common metric and customary units of surface (area) measure.

Metric Unit	Equivalent Customary Unit	Customary Unit	Equivalent Customary Unit
1 m^2	1550 in.^2	1 in.^2	6.452 cm^2
1 dm^2	15.50 m^2		0.0645 dm^2
1 cm^2	0.155 m^2		0.0929 m^2
1 mm^2	$0.001 \ 55 \text{ m}^2$	1 ft^2	9.29 dm^2
			929.03 cm^2
		1 yd^2	0.836 m^2

FIGURE 35-3 Comparative values of common metric and customary units of surface measure

RULE FOR CHANGING METRIC UNITS OF SURFACE MEASURE TO CUSTOMARY UNITS (SQ m, SQ dm, SQ cm TO SQ IN., SQ FT, OR SQ YD)

• Determine the number of metric units of surface measure in one of the required customary units.

• Divide the given metric units by this number.

• Express the quotient in terms of the required customary unit.

EXAMPLE: The area of a sheet of brass is 11.148 square meters. Express this value in units of surface measure (sq ft).

Step 1   Determine the number of metric units of surface measure (m^2) in one of the required customary units (sq ft).

$$0.092 \ 9 \text{ m}^2 = 1 \text{ sq ft}$$

Step 2   Divide the given number of metric units (11.148) by the number of square meters in 1 sq ft (0.092 9).

$$\frac{11.148}{0.092 \ 9} = 120$$

Step 3   Express the quotient (120) in terms of the required customary unit (sq ft).

$$11.148 \text{ m}^2 = 120 \text{ sq ft } (120 \text{ ft}^2) \quad \text{Ans}$$

E. CONVERTING VOLUME AND LIQUID MEASUREMENTS

Figures 35-4 and 35-5 provide values for converting common units of volume and liquid measure in SI metric and customary unit systems.

Metric Unit	Equivalent Customary Unit Value (Approximate)	Customary Unit	Equivalent Metric Unit Value (Approximate)
1 cu meter (1 m^3)	35.314 cu ft (ft^3)	1 cu in. (in^3)	16.387 cm^3
1 cu dm (1 dm^3)	61.023 cu in. ($in.^3$)	1 cu ft (ft^3)	0.028 3 m^3
1 cu cm (1 cm^3)	0.061 cu in. ($in.^3$)	1 cu yd (yd^3)	0.764 6 m^3

FIGURE 35-4 Values of common units of volume measure

RULE FOR CHANGING METRIC UNITS OF VOLUME MEASURE TO BRITISH (CUSTOMARY) UNITS (m^3, cm^3, dm^3 TO CU IN., CU FT, OR CU YD)

- Determine the number of British (customary) units of volume measure in one given metric unit.
- Multiply the given metric unit by this number.
- Express the product in terms of the required British (customary) unit of volume measure.

EXAMPLE: Change 17.3 cm^3 to cu in.

Step 1 Determine the number of British units (cu in.) in one given metric unit (cm^3). **1 cm^3 = 0.061 cu in.**

Step 2 Multiply given metric unit (17.3 cm^3) by this value .061. **17.3 × 0.061 = 1.055**

Step 3 Express the product in terms of the required British units. **17.3 cm^3 = 1.055 cu in. (1.055 $in.^3$)   Ans**

British units of volume measure may be changed to metric units by multiplying the given unit by the number of metric units in one British unit. For example, to change 3.2 cubic feet to cubic meters, simply multiply by (0.028 3), the metric unit equivalent of one cubic foot. The product (0.090 56) is in terms of metric units (m^3). Thus, 3.2 ft^3 = 0.090 6 m^3.

Metric Unit	Equivalent Customary Unit Value (Approximate)	Customary Unit	Equivalent Metric Unit Value (Approximate)
1 liter (ℓ)	1.057 qt	1 qt	0.946 ℓ
	61.023 in.^{3}	1 gal	3.7085 ℓ

FIGURE 35-5 Values of common units of liquid measure

RULE FOR CHANGING TO METRIC AND BRITISH UNITS OF LIQUID MEASURE

- Determine the equivalent value of one British or metric unit from a comparison table.
- Multiply the given value by the equivalent value in the desired unit.
- Express the product in the desired unit.

EXAMPLE: How many liters are there in 3 quarts?

Step 1   Determine the equivalent value in liters of one quart.   **1 qt = 0.946 liters**

Step 2   Multiply the (3) quarts by (0.946).   **3 × 0.946 = 2.838**

Step 3   Express the product as liters.   **3 qt = 2.838 liters**   Ans

ASSIGNMENT UNIT 35   REVIEW AND SELF TEST

PRETEST   *The Review and Self-Test items that follow may be used as a pretest. Pretests are designed to measure a student's beginning level of mathematical skills competency and to determine the starting point of instruction.*

POSTTEST   *The Review and Self-Test items may also be applied as a post test. Post tests are planned to establish the student's level of mathematical skills competency after instruction.*

A. Soft and Hard Conversions: SI and Customary Measurements

1. Use soft conversion to change linear dimensions A, B, and C to equivalent SI metric mm values. Round off each converted value to three decimal places.

Part	Linear Measurement	Decimal Value (.001″)	Equivalent mm Value (.001 mm)	Equivalent (.01 mm) Measurement to (.001″) Precision
A	$7\frac{7}{8}''$			
B	$4\frac{13}{68}''$			
C	$\frac{31}{32}''$			

2. Round off each metric dimension to two decimal places to establish a similar degree of accuracy to each three-place customary decimal measurement value.

3. Describe briefly the effect of hard conversion on the interchangeability of manufactured parts.

4. Determine the hard conversion SI dimensions that a designer might use for parts A, B, and C in problem A. 1.

5. Change each given linear dimension to the system and unit of measurement as specified.
   a. 200 cm to m   c. 5.14 in. to cm   e. 15.3 m to in.   g. 50.8 mm to in.
   b. 575 cm to dm   d. 6.5 yd to m   f. 12.7 cm to in.   h. 70 dm to ft

6. Convert surface measurements (a through h) to equivalent measurements as specified.
   a. 3,100 sq in. to m^2 d. $3\frac{1}{2}$ sq in. to cm^2 g. 3.2 dm^2 to in.^{2}
   b. 62 sq in. to dm^2 e. 9 sq ft to m^2 h. 725.6 cm^2 to in.^{2}
   c. 127.875 sq in. to dm^2 f. $3\frac{1}{2}$ sq yd to m^2

7. Convert volume or liquid measurement (a through h) to equivalent measurements as specified.
   a. .75 dm to m^3 c. 4.5 m^3 to yd^3 e. 17 cℓ to liters (ℓ)   g. 3.171 qt to ℓ
   b. 10 m^3 to ft^3 d. 3.6 ft^3 to m^3 f. 122.046 in.^{3} to ℓ h. 31.5 gal to ℓ

8. Establish the weight of each machine (A through E) in its metric equivalent. Round off each value to one decimal place. Give the metric weight of the five machines.

2,713 lb

612 lb

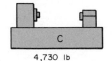
4,730 lb

1,819 lb

12,674 lb

B. Applications of Conversion Tables to Physical Science Problems

1. Identify three specific items of information that are contained in conversion tables of standards of different units of measure.

2. Provide the following technical information for three derived units of measure for the three categories of physical science measurements: mass (A), electric current (B), and time (C).

a. Indicate the nature or area of the measurement.

b. Give the symbol of the selected unit of measure.

c. State the formula for the derived unit of measure.
   Note. The answer to A.1 under mass provides an example.

	Base Units of Measure								
	A. Mass (kilogram, kg)			B. Electric Current (ampere, A)			C. Time (second, s)		
	Derived Units of Measure								
	Nature of Measurement (a)	Symbol (b)	Formula (c)	Nature of Measurement (a)	Symbol (b)	Formula (c)	Nature of Measurement (a)	Symbol (b)	Formula (c)
1	force	N	$N = kg \cdot m/s^2$						
2									
3									

3. Give the reciprocal factor to use to convert the three required unit values (A, B, C, and D) to the given value. The table provides the initial conversion information to change from the given to the required unit value.

	Measurement Values		Conversion Factor and Process	Required Reciprocal
	Given Unit Value	Required Unit Value		
A	in.	mm	25.4 (x)	
B	kg/s	lb · mass/min	0.007 560 (÷)	
C	psi	pascal (Pa)	6 894.757 (x)	
D	grains (g)	kilograms (kg)	15 432.9 (÷)	

4. Use the table of conversion factors for SI metrics in the Appendix.

a. Record the factors for converting the given A, B, C, and D quantities in the following table to the required SI units.

b. Compute the required unit values.

	Category	SI Metric Units		Conversion Factor	Converted Value
		Given	Required		
A	Density	10^{12} g/cm^3	kg/m^3		
B	Energy	75 kW·h	J		
C	Pressure	10^2g (force)/cm^2	N/m^2		
D	Volume	$7\ell \cdot 10^6$	m^3		

5. Compute (a) the equivalent SI metric pressures of systems A and B. (b) Determine the total force exerted by each piston to the nearest two decimal places.

$$Pa = psi\ (6\ 894.757) \qquad kg(force)/m^2 = Pa\ (9.806\ 650)$$

Piston System	System Pressure	Equivalent SI Metric Pressure (a)	Piston Area	Total Piston Force (b)
A	10 psi	Pa	10 in.^{2}	
B	64 Pa	kg (force)/m^2	12.2 m^2	

Section 7
Unit 36 Achievement Review on SI Metrics, Measurement, and Conversion

OBJECTIVES OF THE UNIT

This achievement review serves as an overall test for Section 7. The unit is designed to measure the student's/trainee's ability to

- *Understand how base and derived units of measure are used and the procedures of adoption into the SI metric system.*
- *Identify both advantages and disadvantages of SI metric units of measure as compared with customary units.*
- *Use conversion factors and processes to translate given quantities in customary units to SI metric values.*
- *Read direct metric linear measurements with line-graduated steel rules and drafting room scales.*
- *Read measurements on outside, inside, and depth micrometers, vernier calipers, and height gages.*
- *Interpret angle readings on the vernier bevel protractor.*

UNIT 32. SI METRIC UNITS OF MEASUREMENT

A. Historical Perspective on Standards and Units of Measure

1. State three reasons why international units of measure and standards change slowly but continually.

B. Advantages of SI Metrics

1. a. List five of the seven base units in SI metrics.
   b. Identify the physical property that each unit measures.

SI Base Unit					
Physical Property					

2. Give three advantages of SI metrics over other systems of weights and measure.

C. Scientific Notation System

1. Translate each of the four given values to its equivalent SI value in the required unit.

	Given Quantity and Unit	Required Unit	Quantity SI and Abbreviation
A	1 250 grams	kilograms (kg)	
B	2 500 000 volts	megavolts (MV)	
C	1.75 meters	millimeters (mm)	
D	9 835 800 000 hertz	gigahertz (GHz)	

2. Cite three rules for writing SI metric units of measure.

UNIT 33. PRECISION MEASUREMENT: METRIC MEASURING INSTRUMENTS

A. Direct Linear Measurements in the Metric System

1. Measure the diameter of each of the six bored holes (A through F) with an inside caliper. Transfer each measurement to a rule. Record each diameter to the nearest millimeter.

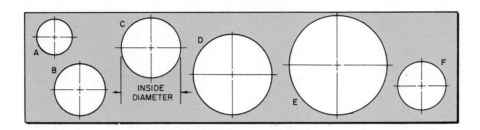

2. Read measurements Ⓐ through Ⓕ on the 1:50 metric drafting scale. State each measurement in terms of meter and fractional decimal values.

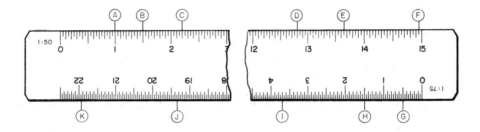

3. Read measurements Ⓖ through Ⓚ on the bottom 1:75 metric drafting scale. Give each reading in terms of its meter/centimeter value.

4. Determine the linear dimensions on the vernier micrometer setting Ⓐ, Ⓑ, and Ⓒ. (*Note*. The 0.1 and 0.05 mm sleeve graduations in the illustration are enlarged for easier reading.)

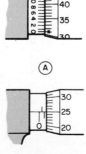
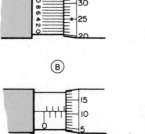
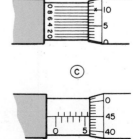

5. Read the three metric vernier caliper settings as indicated at (A), (B), and (C).

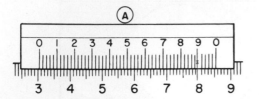

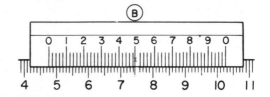

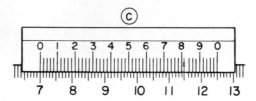

6. Read the metric vernier height gage measurements displayed at (A) and (B).

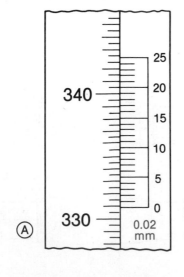

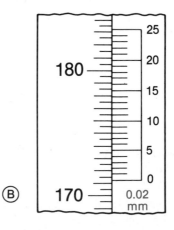

7. Determine the metric depth gage measurements (0.02 mm) shown at Ⓐ and Ⓑ and the inside micrometer measurements using the measuring rods as indicated at Ⓒ and Ⓓ.

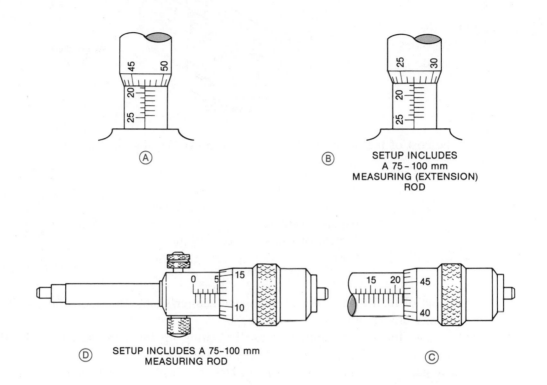

Ⓐ

Ⓑ   SETUP INCLUDES
A 75 – 100 mm
MEASURING (EXTENSION)
ROD

Ⓓ   SETUP INCLUDES A 75–100 mm
MEASURING ROD

Ⓒ

B. Reading Angles with the Vernier Bevel Protractor

1. Determine the vernier bevel protractor readings displayed at Ⓐ and Ⓑ.

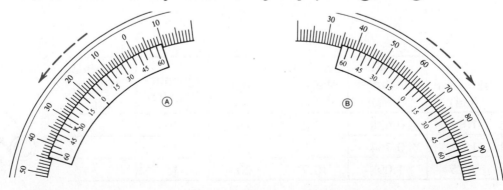

Ⓐ

Ⓑ

C. Computed Measurements in the Metric System

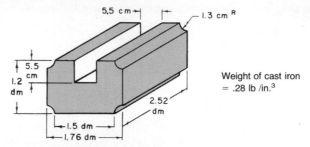

1. Determine the weight of 10 cast iron blocks to the nearest pound.

Weight of cast iron = .28 lb /in.³

2. Compute the cost of the bronze castings, correct to two decimal places.

Outside Diameter	1.2 dm
Inside Diameter	2.54 cm
Height	3.6 dm
Number Required	12
Weight of Brass	.26 lb/in.³
Cost	$3.48/lb

3. a. Secure a set of metric gage blocks or a manufacturer's table to establish the series and sizes of gage blocks in the set.
   b. Select a gage block combination to use in setting up for a measurement of 70.824 mm. (*Note.* Use a 1.5 mm wear block at each end.)
   c. List each gage block in the position (order) in which a craftsperson would set up (wring) the combination.

4. Compute the nominal weight in kilograms of each quantity of thermoplastic pipe according to the sizes, weights, and quantities given.

5. Determine how many metric tons of thermoplastic piping is required, correct to two decimal places.

	Outside Diameter (cm)	Wall Thickness (cm)	Nominal Weight (kg/m)	Standard Length (m)	Quantity Required	Nominal Weight (kg)
A	10.2	0.638	2.2	16	36	
B	15.2	0.754	3.8	24	12	
C	30.5	1.09	10.7	20	15	

6. Determine the weight in kilograms of the 15 bronze bushings described in the drawing. Round off the answer to the nearest whole kilogram value.

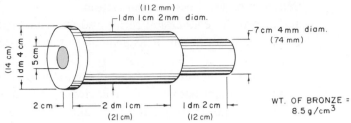

UNIT 34. BASE, SUPPLEMENTARY, AND DERIVED SI METRIC UNITS

1. Make two brief statements about the need for derived units of measure in SI metrics.
2. a. Name any one of the units of measure in SI metrics.
   b. Give the base unit for the group or category.
   c. Name four recommended decimal multiple and submultiple units and give the symbol for each unit.

UNIT 35. CONVERSION: FACTORS AND PROCESSES IN METRICATION

1. Convert each of the temperature measurements (A, B, C, and D) to its equivalent value in each of the two other temperature measurement units.

	SI Metric Temperature		Customary Temperature
	Celsius (°C)	Kelvin	Fahrenheit (°F)
A	− 10		
B	40		
C		1 753.15	
D	3.43×10^3		

2. a. Compute the equivalent Fahrenheit or Celsius degree temperatures that are missing in the table for soft-solder alloys A and B.

Soft Solder Alloy	Melting Range			
	Completely Solid		Completely Liquid	
	°F	°C	°F	°C
A	361		460	
B		270		312.22

b. Give the temperature span (range) between the completely solid and completely liquid states of the two alloys in (1) degrees Celsius and (2) Fahrenheit degrees.

3. Secure a table of conversion factors.
   a. Record the conversion factor and mathematical process for values A, B, C, and D.
   b. Convert the given measurement to the required measurement, correct to two decimal places.
   c. Give the reciprocal and mathematical process to use in converting the computed measurement back to the given measurement.
   d. Recompute the original given value to the nearest whole number.

Measurements		Conversion Factor and Process	Computed Measurement	Reciprocal and Mathematical Processes	Recomputed Given Value
	Given	Required				
A	465 mm	in.				
B	150 kg	lb (mass) avoirdupois				
C	125 Btu/s	kW				
D	240 km/h	m/s				

4. Compute the required volume (mass) in systems A, B, C, and D. Use the two formulas as indicated.
   a. State the conversion factor in each case.
   b. Compute and record each volume in the unit indicated in the table, correct to two decimal places.

	Volume (Mass)		Conversion Factor and Process
A	4 313.80 kg/m^3	lb (mass)/gal (US)	
B	kg/m^3	22.6 lb (mass)/gal (US)	
C	kg/m^3	32 lb (mass)/ft^3	
D	7 220 kg/m^3	lb (mass)/ft^3	

$$\text{lb (mass)/gal (US)} = \frac{\text{kg/m}^3}{119.8264}$$

$$\text{kg/m}^3 = \text{lb (mass)/ft}^3 \times 16.0185$$

PART THREE

Fundamentals of Electronic Calculators

CALCULATORS: BASIC AND ADVANCED MATHEMATICAL PROCESSES

Unit 37 Four-Function Calculators: Basic Mathematical Processes

OBJECTIVES OF THE UNIT

After satisfactorily studying this unit, the student/trainee will be able to

• *State the advantages of electronic calculators over conventional methods of solving mathematics problems.*

• *Understand algebraic (logic) entry, algebraic rules of order, and important components and general operation of basic electronic calculators.*

• *Round off values of significant and nonsignificant figures based on a knowledge of precision and accuracy meanings.*

• *Solve general and practical problems involving the basic arithmetical processes of addition, subtraction, multiplication, and division and other problems requiring chain processes.*

PRETEST   *Use the Review and Self-Test items provided in the Unit Assignment to establish the level of mathematical skills competency and to determine the starting point of instruction.*

A *calculator* is a computing machine that performs either basic arithmetical or advanced mathematical processes. All calculators may be used to compute the four basic processes of addition, subtraction, multiplication, and division. There are, however, larger and more complicated calculators. These models are identified as engineering and scientific calculators, advanced business computers, or electronic slide rules.

A. CHARACTERISTICS OF ELECTRONIC CALCULATORS AND TERMINOLOGY

Calculators that are small and pocket-sized are classified as "hand-held" or "pocket calculators." Larger calculators that are operated from a desk are called "desk-top" models. Some calculators are battery-operated; others depend on electric power, or a combination. Since calculators consist of modern, compact integrated electronic circuits that perform the mathematical operations by processing keyboard input to an output display (Figure 37-1), they are also called *electronic calculators*. Modern electronic calculators have many advantages over the older forms of calculators, slide rules, and adding and other business machines. These advantages include

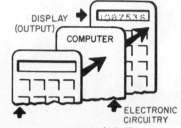

FIGURE 37-1

- Portability: lightweight, small size, comparative inexpensiveness
- Higher calculating speed and greater accuracy
- Performance in carrying out calculations and processes that are not possible by other machines.

Every calculator has a *keyboard*. The keyboard consists of a number of keys. There are ten keys for the basic numerals: 0, 1, 2, 3, 4, 5, 6, 7, 8, and 9. In addition, there is a decimal point and another set of keys for instructions relating to mathematical processes. The numerical keys, when pressed, enter the values to be used in the computation. The mathematical keys indicate the processes ($+$, $-$, $\times$, etc.) that are to be performed.

The mechanical operations of the keys provide the *input*. The computer component of the calculator performs the mathematical processes as directed by the input. The answer appears as an easy-to-read numerical value in the *display*.

The four basic arithmetic processes (*functions*) are addition, subtraction, multiplication, and division (Figure 37-2). These processes are required in more than 90% of all calculations. The electronic calculator that performs these

FIGURE 37-2 Four basic mathematical function calculator (*Courtesy of Sharp Electronics*)

processes is identified as a *four-function* type. This four-function calculator can also be used to perform higher mathematical processes. This is accomplished by breaking down the problem into a form that requires the four basic processes. The four-function calculator usually includes a percent (%) key to perform this particular arithmetical process.

Programmable Calculators

Programmable (scientific/engineering) calculators contain a *memory* to store numbers and *instructions* for performing a sequence of operations. The sequence in which the instructions are to be executed on the numerical values forms a *program* for solving the problem.

The program is stored in the calculator. Problems are entered and all mathematical processes are performed electronically according to the specified sequence. The program may be repeated any number of times using different numerical values. At the completion of the problem the correct answer is displayed. There is no further operator input into the problem.

Programmable calculators perform other sophisticated processes such as *decision making* and *branching*. Under specified conditions quantities are analyzed. Certain numerical values are then automatically processed by branching through a different program. Programmable calculators have the capacity to execute a program that is stored in a memory of the calculator. A depth study is made in Unit 39 of scientific calculator operations, functions, and advanced applications (Figure 37-3).

FIGURE 37-3 Handheld engineering and scientific calculator (*Courtesy of Sharp Electronics*)

Algebraic Entry

The sequence in which the numerals of a problem are entered on the calculator is referred to as the *algebraic (logic) entry*. This means that a problem is solved according to standard *algebraic rules of order*. Thus, the calculator uses the same logical sequence in which an individual solves a problem. The keys on an algebraic calculator are identified as $+$, $-$, $\times$, $\div$, and $=$.

Numerical values and mathematical processes are entered on the calculator in the sequence in which they occur in the problem. Basic rules of order must be followed. For example, all of the multiplication and division processes in the problem must be performed first. Then, the addition and subtraction calculations are completed as required.

The advantages of this system are simplicity and consistency. Its disadvantages lie in the fact that the individual must arrange the sequence of processes, write down and remember intermediate results, and apply these values later in the solution of the problem.

Some of the more advanced calculators use another form of algebraic entry. Engineering calculators are able to perform a series of *hierarchy* operations. Internal storage registers eliminate intermediate storage as well as the need to remember the rules of order. Information is entered as an algebraic entry in the same sequence in which the problem is to be solved. The correct answer appears when the equals ($=$) key is depressed.

Arithmetic Entry

Computations for addition and subtraction are carried on differently on the arithmetic calculator. The arithmetic calculator is distinguished by its $+=$ key and $-=$ key. The $+=$ key is pressed after all positive numbers. The $-=$ key is pressed after all negative numbers or if the numbers are to be subtracted. For instance, if 27 is to be subtracted from 85, the key sequence on the arithmetic calculator is $85 + = 27 - = 58$.

The procedures for multiplication and division are the same on both the algebraic and the arithmetic calculators.

True-Credit Balance

A calculator is said to have a *true-credit balance* when the display indicates whether the resulting numerical value is positive or negative. Usually, a negative result or difference on a true-credit balance calculator is displayed with a minus sign.

Registers

A *register* is a memory circuit. It is used to remember multidigit numbers. Four-function calculators have at least three *operating registers*. These registers are used to enter the numbers of a problem, display the numbers, and carry out the arithmetical processes. The most significant register is known as the *accumulator* or *display register*. The numbers that the operator enters via the keyboard are stored in the accumulator register. The numbers appear on the display register (Figure 37-4).

When the arithmetical process is entered on the keyboard, followed by the next value in the problem, the first

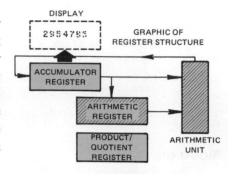

FIGURE 37-4

number is transferred from the accumulator register into the *arithmetic register*. When the equals key is pressed, the numbers (which are stored in the accumulator register and the arithmetic register) are transmitted to the *arithmetic unit*. Here the values are added or subtracted. The addition or subtraction result is stored in the accumulator. It is also displayed.

Since multiplication and division processes are shortened methods of adding and subtracting, a third register known as the *product/quotient register* is used to give the calculator additional memory storage.

Memory: Processing Store/Recall Data

All calculators have registers (memories) that store values to be used in a problem. The registers are fixed because they are permanently tied into the calculator circuitry. The registers are adequate for everyday applications.

Other calculators have greater capacity with long multidigit numbers that normally require complex calculations. Auxiliary memory is provided in such calculators to permit the storage and later recall of intermediate calculation results.

Separate keys are used to control the auxiliary memory features. A **CM** or **MC** key clears all numbers from the memory register, returning to zero. Pressing the **RM** or **MR** key recalls and transfers the value in memory to the accumulator register. The numerical value is displayed and is ready to be used in some computation without clearing or disturbing the memory accumulation.

The sum or difference of the memory content and the content of the accumulator register may be computed by pressing the appropriate **M+** or **M−** key. The sum or difference calculation is then contained in the memory. A number is stored in memory by first clearing the memory with the **CM** key. The **M+** key is pressed to transfer the accumulator contents to the memory. The **M−** key subtracts numbers in display from memory. On some calculators, when memory is in use, a decimal place appears in the furthest left ("left-most") digit space. Additional information on memory processes is presented in Unit 39 in relation to scientific calculators.

Clear, Clear-Entry, and Clear Display Key Functions

These keys are used to clear or erase the results of previous calculations and to reset the calculator. The clear **C** key is pressed on many calculators to reset the contents of the accumulator display register and the arithmetic register to zero. The circuitry of the calculator is then set to start a new calculation.

The clear-entry **CE** or clear-display **CD** key is used to correct number-entry errors without disturbing any previous part of the calculating sequence. The **CE** key is pressed to erase an error and to enter a correct new value. On some models the clear-error and total-erase processes are performed by a key marked **C/CE**. An entry mistake is corrected by pressing the **C/CE** key once and entering the correct value. This action does not disturb any previously recorded data or calculations. The **C/CE** key is pressed twice to erase all previous data and to clear the machine completely.

Calculator Display

The numerical readout on a calculator is found in a lighted *display*. The display serves three main purposes.

- Answers to problems are presented as a series of numerical digits in a readout display.
- Intermediate numbers used in a calculation are entered on the keyboard and presented on the display as a visual check.
- Information is fed back to a calculator operator through the <u>display</u>.

The number of digits in a calculator display varies. Six-digit displays are found in the less expensive models. Twenty or more digits are used in scientific calculators where greater precision is required. A practical common electronic calculator has an eight-digit display. This calculator has the capacity to display any number from 00 000 000 to 99 999 999. Fractional values are displayed by using a decimal point preceding the appropriate digit. Calculators are available in a variety of lighted color-digit displays. The displays operate on extremely small quantities of power.

Some accounting, scientific, and other business applications require a printed record. Calculators are designed with impact ribbon, heat-sensitive mechanical printers, ink injectors, and other mechanisms to meet such requirements and to produce a record tape.

B. SIGNIFICANT FIGURES, PRECISION, AND ACCURACY

Precision refers to exactness in defining a quantity. Precision is determined by the smallest increment that can be distinguished in the change of a quantity. When a higher degree of precision is required, there must be greater *resolution* and number of *significant figures* in a problem result.

A linear dimension like 3.01″ may be measured directly on a steel rule to a precision of roughly one one-hundredth part of one inch. If a precision of .001″ or 0.0001″ or 0.0254 mm or 0.00254 mm is required, measuring instruments with capability to measure to these degrees of precision are needed. In either instance, the number of significant digits in a required dimension is determined by the required precision. In these two examples, the steel rule may not be used to resolve any finer, more precise differences in length. Its limits of precision are reached at ± .01″. The micrometer, by comparison, has capacity to increase the precision from 10 (0.001″ or 0.0254 mm) to 100 (0.0001″ or 0.00254 mm) times. Precision is determined by the number of significant figures in a quantity, particularly those relating to decimal values.

The number of significant digits in whole numbers is determined by counting the number of digits with values from 0 through 9 that appear in the display to the left of the decimal point. For whole numbers and decimal values, all of the digits are counted (Figure 37-5).

At this point a statement should be made about accuracy as contrasted with precision. *Accuracy* is associated with degree of closeness to precision. While a micrometer may be used to measure to a precision of 0.001″ or 0.0001″ (0.2 mm or 0.02 mm), the operator's handling of the instrument may cause an inaccurate measurement to be taken. Accuracy is the relationship between a computed measurement and any variance with the true required measurement. Accuracy is usually stated as a percent of the actual (or calculated) value to the true value. If a measurement is computed to be 7.65 meters and its true length is known to be 7.77 meters, the calculated value is $^{7.65}/_{7.77}$ of 100% or 98.5% accurate.

Significant Digits (Figures)	Examples	
1	0.000 005	
2	98.	or 0.098
3	268.	or 0.000 268
4	3.192	or 0.003 192
5	29 323	or 29.323
6	29.3230	or 0.293230

FIGURE 37-5

C. BASIC MATHEMATICAL PROCESSES ON A FOUR-FUNCTION CALCULATOR

All calculators may be used for the four basic arithmetical processes of addition, subtraction, multiplication, and division. To do this, an individual must know the basic arithmetic principles and their application. The calculator simplifies all of the basic processes, ensures accurate results, and saves time.

Addition

The first of the processes to be discussed is that of addition. Two common methods of carrying on addition are described.

Method 1   *Simulated Example*. Add quantities $A_1 + A_2 + A_3 + A_4$

Step 1   Turn the calculator on.

Step 2   Press the numbered keys to enter the numerical value of A_1. Read the visual display to check that the correct number is recorded.

Step 3   Press the ⊕ key.

Step 4   Press the numbered keys to enter the second numerical value (A_2). Read the visual display for accuracy.

Step 5   Press the ⊜ key. This value is the sum of $(A_1 + A_2)$.

Step 6   Press the ⊕ key.

Step 7   Press the numbered keys to enter the value of A_3. Read the visual display to check the entry.

Step 8   Press the ⊜ key. This is the sum of $(A_1 + A_2) + A_3$. The display shows this sum.

Step 9   Press the ⊕ key.

Step 10   Enter the value of A_4. Check this value on the display.

Step 11   Press the ⊜ key. The display total is the answer, representing the sum of $A_1 + A_2 + A_3 + A_4$.

Note. If whole numbers and decimal values are involved, the decimal key is used to enter the value. The *floating-decimal point* within the calculator automatically indicates the position of the decimal point in the answer.

Method 2   A number of key strokes may be eliminated by a second method.

Example. Add 1 763 (A_1) + 23 825 (A_2) + 197 (A_3) + 192 368 (A_4)

Step 1   Turn the calculator on.

Step 2   Press the numerical keys to enter the value of A_1 of 1 763. Check this number on the display.

Step 3   Press the ⊕ key.

Step 4   Press the numerical keys to enter the value of A_2 of 23 825. Check the display.

Step 5   Press the ⊕ key.

Step 6   Enter the value of A_3 of 197. Again, check this entry on the display.

Step 7   Press the ⊕ key.

Step 8   Enter the value of A_4 of 192 368. Check the display.

Step 9   Press the equals key. Read the answer 218 153 on the display. In this example, **1 763 + 23 825 + 197 + 192 368 = 218 153.   Ans**

Note. All problems should be checked by repeating the steps and comparing the answers.

Subtraction

The same two methods may be used to solve problems involving subtraction. Instead of pressing the ⊕ key, the ⊖ key is used.

Correcting an Error

If a wrong numerical value is entered, two simple steps are required to correct the error.

- Press the **C** (or **CE** key on some models) key once. This removes the last entry.
- Enter the correct value on the keyboard. Check this value on the display. Then proceed with the remaining part of the problem.

EXAMPLE: Add 79 and 67.

	The Display Reads
Step 1 Enter the 79.	79
Step 2 Press the **+** key. Check the display.	79
Step 3 Assume you enter 65 by mistake.	65
Step 4 Press the **C** key.	0
Step 5 Enter the 67. Check the display.	67
Step 6 Press the **=** key. The display quantity is the answer.	146 Ans

FIGURE 37-6

Multiplication and Division

In addition to regular multiplication and division problems, a number of calculators have what is called a *constant feature*. Some calculators have a constant **K** panel switch to enter a required constant with the keyboard. This means that if a constant number is used in a series of multiplication or division problems, the constant is entered once. For example, in the accompanying series of multiplications

$$12 \times 12 \qquad 12 \times 16 \qquad 12 \times 47$$

12 is the constant. Figure 37-7 shows how each value is computed.

Step	Press/Enter	Display Reads	
1	CC (Press C key twice)	0	
2	12 (and the × key)	12	
3	=	144	(Ans to 12 × 12)
4	16	16	
5	=	192	(Ans to 12 × 16)
6	47	47	
7	=	564	(Ans to 12 × 47)

FIGURE 37-7

A similar series of steps is followed in division. For example, if -10 is to be divided three times by 3.2,

$$-10 \div 3.2 \div 3.2 \div 3.2$$

the 3.2 would be a *constant*. If an answer to three significant digits is required, the -0.3051757 would be rounded off to -0.305 (Figure 37-8).

Step	Press/Enter	Display Reads
1	CC	0
2	-10	-10
3	$\div$	-10
4	3.2	3.2
5	$=$	-3.125
6	$=$	-0.9765625
7	$=$	-0.3051757

FIGURE 37-8

Combination or Chain Arithmetical Processes

Many practical problems require a series of mathematical processes. Such series are known as *chain calculations*. They may involve addition, subtraction, multiplication, and division in varying combinations. In chain processes one of the goals is to solve the problem without having to store or write down intermediate values. Another goal is to keep the number of key strokes to a minimum.

There are rules of order to follow in establishing a sequence for carrying out chain calculations.

- Multiplication and division calculations are completed before addition and subtraction. For example, to solve the problem

$$7 + 9 \times 5 - 8 \div 4,$$

carry out the multiplication ($9 \times 5 = 45$) and division ($8 \div 4 = 2$) processes first. Then add $7 + 45 - 2 = 50$ **Ans**

Note. If the rules of order were not followed, an incorrect answer is obtained. For example,

$$7 + 9 = 16; \ 16 \times 5 = 80; \ 80 - 8 = 72; \ 72 \div 4 = 18.$$

- Portions of the problem are grouped before any mathematical processes are started. In the example, the values may be grouped and enclosed in parentheses.

The correct order or sequence for more complicated problems may be clarified by using brackets and parentheses. The operations in the parentheses are performed first. Those within the brackets are carried out next. The division process (which refers to all items within the brackets) is last.

The keys, processes, and display readouts for this problem are shown in Figure 37-9.

$[27 \times (19 + 17) + 68] \div 56$

$[27 \times \quad (36) \quad + 68] \div 56$

$[\quad \quad 972 \quad + 68] \div 56$

$\quad \quad 1040 \quad \quad \quad \div 56 = \boxed{18.571428}$ **Ans**

Step	Press	Display
1	CC	0
2	19	19
3	+	19
4	17	17
5	×	36
6	27	27
7	+	972
8	68	68
9	÷	1040
10	56	56
11	=	18.571428

FIGURE 37-9

ASSIGNMENT UNIT 37 REVIEW AND SELF TEST

PRETEST *The Review and Self-Test items that follow may be used as a pretest. Pretests are designed to measure a student's beginning level of mathematical skills competency and to determine the starting point of instruction.*

POSTTEST *The Review and Self-Test items may also be applied as a post test. Post tests are planned to establish the student's level of mathematical skills competency after instruction.*

A. Characteristics of Electronic Calculators

1. Indicate the one process in column II that correctly relates to each machine or function (column I).

Column I	Column II
Calculator or Function	*Process*
(1) Four-function calculator	a. Contains memory and instructions capability
(2) Display	b. Problem solving in the same logical sequence a person usually follows
(3) Programmable calculator	c. Performs addition, subtraction, multiplication, and division processes only
(4) Program	d. Total-erase process
(5) Algebraic entry	e. Visible colored entry or problem solution
	f. Registers that store problem values
	g. Corrects a number entry error
	h. Sequence for executing instructions entered in a computer

2. State the functions performed by the (a) **CM**, (b) **RM**, and (c) **M+** memory keys.

3. Indicate (a) how a calculator may be reset to zero and (b) a method of correcting an entry error.

4. (a) Describe what an eight-digit display means. (b) Give a numerical example of the capacity of this model calculator.

B. Significant Figures, Precision, and Accuracy

1. Give four examples of whole quantities and four of mixed number quantities. The quantities must contain the required significant digits given in the table for A, B, C, and D.
2. Define (a) precision and (b) accuracy.

	Required Significant Digits	Quantity Examples	
		Whole Number Values	Mixed Number Values
A	1		
B	3		
C	5		
D	7		

C. Basic Arithmetical Processes: Addition, Subtraction, Multiplication, Division

1. Add the linear measurements given in the table with a calculator. List (a) the sequence of steps, (b) the key(s) and values to be entered, and (c) the display readings. Mark the answer with the appropriate unit of measurement.

	29 375. Kilometers (km)
Problem	7 645.2 km
(Add)	987.96 km
	1 969.7 km

Step (a)	Keyboard Entry (b)	Display Reads (c)

2. Add the four quantities given in the table with a calculator. Indicate (a) the sequence of steps, (b) keys to be pressed or value to be entered, and (c) the display reading.
   Between the second and third quantity introduce an error. Include the steps to show the wrong entry, the display, and how the error is corrected.

Problem	797.75 RPM
(Add)	8 756.90 RPM
	+95 682. RPM
	439.6 RPM

Step (a)	Press (Keyboard) (b)	Display Reads (c)

3. Set up a problem with a quantity that is to be divided three times by a constant value. State the problem, the keyboard inputs, and the display readout for each step.

Problem

Step	Keyboard Input	Display Readout

4. Solve the problem, using the rules of order. Round off the decimal value to two significant places.

$$27 + 19.2 \times 31.3 \div 79$$

Unit 38 Four-Function Calculators: Advanced Mathematical Processes

OBJECTIVES OF THE UNIT

After satisfactorily studying this unit, the student/trainee will be able to

- *Understand how a calculator truncates values greater than the capacity of the instrument as underflow/overflow conditions.*
- *Use the calculator to convert fractions and decimals.*
- *Add and subtract numbers that have a greater number of digits than the capacity of the calculator.*
- *Deal with algorithms in solving calculator problems.*
- *Raise numerical values to higher powers and calculate reciprocals.*
- *Calculate square, cube, and higher root problems using a four-function calculator and use rounded-off values of π.*

PRETEST   *Use the Review and Self-Test items provided in the Unit Assignment to establish the level of mathematical skills competency and to determine the starting point of instruction.*

Persons who regularly deal with financial, business, engineering, scientific, slide rule, or other advanced mathematical processes usually perform them on more expensive, special calculators. The electronic circuitry is more complex in these special calculators and there is additional built-in capacity and accuracy. Also, additional keys are included in order to carry on many combinations of mathematical functions.

It is possible, however, to perform higher mathematical processes involving algebraic, trigonometric, or geometric calculations using the four-function calculator. Step-by-step procedures are described in this unit for performing advanced mathematical processes.

Rounding Off Values

Regardless of whether a value is rounded off by conventional mathematical computing processes or on a calculator, error results. For most practical purposes, dimensions and values that are rounded off to two, three, and four decimal places are well within the usual range required. Parts produced within dimensional tolerances of 0.001″ and 0.0001″ (0.2 mm and 0.02 mm) are considered to be practical to measure and may be reproduced at reasonable cost. Movable units, machined to such tolerance limits, will mesh and work properly.

The degree of required precision should be determined first, before the solution of a mathematical problem. The precision will then establish the number of significant figures in the answer and at what point the computed value may be rounded off.

The calculator also *truncates* values. For example, the display of the decimal value for ⅓ is *truncated*. This means the display value on an eight-digit calculator reads .33333333, rounded off by eliminating the least significant digits. While this degree of accuracy in a final result is acceptable in most everyday problems, added care must be taken in making a long series of calculations. When values are rounded off after each calculation, the *cumulative* error may result in an inaccurate measurement.

A. OVERFLOW AND UNDERFLOW CALCULATOR CONDITIONS

When the numerical capacity of a calculator is exceeded, a condition of *overflow* exists. Overflow occurs when a quantity greater than the number of digits in the calculator is fed into it or when the result of a calculation exceeds the calculator capacity. This overflow condition is indicated in the display by the appearance of the letter **E** or **C** in the furthest left position of the display.

The overflow may be cleared by dividing by 10. This is repeated enough times to bring the decimal point into the display. Calculations are then continued. The result will be multiplied by 10^N. N represents the number of times the overflow was divided by 10.

Underflow denotes a condition where, again, the capacity of the calculator is exceeded. In this instance, the least significant digits may be lost. For example, in an eight-digit calculator, the display would register zero if the keys were pushed to record a value like **.000 000 003 256.**

B. CONVERSIONS OF FRACTIONS AND DECIMALS

Calculators are used with whole numbers or decimal parts. Sometimes, it is necessary to convert a fractional quantity to its equivalent decimal value. This process is easily accomplished by dividing the numerator by the denominator. Where the floating decimal is incorporated into the calculator, the decimal is automatically located. For instance, to convert the fraction $\frac{7}{64}$ to the equivalent decimal value, divide the numerator 7 by 64 = 0.109375.

To convert a decimal quantity to its fractional value, determine the number of decimal digits in the fractional part. A quantity like 1.104 indicates that there are 104 thousandths in the fractional part. So, the 1.104 quantity, expressed as a mixed number, $= 1\frac{104}{1\,000}$. The fraction is further reduced by dividing by the greatest common divisor (8). Expressed in its lowest terms, the $1\frac{104}{1\,000} = 1\frac{13}{125}$.

C. DOUBLE-PRECISION CALCULATIONS

Addition of Numbers with Digits Exceeding Calculator Capacity

Occasionally, it is necessary to add numbers that have more significant digits than the capacity of the calculator. Two numbers, like

140 569 074 279.   and

9 435 829 308 695.

can be added on an eight-digit calculator by following these steps.

Step 1   Align the decimal points.

Step 2   Split each number into two parts: most significant digits and least significant digits. Keep each set of digits properly aligned. The number of least significant digits should be one less than the number of digits in the calculator display.

Most Significant Digits	Least Significant Digits
140 56	9 074 279.
9 435 82	9 308 695.
① Added ⟵----------- Carried⟵----①	8 382 974.
over	

Step 3   Add the least significant digit values. Record the answer. Note that there are eight digits in this answer. The eighth place digit value ① is carried over and added to the most significant digit value in the furthest right digit column.

140 56

+9 435 82

+   ①   (carryover)

9 576 39

Step 4   Combine the sums of the two parts. The sum of the two combined display readings of **9576398382974** may be more easily read and accurately used when it is written as **9,576,398,382,974.** In SI metrics, the sum is given in this form: **9 576 398 382 974.**

Subtraction of Numbers with Digits Exceeding Calculator Capacity

Large quantities may be subtracted in a similar manner. The decimal points are aligned. The values are placed in two columns (1) the most significant part and (2) the least significant part. There must be one less number of digits than the capacity of the calculator in the least significant part. The values (1) and (2) are then subtracted.

It should be noted that if the top number is less than the lower number, one (1) must be *borrowed* the same as in simple arithmetic. When (1) is borrowed in the last (left) digit position of the least significant part, the right (and first) digit in the most significant part is reduced by (1). After the two parts are subtracted, they are combined to obtain the answer.

EXAMPLE:

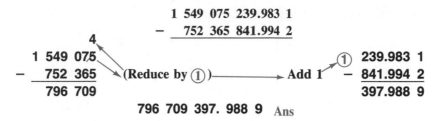

$$
\begin{array}{r}
1\ 549\ 075\ 239.983\ 1 \\
-\quad 752\ 365\ 841.994\ 2
\end{array}
$$

796 709 397. 988 9 Ans

D. ALGORITHMIC PROCESSES APPLIED TO FOUR-FUNCTION CALCULATORS

An *algorithm* is a step-by-step routine used in solving a problem. It is this same routine that is used internally in calculators. The electronic circuitry within a calculator performs three ·basic processes to solve or implement any mathematical problem. The circuitry can perform only addition, subtraction, and shifting. The term *shifting* means moving a number one or more digits to the right or left in relation to another number. As stated many times, multiplication and division processes require successive addition and subtraction, respectively.

In summary, an algorithm is a set of mathematical rules, numerical procedures, or logical decisions. It provides variable input data to a calculator. Once an algorithm is mastered, advanced problems that would involve a single key on a sophisticated calculator may be easily performed on the four-function calculator.

Step	Keyboard Entry	Display Readout
1	CC	0
2	16	16
3	÷	16
4	=	1
5	=	0.0625

FIGURE 38-1

The *reciprocal* interchanges a numerator and denominator. This interchanging is a valuable technique in chain processes. Since division is involved in finding a reciprocal, the value (1) is divided by a designated value. For instance, the reciprocal of 16 is 1 divided by 16 = 0.0625. Another procedure for finding the reciprocal of 16 is illustrated by the five steps shown in Figure 38-1.

When used in a chain process, the algorithm (Figure 38-2) applies. For example, find the reciprocal of (275.392 75 ÷ 17.687).

Step	Keyboard Entry	Display Readout
1	CC	0
2	275.392 75	275.392 75
3	÷	275.392 75
4	17.687	17.687
5	=	15.570 348
6	÷	15.570 348
7	=	1
8	=	0.064 224 6

FIGURE 38-2

Raising Numerical Values to Higher Powers

One of the simplest problems of raising a number to a higher power is squaring the number. This process requires the use of the exponent (^{2}) with the number. Squaring problems are computed on the calculator the same as in ordinary multiplication. The square of 6.7 is 6.7 $\times$ 6.7 = 44.89.

There are two common algorithms that may be used when a number (N) must be raised to a higher power like the (nth) power. The first sequence is practical when the value of the power is smaller than ten. The number to be raised to a given power is entered on the keyboard. Then the multiplication ✖ key is pressed. This step is followed by pressing the equals ═ key one less time than the required power. The final value represents N^n, the quantity (N) raised to the required (^{n}) power.

If there is a chain calculation involving the raising of a quantity to a power, the problem should be restated in the sequence in which the combined mathematical processes are performed. The example eliminates going back to parts of a problem.

EXAMPLE:

$$\frac{(17.96 + 8.53)\ 6.2^3}{(7.23 \times 3.45)}$$

Restated as (17.96 + 8.53) $\times$ 6.2 $\times$ 6.2 $\times$ 6.2 $\div$ (7.23 $\times$ 3.45), the problem is keyed as a chain process into a calculator.

Raising a Number to a Very High Power

Raising a quantity to a very high power by using the first method is cumbersome and subject to possible error of a lost count. A simple algorithm to follow requires the breaking down of the high power into smaller factors. Assuming a quantity N is to be raised to the 180th power $(N)^{180}$, the 180 is broken down into smaller factors, like 3 $\times$ 3 $\times$ 4 $\times$ 5.

EXAMPLE: Raise N (value 1.3) to the 45th power.
$$(1.3)^{45} = [(1.3^3)^3]^5.$$

- Enter N (1.3).

- Press ⊗ key.

- Press ⊜ key one less time than the value of first power.

- Press ⊗ key.

- Press ⊜ key one less time than the value of the second power.

- Press ⊗ key.

- Press ⊜ key one less time than the value of the third power.

- Read the display.

Step	Keyboard Entry	Display Readout
1	CC	0
2	1.3	1.3
3	×	1.3
4, 5	=, =	1.69 2.197
6	×	2.197
7, 8	=, =	4.826 809 10.604 499
9	×	10.604 499
10	=	112.455 539
11	=	1 192.533
12	=	12 646.215
13	=	134 106.77

FIGURE 38-3

The thirteen steps shown in the table may be reduced if the calculator has a key for squaring x^2. After the required power has been factored, each factor may be further broken down. For example, $N^4 = N^2 \times N^2$; $N^3 = N^2 \times N$; $N^8 = N^2 \times N^2 \times N^2$. The required quantity is factored by using the x^2 key in combination with the × key to produce the equivalent exponential factor value.

E. CALCULATING SQUARE ROOT ON A FOUR-FUNCTION CALCULATOR

Square root problems are easily solved on calculators that have a square root function key. The required quantity for which the square root is needed is entered on the keyboard. The √ key is pressed. The display readout is the answer.

The process is more complicated on the four-function calculator because a trial-and-error procedure and a formula are used. An estimate is made of the square root of the required quantity. Suppose the square root of 97 is required. An estimate of 10 is made. Substitute the first estimate (10) for E and 97 for N in the formula.

$$\sqrt{R} = [(N \div E) + E] \div 2$$
$$= [(97 \div 10) + 10] \div 2$$

$$= 9.85$$

$\sqrt{R}$ = square root value

N = number for which the square root is required

E = estimated square root of N

- The approximate root (9.85) is now squared. The result 97.022 5 is too inaccurate. So, the process is repeated. This time the second estimated new root of 9.85 is substituted.

$$\sqrt{R} = [(N \div E) + E] \div 2$$
$$= [(97 \div 9.85) + 9.85] \div 2$$
$$= 9.848\ 857\ 5$$

- Again, squaring the resulting third estimated root of 9.848 857 5 gives a product of 96.999 994. With this degree of precision, the third estimate is accepted as the answer. Normally, a four-place decimal value is sufficiently accurate.
- If the last estimate does not produce a product within the limits of accuracy required in the problem, the estimating procedure is repeated one or more times. When still higher degrees of accuracy are needed, a calculator with a greater number of digits or built-in square root capacity should be used.
- The percent of error may be calculated by using a formula and substituting values.

$$\frac{(N) - (R)^2}{(N)} \times 100 = \text{Percent Error}$$

For example, if the second estimated root of 9.85 is squared and the product (97.022 5) is substituted for R^2

$$\frac{97 - 97.022\ 5}{97} \times 100 = -0.023\ 19\% \text{ Error}$$

F. CALCULATING CUBE ROOT AND HIGHER ROOTS ON A FOUR-FUNCTION CALCULATOR

Another algorithm is needed to obtain the cube and higher roots. While a similar procedure is followed as for determining square root, a different formula is used.

$$R^x = [(N \div A^{n-1}) + (n - 1)A] \div n$$

R^x = cube root or higher root
N = quantity for which root is to be computed
n = order of the root
A = approximate value of the nth root

As an example of the steps and entries that are made on a calculator, assume the cube root of 500 is required (Figure 38-4). The root value should be within .001% of accuracy. The estimated value (A) of the nth (^{3}) root is 8.

Formula		Step	Keyboard Entry	Display Readout
Item	Quantity			
		1	CC	0
$[N$	500	2	500	500
$\div$		3	$\div$	500
(A^{n-1})	(8^{3-1})	4	64	7.8125
$+$		5	$+$	7.8125
$(n-1)A]$	$(3-1)8$	6	16	23.8125
$\div$		7	$\div$	23.8125
n	(3)	8	3	3
$=$	R^x	9	$=$	7.9375

FIGURE 38-4

The cube of $7.9375 = 500.0935$. If the error of .0935 is too great, the process may be repeated. In this case, the second estimated root value should be slightly smaller. For closer precision, an estimated value smaller than the first estimate of 7.9375 (like 7.937) may be used $(7.937)^3 = 499.999$. This value is accurate to within 0.0002%.

While a cube root example was used to show the steps involved with a four-function calculator, higher root values may be computed using the same formula and processes.

G. ROUNDED PI (π) VALUES

Throughout the text, pi (π) has been used in many different types of practical mathematical problems. Pi is an *irrational number* whose value is 3.141 592 653 589 793. . . . It is obvious that most calculations are not this precise. So, the exact value is rounded off to a required number of decimal places. Values of 3.1416, 3.142, and the fractional value $\frac{22}{7}$ are widely used for most everyday problems. The highest degree of accuracy possible using a four-function, eight-digit calculator is to round the value of π to 3.141 592 7.

ASSIGNMENT UNIT 38   REVIEW AND SELF TEST

PRETEST   *The Review and Self-Test items that follow may be used as a pretest. Pretests are designed to measure a student's beginning level of mathematical skills competency and to determine the starting point of instruction.*

POSTTEST   *The Review and Self-Test items may also be applied as a post test. Post tests are planned to establish the student's level of mathematical skills competency after instruction.*

A. Overflow, Underflow, and Rounded-Off (Truncated) Values

1. Define (a) overflow and (b) underflow. Give a numerical example in each case.

2. State how rounding off errors in high precision calculations result from truncated display values.

3. Subtract the three numerical quantities, using an eight-digit calculator. Round off the metric measurement to five significant digits.

8 957.625	mm
− 2 868.732 9	mm
− 4 927.843 42	mm

B. Conversions of Fractions and Decimals

1. Compute the missing fractional, decimal, or metric (mm) equivalent values from the information provided for measurements A through F. Round off decimal values to four places.

Kind of Measurement	Equivalent Measurements					
	A	B	C	D	E	F
Fractional (")	$\frac{1}{16}$	$\frac{1}{64}$				
Decimal (")			0.125	0.3125		
Metric (mm)					37.7	108.346

C. Double-Precision Calculations

1. Add the following quantities with a calculator having fewer digits than any of the three quantities. (a) Arrange the quantities into two columns of most significant digits and least significant digits. Then proceed to find the sum. (b) State the answer in megahertz (MHz) to four decimal places (least significant digits).

765 372 987.37   Hz
8 849 235.967 89 Hz
26 795 869.758 7   Hz

D. Algorithmic Processes Applied to a Four-Function Calculator

1. Describe briefly the importance of an algorithm in using a calculator.

2. a. Calculate the reciprocal of the four quantities given. List the display readouts.
   b. State the answer, rounding each decimal value to four significant digits.

	Quantity	Reciprocal Value	
		Display Readout (a)	Rounded Answer (b)
A	18		
B	$\frac{1}{7.4}$		
C	$\frac{176.33}{16\ 206.4}$		
D	$\frac{422.17 \times 16.7}{14\ 692.9}$		

3. a. Raise each quantity (A through H) to the required power. Record the display readout.
   b. Indicate each measurement by rounding off the decimal display values to two significant digits, unless otherwise indicated.

	Quantity	Power Value	
		Display Readout (a)	Measurement (b)
A	8^6		
B	7.2^3		
C	$8.4^2 \times 10.6^2$		
D	$10.04^3 \times 16.2^2$		
E	2.2^{12}		
F	0.46^{15}		
G	π^4 Use $\pi = 3.1416$		
H	α^6 Use $\alpha = 2.7183$		

E. Calculating Square and Higher Roots on a Four-Function Calculator

1. Compute the square root and the higher roots for quantities A, B, C, and D. Note the required degree of precision.
   a. Give the display readout of each final root value.
   b. Round each root value to four decimal places.

Quantity	Required Precision	Root Value	
		Display Readout (a)	Required Dimension (b)
A $\sqrt{86}$	$\pm.001$		
B $\sqrt{\dfrac{(6.2 \times 9)}{1.74}}$	$\pm.01$		
C $\sqrt[3]{(5.07 \times 4.02)^2}$	$\pm.02$		
D $\sqrt[6]{716.88}$	$\pm.005$		

Unit 39 Scientific Calculators: Operation, Functions, and Programming

OBJECTIVES OF THE UNIT

After satisfactorily studying this unit, the student/trainee will be able to
- *Interpret the range of a scientific calculator with respect to memory usage, basic mathematical functions, statistical functions, and advanced algebraic, trigonometric, and geometric functions.*
- *Identify each key and key sequence function and read display register data.*
- *Establish the calculator mode; clear a scientific calculator; isolate numerical expressions; and enter data.*
- *Evaluate and solve practical problems involving combinations of basic mathematical processes.*
- *Program a scientific calculator for statistical (central tendency) measurements of mean, standard deviation, and variance functions.*

PRETEST   *Use the Review and Self-Test items provided in the Unit Assignment to establish the level of mathematical skills competency and to determine the starting point of instruction.*

A. INTRODUCTION TO THE SCIENTIFIC/ENGINEERING CALCULATOR

More advanced problems encountered by technicians, engineers, and scientists, who formerly used slide rules, are now solved with sophisticated and powerful electronic hand calculators (Figure 39-1). While each calculator incorporates different design features, similar mathematical processes are performed.

A representative *scientific calculator,* designed for data entry by *algebraic notation,* is used as a model in this unit. According to the *Algebraic Operating System* (AOS)^{TM}, each number and function is programmed into the calculator. Thus, each entry follows the same mathematical sequence in which each function and quantity is written. Intermediate results are obtained by pressing the *equals key* $=$

Calculators in this family or level have the ability to accept entries in *scientific notation* (or *powers of ten*). The significant number of digits in internal calculator capacity is eleven. However, eight digits are displayed. The last number displayed for most functions is generally rounded off to ± 1 for the eighth-digit value.

FIGURE 39-1 Representative model of a scientific/engineering electronic calculator *(Courtesy of Texas Instruments, Inc.)*

First and Second (Alternate) Function Keys

Some calculator keyboards have keys that provide a *second (alternate) function.* The first function is printed on each key. The second function appears above a key. The drawing (Figure 39-2) shows one row of second function keys on a calculator as designed by one manufacturer. The first function keys

sin	cos	tan	π	1/x
1	2	3	×	÷

FIGURE 39-2 Sample row of keys with identification of first and second (alternate) functions.

1, **2**, and **3** are numeric; the following **✕** and **÷**, arithmetic. The alternate (second) functions of sin, cos, and tan relate to trigonometry. The **π** key is used for circle and angle measurement and the **1/x** key for reciprocal functions. First and second function keys may be positioned horizontally or vertically, depending on the design of the calculator.

Particular attention was given in preceding units to the use of a simple four-function calculator. The applications related principally to basic arithmetic functions of addition, subtraction, multiplication, division, and combinations of these processes.

Identification and Functions of Keys

Each key on the representative model as illustrated in Figure 39-2 is identified by name, function, and processing practices. Similar information with procedures follow in later units for those key and key sequences that apply to algebra, trigonometry, and geometry. This unit builds upon the skills developed with a four-function calculator.

B. BASIC CALCULATOR OPERATIONS

A *power-on condition* is provided in the model scientific calculator by pressing the **ON/C** key. A ⎸ 0 ⎹ *display register* indicates that the calculator is *clear* of *all pending operations and statistical registers*. Some calculators are designed so that *user memory* is not cleared when power is removed from the calculator by the **OFF** key.

Mode Indicator

Scientific/engineering calculators use a *mode indicator* in the display. During short periods of a calculation, the abbreviation DEG (degree), RAD (radian), and GRAD (grad) may appear. These letters are *angular mode indicators*. Similarly, STAT in the register indicates the computer is in the *statistical mode*. The statistical mode indicator may be turned off, the STAT register cleared, and the calculator set for normal calculations by pressing the **2nd** **CSR** *clear statistical register key sequence*.

Data Entry

- The **·** *decimal point key* permits a fractional part of a number to be entered. Each digit value in the whole number is entered. The decimal point then *floats* with the additional numeric values added at each digit position. On this model (Figure 39-1), a decimal with a maximum of eight digits may be entered.

- *Numeric Keys* **0** through **9**, like the earlier four-function calculator, enter values of 0 through 9 as appropriate digits.

- The **+/−** *change sign key* is pressed after a number is entered in a calculator to change the sign of the number on display.

- The constant value of (π) pi (rounded to 3.1415927 on the display) is entered by pressing the **2nd** **π** *pi key sequences*. The *internal calculator value of pi* in the model is correct to eleven digits.

- The **K** *constant key* is used to store a number and its associated operations for repetitive calculations.

- Values up to eleven digits may be entered from the keyboard. A value like 5232.7684713 is entered as the sum of two numbers (5232 plus .7684713).

Enter (Value)	Press (Function)	Display
5232	+	5232
.7684713	=	5232.7685

Clearing the Calculator

The **ON/C** *clear entry/clear key* is pressed before any function or operation key to remove an incorrect entry from the display. The display, constant, and all pending operations are cleared when the **ON/C** key is pressed after an operation key or a function key or if the **ON/C** key is pressed twice. It should be noted that the **ON/C** key does not affect the user memory or statistical registers.

An incorrect number entry may be cleared by pressing the **ON/C** key before any non-number key. This action does not affect any calculations that are in progress.

C. ISOLATING NUMERICAL EXPRESSIONS

Parentheses serve the important function of grouping *(isolating)* particular mathematical values and functions. The closing of parentheses () indicates that all necessary information is complete. Each group is isolated so that the numerical values and required mathematical processes may be evaluated and performed on the calculator just as they are written.

EXAMPLE: Evaluate 6 × (7 + 4) ÷ (7 − 2).

Numerical Entry	Key or Key Sequence	Display	Notes
6	× (	6	The 6 × is stored
7	+	7	(7 + is stored
4	)	11	(7 + 4) is stored
	÷	66	6 × (11) is evaluated and the operation is performed
	(	7	
7	−	5	(7 − is stored
2	)		(7 − 2) is evaluated and then divided into
	=	13.2	6 × (7 + 4)

D. MEMORY FUNCTIONS AND USAGE

Scientific calculators are designed with *memory storage*. Data and operations may be stored in the *memory register* even when the calculator is turned off. Operations in progress are not affected by the use of memory. The advantage of memory is that complicated computations and high powers of numbers and multiple calculations may be performed once, stored, and recalled any number of times.

Memory Store

The STO *memory store key* is used to store a displayed value in the memory. When the key is pressed, the new entry replaces any previous value or quantity in memory.

Memory Recall

Contents of the memory register are recalled into display by pressing the RCL *memory recall key*. The content of the memory register is not affected.

EXAMPLE: Store and recall the value 46.375.

Numerical Entry	Key or Key Sequences	Display
46.375	STO	46.375
	OFF ON/C	0
	RCL	46.375

Adding a Display Value (Sum) to Memory

The SUM *sum to memory key* permits the display value to be added to the content in memory register. The number on display or any calculation in progress is not affected by this key. When pressed, the SUM key accumulates the quantities from a series of independent calculations.

EXAMPLE: Store the sum to memory of the calculations:

$$(16.36 \times 4.2) + (12.974 + 7.286) + (3.14156 - 2.1347 + 6.42)$$

Numerical Entry	Key or Key Sequences	Display	Sum to Memory
16.36	×	16.36	0
4.2	= STO	68.712	68.712

	+ (		
12.974	+	12.974	68.712
7.286	) SUM	20.26	88.972
	+ (		
3.14156	−	3.14156	88.972
2.1347	+	1.00686	88.972
6.42	) SUM	7.42686	96.39886
	RCL	96.39886	96.39886 Ans

Memory Exchange

The content of the memory is exchanged for the display value by pressing the **EXC** *exchange key*. This key is used to store numbers or a result and to recall such values for comparison calculations or further computations.

E. SQUARING NUMBERS

The **x^2** *square key* is used to calculate the square of the number entered into the display.

EXAMPLE: Square the value 27.435.

Numerical Entry	Key to Press	Display	
27.435	x^2	752.67922	Ans

F. STATISTICAL FUNCTIONS

Data related to quality control in manufacturing, cost analyses, sales and marketing, and other business activities are usually evaluated by statistical methods. Such statistics show central tendency. Such common terms as mode, median, standard deviation, and curves of distribution were described and applied earlier. Values obtained were computed ''long hand'' using basic mathematical processes.

The scientific electronic calculator is designed for effectively evaluating data and calculating *mean, standard deviation,* and *variance* for a total population or a sample of the population.

The *second function keys* for statistical functions appear on the keyboard as **2nd** $\bar{x}$ for the *mean key;* **2nd** **σn** for the *population standard deviation key sequence;* and **2nd** σn-1 for the *sample standard deviation key sequence*. The *variance key sequences* appear on the keyboard as the *population variation key sequence*, which includes three keys: 2nd, σn, x^2; the *sample variation key sequence* involves pressing the 2nd, σn-1, x^2 keys. The keys and sequential data are internally programmed according to the appropriate mathematical processes and data-gathering formulas.

Statistical Data Entry and Removal

The term *population* relates to a large quantity of values or sets of items. A *sample* or *sampling* represents a smaller portion selected from the population. The calculator is set in the *statistical mode* by the Σ+ *sum plus key* after the first entry is made. The mode is identified as ⌐ STAT ¬ on the display.

Resetting the Calculator for Regular Functions

The statistical registers, the STAT indicator, and the calculator may be reset for regular calculations and functions by pressing the 2nd CSR keys.

The 2nd Σ−1 *sum minus key sequence* is used to remove any unnecessary data points. After removal, the X register displays the current number of data points.

G. PROGRAMMING FOR STATISTICAL MEASUREMENTS OF CENTRAL TENDENCY

Calculating Mean, Standard Deviation, and Variance

Once the calculator is clear and set in the statistical mode, data points are entered by pressing the Σ+ *sum plus key*. If an error is entered, the data point entry is removed by pressing the 2nd Σ− keys. This action produces an ⌐ ERROR ¬ display and causes the statistical registers to be cleared. After clearing and all new data are entered, the *mean value* is computed and displayed by pressing the 2nd $\bar{x}$ keys. Similarly, the 2nd σn keys are pressed for *standard deviation* and the x^2 *square key,* to obtain the *variance*.

It is important to remember that the 2nd CSR keys are pressed before entering data points for arithmetical calculations.

EXAMPLE: Quality control inspection of a total population run of ground bearing races produces quantities of parts that meet dimensional and form requirements as given in the table. Calculate the mean, standard deviation, and variance.

Category	A	B	C	D	E
Accepted Quantities	990	950	894	943	996

Numerical Entry	Key Sequence (Programming)	Display	Notes and Functions
	ON/C		Applies power Clears calculator
	2nd CSR	0	Clears the statistical register Clears the STAT indicator Resets calculator for manual operation
990	Σ+	1 STAT	Enters first data point
950	Σ+	2 STAT	Enter, add, and store second data point in statistical memory
894	Σ+	3 STAT	Enter, add, and store third data point
997	Σ+	4 STAT	Enter, add, and store fourth data point *Note.* Incorrect entry
997	Σ−	3 STAT	Remove fourth entry
943	Σ+	4 STAT	Enter, add, and store correct fourth data point
996	Σ+	5 STAT	Enter, add, and store fifth data point
	2nd $\bar{x}$	954.5 STAT	Mean*(average number of parts that satisfactorily meet standards) Ans
	2nd σn	36.86516 STAT	Standard deviation Ans
	x^2	1359.040 STAT	Variance Ans *Values are rounded to the whole number in the display

H. KEYS FOR PROGRAMMING ALGEBRAIC, TRIGONOMETRIC, AND GEOMETRIC FUNCTIONS

Figure 39-3 identifies the remaining keys on the model. The keys in Figure 39-3 are grouped according to functions in each branch of mathematics and do not appear in the position occupied on the calculator.

Algebraic Functions		Trigonometric and Geometric Functions			
Key	**Applications**	**Key**	**Applications**	**Key**	**Applications**
1/x	Reciprocals	$\log$	Common logarithm	sin	Trigonometric functions
$x!$	Factorial	$\ln x$	Natural log	cos	
$\sqrt{x}$	Square root	e^x	Natural antilog	tan	
Y^x	Powers of numbers	DRG	Degree/Radian/ Grade mode		
$\sqrt[x]{Y}$	Roots of numbers				
EE	Enter exponent				
INV	Inverse				
K	Constant				
%	Percentage				

FIGURE 39-3

ASSIGNMENT UNIT 39   REVIEW AND SELF TEST

PRETEST *The Review and Self-Test items that follow may be used as a pretest. Pretests are designed to measure a student's beginning level of mathematical skills competency and to determine the starting point of instruction.*

POSTTEST *The Review and Self-Test items may also be applied as a post test. Post tests are planned to establish the student's level of mathematical skills competency after instruction.*

A. Scientific Calculator Operation

1. State two important differences between a simple four-function calculator and a scientific calculator.
2. Identify (a) two mode indicators and (b) the display readout(s) for each indicator.
3. a. Tell how to remove an incorrect entry.
   b. Tell how to clear the display and any constant and/or pending operation on a scientific calculator.

4. Identify the key sequence to the following operations:
   a. Clearing the statistical registers, turning off the STAT display indicator, and setting the calculator for normal operation.
   b. Entering the value of (π) pi.

B. Isolating Numerical Expressions: Chain Calculations

1. List the scientific calculator process by (a) numerical entry, (b) key(s), (c) display, and (d) notes for solving the following problem.

$$3.1416 \times (8.2 + 5.73) \div (9.84 - 3.652).$$

2. Solve problems (a) and (b). Round off the final value to two decimal places.
   (a) $(4.264 + 2.40 - 0.96) \times (13.568 \div 2.6)$
   (b) $(79.8 - 13.47 \times 0.22) \div (12.9 + 6.7 - 3.142)$
3. Compute the weight of metal required to stamp and form parts A, B, and C to the nearest pound.
   Note. The weight per piece is equal to the mass (volume) of each part multiplied by the weight factor given in the table. The scrap allowance is added to the length of each part stamped.

Parts	Quantity Required	Surface Area		Thickness	Metal	Weight Factor	Scrap Allowance
		Length	Height				
A	10,000	2.00"	2.00"	0.03196"	Brass	0.306	0.25"
B	5,275	3.40"	1.50"	0.05082"	Steel	0.284	0.35"
C	20,625	5.687"	3.25"	0.02535"	Copper	0.322	0.032"

4. Calculate the change in overall length (in inches) of current carrying conductors A and B for the (a) lowest and (b) highest temperatures recorded in the table. Round off the final answers to three decimal places.

Conductor	Metal	Length of Conductor at 25°C	Coefficient of Linear Expansion (inches/foot)	Temperature Range	
				Low (a)	High (b)
A	Pure copper	1,000 ft	0.000 009 2	−40°C	50°C
B	Copper alloy	1,000 ft	0.000 009 9	−40°C	50°C

C. Processing Reciprocals and Pi Values

1. Give the processes involved in determining the reciprocal of a physical measurement like 25.4 millimeters. Indicate the (a) numerical entry, (b) name and symbol of the key, and (c) the display.

2. Calculate the pitch of the following national standard screw threads, correct to three decimal places.

   (a) $\frac{1}{2}$-13 UNC   (b) 1"-12 UN   (c) 2-56 UNF

   Note. The pitch is equal to $\dfrac{1}{\text{No. threads/inch}}$.

   The number of threads per inch is 13, 12, and 56, respectively.

3. List each (a) entry, (b) name and sequence of keys, and (c) display in calculating the mass (volume) of the cored bronze casting.

Core diameter = 17.78 cm diam
Outside diameter = 27.94 cm diam
Length = 20.32 cm

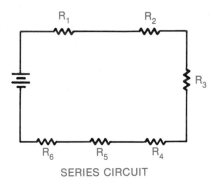

Note. Mass (volume) = $(\pi) \times (\text{radius})^2 \times \text{length} - (\pi) \times (\text{radius})^2 \times \text{length}$

D. Calculating Statistical Measurements

1. Find the total resistance (R_T), mean, standard deviation, and variance of series circuits A and B. Round off each answer to three decimal places.

   Note. $R_T = \Sigma$ of $R_1 + R_2 + R_3 + R_4 + R_5 + R_6$

Series Circuit	Resistance in Ohms					
	R_1	R_2	R_3	R_4	R_5	R_6
A	5.00	12.50	6.72	12.97	32.63	17.26
B	41.70	26.0	38.0	46.42	18.40	9.20

SERIES CIRCUIT

2. a. Read and record the revolutions per minute (RPM) as displayed on each of the five tachometer test instruments.
   b. Calculate the (1) mean of the high speed turbine engine speeds, (2) the standard deviation, and (3) the variance.

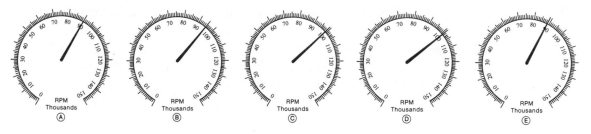

Unit 40 Section 8: Achievement Review on Basic and Scientific Calculators

OBJECTIVES OF THE UNIT

This achievement review serves as an overall test for Section 9. The unit is designed to measure the student's/trainee's ability to

- *Describe the characteristics, advantages, and selected features of an electronic calculator.*
- *Deal with applications of significant figures with relation to precision and accuracy.*
- *Work with a four-function calculator to solve typical arithmetic and chain process problems.*
- *Apply algorithms with a four-function calculator to solve practical problems involving the substitution of values in formulas and chain processes.*
- *Isolate values for separate mathematical processes.*
- *Identify keys and functions of a scientific calculator.*
- *Calculate statistical measurements using a scientific calculator.*

SECTION PRETEST/ POSTTEST

The Review and Self-Test items that follow relate to each Unit within this Section. The test items may be used as a Unit-by-Unit pretest and/or Section post test.

UNIT 37. FOUR-FUNCTION CALCULATORS: BASIC MATHEMATICAL PROCESSES

Characteristics of Electronic Calculators

1. State four advantages of modern electronic calculators over older calculating machines and devices.
2. a. Describe the purpose served by the registers of a calculator.
   b. State the function of three operating registers in a four-process calculator.
3. Cite two purposes that are served by the calculator display.

Significant Figures, Precision, and Accuracy

1. Tell and illustrate how the number of digits in a least significant decimal value indicates the precision of a quantity.
2. State a problem with three mixed numbers having an overall total of 12 digits and requiring addition and subtraction processes. Show how the three quantities may be grouped for an eight-digit calculator for double-precision calculations.

The Four-Function Calculator and Basic Arithmetical Processes

1. Solve the problem. (a) Number the steps and indicate the (b) keyboard entries and (c) display readings. Round the answer to six significant digits and mark the answer with the appropriate unit of measurement.

Problem	Step (a)	Keyboard Entry (b)	Display Readout (c)
865.7 volts			
+1 720. volts			
− 440.80 volts			
− 720.095 volts			

2. Add and subtract the quantities given in the problem. Use an eight-digit calculator.
   a. Show how the quantities are arranged in two columns.
   b. Label and give the answer corrected to eight significant digits.

$$
\begin{array}{rl}
 & 4\,3\,5\,8\,7\,2.5 \qquad\ \text{cycles} \\
+ & 8\,9\,4\,2\,5.7\,6 \qquad\quad " \\
+ & 4\,9\,8\,7.5\,3\,7 \qquad\quad " \\
- & 2\,0\,9.6\,7\,9\,6 \qquad\quad " \\
- & \underline{3\,2\,6\,8.7\,8\,2\,5} \qquad " \\
\end{array}
$$

3. Multiply the quantity of 208 volts by the constant value 2.67 for three times. Use a hand calculator and give the answer correct to three decimal places.

UNIT 38. FOUR-FUNCTION CALCULATORS: ADVANCED MATHEMATICAL PROCESSES

1. Determine the ignition pulse (distributor point) frequency (ipf) for the two- and four-stroke cycle engine specifications given in the table.

   Use the formula, $\text{ipf} = \dfrac{(n) \times (\text{RPM})}{(X)}$

   n = number of cylinders
   X = 120 for 4-stroke cycle engines
   = 60 for 2-stroke cycle engines

   Note. Round off the pulses per second (ips) to one decimal place.

	Engine		Specification	
	Cylinders (n)	Crankshaft (RPM)	Stroke Cycle	Ignition Impulse Frequency (ipf)
A	8	2 000	4	
B	12	1 625	4	
C	6	1 375	2	
D	4	1 840	2	

2. The rise time (t_r) and the cutoff frequency (f_2) of a circuit may be computed by the following formulas.

$$ t_r = \sqrt{(c^2) - (a^2 + b^2)} \qquad f_2 = \frac{0.35}{t_r} $$

Use a calculator to compute the (t_r) and (f_2) values for the A and B circuit conditions given in the table. Round off megahertz values to two decimal places.

	Rise Time (in nanoseconds)				Upper Cutoff Frequency (f_2) in mega-hertz (MHz)
Circuit	Oscilloscope (a)	Square Wave (b)	Pulse (c)	Circuit (t_r)	
A	40	20	110		
B	32	18	90		

UNIT 39. SCIENTIFIC CALCULATORS: OPERATION, FUNCTIONS, AND PROGRAMMING

1. (a) Name and (b) state the function of each key or key sequence as represented by items 1 through 8.

	Key or Key Sequence	Name (a)	Function (b)		Key or Key Sequence	Name (a)	Function (b)
1	1/x			5	2nd CSR		
2	Σ+			6	+/−		
3	SUM			7	2nd π		
4	( )			8	x^2		

Chain Calculations: Isolating Values for Separate Processes

1. Solve the two problems (a and b) by isolating the bracketed values for separate mathematical processes. Round off the final value to three decimal places.
   a. $(\pi) \cdot (6.50)^2 - (\pi) \cdot (4.75)^2 \times (12.25 + 2.50 - 0.92)$
   b. $(72.4 + 26.37 - 14.67) \div (\pi \times 17.2) \times (14)^2 - (16.4 - 10.26 - 14.38)$

Calculating Statistical Measurements

1. a. Read the 0.01 mm metric dial indicator readings for five sample parts.

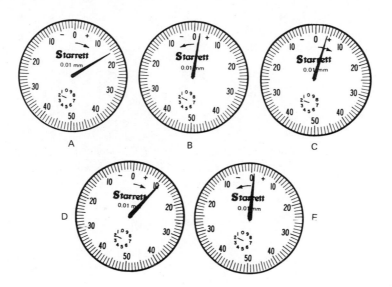

b. Calculate the (1) mean measurement, (2) standard deviation, and (3) variance.

Courtesy of the L. S. Starrett Company

PART FOUR

Mathematics Applied to Consumer and Career Needs (with Calculator Applications)

FUNCTIONAL CONSUMER AND CAREER MATHEMATICS

Unit 41 Mathematics Applied to Money Management and Budgeting

---

OBJECTIVES OF THE UNIT

After satisfactorily studying this unit, the student/trainee will be able to
* *Make calculations that involve the addition, subtraction, multiplication and division of money quantities either by conventional mathematical methods or by using a calculator.*
* *Set up plans and prepare forms for money management and budgeting.*
* *Prepare a trial budget, an actual budget, and a control system to meet personal needs.*

PRETEST   *Use the Review and Self-Test items provided in the Unit Assignment to establish the level of mathematical skills competency and to determine the starting point of instruction.*

A. CONCEPT OF THE MONEY SYSTEM

Money transactions are made daily in all branches of business, industry, and the home. The money system in the United States is based on the dollar as a unit. All fractional parts of the dollar are expressed in the decimal system. Thus, all amounts from one cent to ninety-nine cents may be written as decimals.

Two symbols, the dollar ($) and the cent (¢) signs are used with money. In writing a sum like ten dollars, the ($) sign is placed in front of the number, for example, $10 or $10.00. An amount like twenty-five cents may be written with the symbol (¢) as 25¢ or as the decimal ($.25). Neither the ($) sign nor the decimal point is used with the cent sign (¢).

The decimal point separates the dollar values from the fractional parts of the dollar. The value of a few common digits in money calculations is illustrated.

Number of Tenths of a Dollar (dimes)

Number of Units (cents)

Number of Tenths of One Cent (mills)

$94,235.743

Number of Dollars

Number of Ten Dollars

Number of Hundred Dollars

Number of Thousand Dollars

Number of Ten–Thousand Dollars

B. CALCULATIONS INVOLVING ADDITION, SUBTRACTION, MULTIPLICATION, AND DIVISION OF MONEY

Since the money system is a decimal one, it is possible to add, subtract, multiply, and divide the numbers representing different amounts in the same way as any other decimal. Care must be taken to keep the units in each number in columns, one under another, to point off the correct number of places in an answer, and to check each step.

C. MATHEMATICS APPLIED IN MONEY MANAGEMENT AND BUDGETING

Effective money management requires the setting up of a system of records, maintaining them in a regular and orderly manner, and controlling and managing income and expenses through a budget. All essential information such as bills, payments, receipts, and cancelled checks, should be recorded in a book or ledger and kept in properly labelled envelopes or files. This requires putting valuable papers, bonds, savings account passbooks, insurance policies, tax records, and other valuable documents, in a safe deposit box at a bank or in a fireproof container.

The methods described in this unit for setting up and managing a money management plan permit adaptation to meet particular individual needs. The plan may be programmed into a personal computer where the data may be stored, retrieved, and changed as needed.

Money management forms are commercially available as horizontally ruled sheets with vertical columns that may be inserted into and removed from a loose-leaf binder. The standard forms are easily adapted for personal use.

Fixed and Variable Items

In its simplest form, money management requires the planning of a budget and then managing how money (income) will be used to meet planned expenditures. The expenses in a budget are usually grouped as *fixed* or *variable* (flexible). The *fixed expenses* relate to bills that are due and payable on a regular weekly, monthly, quarterly, semiannual, or other schedule. Examples of fixed income items are federal and state taxes on earned income, property taxes, monthly payments on a house, insurance premiums, automobile license fees, student or professional association dues, to name a few.

Variable (flexible) expenses represent those that may be changed. Such expenses permit flexibility in adjusting a budget to meet emergency conditions or other changes in earnings or expenses. Expenditures for flexible items cover recreation and hobbies, gifts, entertainment, nonessential clothing, additional furniture, dining out, and any items that may be included, changed, or dropped from a budget.

D. SETTING UP A BUDGET FOR A MONEY MANAGEMENT PLAN

√ Preparing the Income Part of a Trial Budget

Step 1   List the main categories for all items of income on a ruled worksheet.

Step 2   Set up *amount* and *schedule* columns to indicate how much and when the income (receivable) items are due.

Step 3   Label the columns for recording the monthly and annual income totals on the worksheet.

Step 4   Add the monthly income entries vertically for each item category to obtain the *grand total;* horizontally, for the *annual income by category*.

Step 5   Add the annual income entries for all categories vertically to determine the *gross income*.

Note. Total wage, salary, or other income should be recorded without payroll deductions. Deductions will be reported as expenditures.

EXAMPLE:

✓ Preparing the Expenditure Part of a Trial Budget

Step 1   Set up a worksheet listing major categories of expenditure under two headings: (A) *fixed* and (B) *variable (flexible)* expenditures.

Step 2   List the major categories and the main items under each category in a vertical column.

Step 3   Label a series of vertical columns for each month and an annual total column.

Step 4   Total the weekly expenses for each fixed item for each month. Record (post) each amount. Monthly expenses may be averaged by multiplying the weekly amount by $4\frac{1}{3}$.

Step 5   Total the monthly columns horizontally to get the annual expenditures for each major *fixed item*. Add the column entries vertically to get the monthly expenses and the grand total for the year.

Step 6   Continue to list the major categories and main items for all *variable expenses* for each week. Total these for the month and year.

Step 7   Add the grand totals for all fixed and variable expenses. Record these amounts by the month and as the overall *gross amount*.

EXAMPLE:

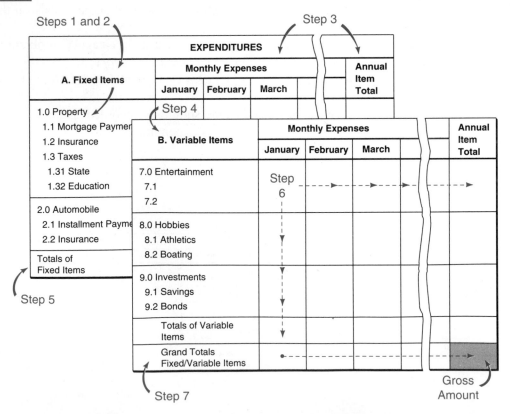

✓ Setting Up the Actual Budget

Step 1   Compare the gross amounts for income and expenses. If the income is less than the projected expenses, set up priorities and reduce the variable expenditure items first.

Step 2   Prepare the budget after all adjustments are made and the income and expenses are in balance.

✓ Managing the Budget: Control System

Step 1   Develop a set of weekly income and expense control cards for the year. These cards should include three parts as shown in the example that follows: *balance, income,* and *expense.*

Step 2   Total the weekly income and weekly expenditures. Add the income to the *starting balance.* Then, subtract the expenditures. Record this difference as the *new balance.*

    Note • A budget review should be done at least a month in advance to ensure that there is an adequate reserve to cover all bills.

           • Where an expense exceeds the balance, record the amount as a negative one: for example, −$50 or ($50).

Step 3   Review, recompute, and adjust the budget whenever there is any significant change in either income or expenditures.

EXAMPLE:

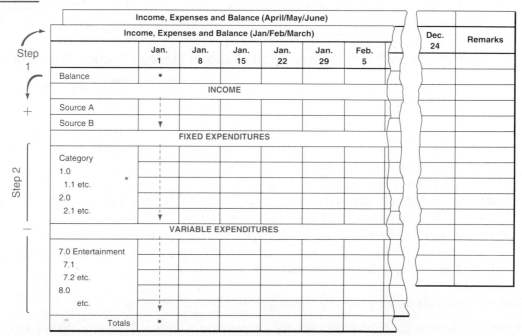

ASSIGNMENT UNIT 41 REVIEW AND SELF TEST

PRETEST
The Review and Self-Test items that follow may be used as a pretest. Pretests are designed to measure a student's beginning level of mathematical skills competency and to determine the starting point of instruction.

POSTTEST
The Review and Self-Test items may also be applied as a post test. Post tests are planned to establish the student's level of mathematical skills competency after instruction.

CALCULATOR APPLICATIONS
The problems in this Unit may be solved and/or checked either by conventional mathematical processes or by using a calculator. Each answer displayed on the calculator register is rounded off to the required number of decimal places.

A. Concept of the Money System (General Applications)

1. Write the amounts A to G in words.

A	B	C	D	E	F	G
$1.12	$20.07	$2,692.75	$0.02	37¢	2.37\frac{1}{2}$	46$\frac{1}{4}$¢

2. Check each amount and indicate those that are incorrectly written.

A	B	C	D	E
.29¢	$0.17¢	$124.04¢	$6,292.16	57¢

B. Addition, Subtraction, Multiplication, and Division of Money (Practical Problems)

1. Find the cost of 30 lengths of steel as shown by the sketch.

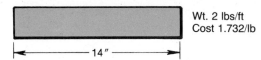

Wt. 2 lbs/ft
Cost 1.732/lb

|← 14″ →|

2. Twenty castings cost $189.40. If the cost per pound is $3.30, how much does each casting weigh and cost? Round the weight and cost to two decimal places.
3. How much coolant is required and what is the cost to fill a machine reservoir $\frac{3}{4}$ full of a ready-mix preparation that sells for $5.50 per gallon? The reservoir measures 2′ wide × 3′ long × 8′ deep. (Use 1 gallon = 231 cubic inches).
4. The labor and material costs for a job total $2,701.12. The labor costs are 68$\frac{1}{2}$% of this amount; the material is 31$\frac{1}{2}$%. Determine, correct to two decimal places, what the materials cost and what charges were made for labor.

C. Mathematics Applied to Money Management and Budgeting

Problems 1 through 6 relate to (a) the preparation of a trial budget, (b) an actual or simulated working budget, and (c) a system for controlling and managing the budget.

1. Prepare a worksheet for listing major categories of (a) income and (b) fixed and variable expenditure items that are to be included in an actual or a simulated trial budget.
2. Add columns for both income and expenses. Then, record (a) the amounts for exact or anticipated income and expenses and (b) prepare a schedule showing when each item is due and payable.
3. Add still other columns to show monthly and annual income for each item. (a) Compute and record monthly income and expenditures. (b) Add the monthly totals and post the yearly amounts. (c) Add and record the gross annual totals.
4. Develop a balanced budget by establishing priorities and making adjustments until the income and expenditures are equal. Then, prepare the working budget.
5. Develop a system for the control and management of the one-year budget.
6. Compute the percent of the total annual income that each major category of (a) income and (b) expenditures constitutes. Add another column on the budget forms and record these percents.

Unit 42 Mathematics Applied to Purchasing Goods and Credit Financing

OBJECTIVES OF THE UNIT
After satisfactorily studying this unit, the student/trainee will be able to:
- *Apply percentages in computing manufacturing costs.*
- *Solve practical problems in purchasing goods that require the application of single and multiple discounts.*
- *Compute the cost of purchasing on credit.*
- *Make comparisons of credit charges based on calculations of credit rates and different patterns of financing.*

PRETEST   *Use the Review and Self-Test items provided in the Unit Assignment to establish the level of mathematical skills competency and to determine the starting point of instruction.*

A. MATHEMATICS APPLIED TO THE PURCHASING OF GOODS

Determining Manufacturing Costs

Manufacturing costs consist of three basic elements: (1) raw materials costs, (2) labor costs to convert raw materials into marketable products, and (3) overhead costs such as taxes, rent, light, power, equipment depreciation and maintenance, and worker benefits.

In arriving at manufacturing costs, simple addition, subtraction, multiplication, and division are used. Also, percentages and averages are needed for comparison purposes. The manufacturing cost is equal to the cost of raw materials, direct labor, and overhead.

RULE FOR DETERMINING MANUFACTURING COSTS AND PERCENTAGES

- Total the cost of all materials.
- Total all labor charges.
- Determine what items of overhead to include.
- Check by subtracting from the manufacturing costs any one of the three cost figures. The difference is the sum of the other two costs.

EXAMPLE: Determine the manufacturing costs and the percent of each item. The raw materials cost $245, the labor charges are $1,535, and the overhead costs are $450.

Step 1   Add raw materials and labor and overhead costs.

$$\begin{array}{r} \$\ \ 245 \\ 1{,}535 \\ \$\ \ 450 \\ \hline \end{array}$$

Manufacturing cost = $2,230   Ans

Step 2   Divide *raw materials* cost by manufacturing cost to get percent of total.

$$\frac{245}{2{,}230} = 11\%$$

Step 3   Divide the *labor cost* by the manufacturing cost for percent.

$$\frac{1{,}535}{2{,}230} = 69\%$$

Step 4   Divide *overhead costs* by manufacturing costs for percent.

$$\frac{450}{2{,}230} = 20\%$$

$$\begin{array}{r} 11\% \\ 69\% \\ 20\% \\ \hline 100\% \ \ \text{Proof} \end{array}$$

Step 5   Check percent by adding to see if the total is 100%.

Applying Single and Multiple (Successive) Discounts

Manufacturers and wholesalers generally use a *suggested retail price* or *list price* as a guide for determining the *net price* that a retailer pays for a particular product. The difference between the list price and the net price represents a *trade discount* (the retailer's *margin of profit*).

When a product is sold to a consumer at a reduction from the list price, the discount is usually stated as a percent reduction (regular price less 20% discount or 10% off *manufacturer's suggested retail price*) or as a fraction ($\frac{1}{2}$ off list price).

Discounts may be *single* (like a 10% discount) or *multiple;* as for instance, 20%, 10%, and 5%. Multiple (successive) discounts are often given on large quantity purchases because each order can be fulfilled at less cost. When multiple discounts are applied, each successive discount is taken on the remainder.

RULE FOR TAKING A SINGLE DISCOUNT

- Change the percent discount to a decimal.
- Multiply the list price by the discount.
- Point off the decimal places in the answer. The product represents the discount.

EXAMPLE: Find the net cost of an instrument that lists at $630 and discounts at 12%.

Step 1 Change the 12% to its decimal equivalent.	12% = .12
	$ 630
Step 2 Multiply the list price (630) by the discount (.12).	× .12
	$75.60
Step 3 Subtract the discount ($75) from the list price ($630) to get the net price of $555.	$ 630
	−75
	$ 555 Ans

RULE FOR TAKING MULTIPLE DISCOUNTS

- Take the first of the series of discounts as a single discount.
- Subtract the discount from the list price to get a net price.
- Multiply the net price by the decimal equivalent of the second discount.
- Subtract the second discount from the net price.
- Continue these same steps for each discount. The last difference represents the actual cost to the consumer.

<u>E XAMPLE</u>: Determine the cost of 4,200 pounds of metal at $1.45 per pound less 15%, 10%, and 5% discounts.

Step 1	Change the percents to decimal equivalents.	**4,200**
Step 2	Multiply the 4,200 pounds by $1.45 to get the gross cost.	**× $1.45**
		$6,090.00

Step 3	Take the first discount of 15% on $6,090.00.	**$6,090.00** **× .15** **$ 913.50**
Step 4	Subtract the first discount from $6,090.00.	**$6,090.00** **− 913.50** **$5,176.50**
Step 5	Take the second discount of 10% on the remaining $5,176.50.	**$5,176.50** **× .10** **$ 517.65**
Step 6	Subtract the second discount from the $5,176.50.	**$5,176.50** **− 517.65** **$4,658.85**
Step 7	Repeat the last two steps with the third discount of 5%.	**$4,658.85** **× .05** **$ 232.94**
	Note. The $4,425.91 is the actual cost of the metal.	**$4,658.85** **− 232.94** **$4,425.91 Ans**

RULE FOR CALCULATING A SINGLE DISCOUNT PERCENT EQUIVALENT

A short method for solving problems that involve a series of discounts is to calculate and use a *single discount percent equivalent*.

- Subtract the first discount from 100 percent to establish the first *remaining discount percent*.
- Repeat this step to obtain the remaining discount percent for each successive discount.
- Multiply each remaining discount percent in the series. The product represents the single discount percent equivalent.
- Multiply the list price by the single discount percent equivalent. The product represents the invoice price of the manufacturer's product.

EXAMPLE: Calculate the invoice price of an appliance that a manufacturer lists at $882.50. A trade discount of 40% and quantity discounts of 10% and 6% apply.

Step 1 Change each discount to a remaining discount to a decimal equivalent.

First discount 40% $100\% - 40\% = 60$ percent remaining discount

$= .60$ decimal equivalent

Second discount 10% $100\% - 10\% = 90\% = .90$

Third discount 6% $100\% - 6\% = 94\% = .94$

Step 2 Multiply all of the decimal equivalents of each remaining discount percent. The product represents the single discount decimal equivalent.

$$(.60) \times (.90) \times (.94) = (.5076)$$

Step 3 Multiply the manufacturer's list price by this decimal factor. The product is the invoice price of the appliance.

($882.50) $\times$ (.5076) = $447.96 Invoice Price. Ans.

Step 4 Check by applying successive discounts with each remaining balance.

First discount $882.50 \times .40 = 353.00 $(882.50 - 353.00) = 529.50

Second discount $529.50 \times .10 = $ 52.95$ $(529.50 - 52.95) = 476.55

Third discount $476.55 \times .06 = $ 29.59$ $(476.55 - 29.59) = 447.96

$447.96 is the Invoice Price. Ans.

B. MATHEMATICS APPLIED IN CREDIT FINANCING

Buying on an investment plan means that the purchaser is actually using a product or service before completely owning it. In a free enterprise system, many sellers provide credit for customers in order to stimulate and maintain sales. Some have open charge accounts with a grace period, after receipt of the bill, to pay it without credit charges. Others may require the customer to cover the payments by financing through some type of finance organization.

The Cost of Credit

EXAMPLE: The cost of lumber, panelling and hardware totals $1,500. The lumberyard requires a 20% down payment and 12 monthly installment payments at the rate of $1\frac{1}{2}\%$ of the unpaid balance. Determine the additional expense of making installment payments.

Step 1 Subtract the 20% down payment from the total cost of $1,500 to find the amount subject to installment charges.

$1,500 Total cost
$\times$ .20 Down payment (%)
$ 300 Down payment ($)
$1,500 - 300 = $1,200

Step 2 Compute the credit charge on the beginning balance of $1,200. The first month credit charge is $18.

$$\frac{\$100}{12 \overline{)\,1,200}} \text{ Monthly payment}$$

$$\$1,200 \times .015 = \$18$$
first credit charge

Step 3 Subtract the monthly payment on the principal for the second month ($1,200 − 100 = $1,100). Multiply the new principal ($1,100) by the installment charge rate of $1\frac{1}{2}\%$. The second monthly credit charge is $16.50.

$$\$1,100 \times .015 = \$16.50$$
second credit charge

Step 4 Repeat step 3 until all the monthly payments are computed.

Step 5 Total the installment charges for the 12-month period. These equal $117.00.

Month	Credit Charge Principal	Charge
1	1,200	$ 18.00
2	1,100	16.50
3	1,000	15.00
4	900	13.50
5	800	12.00
6	700	10.50
7	600	9.00
8	500	7.50
9	400	6.00
10	300	4.50
11	200	3.00
12	100	1.50
		$117.00 Total Charges

Step 6 Divide the installment charges by the original principal. The quotient is the percent charged annually for buying the hardware items on the installment plan of the dealer. $\left(.0975 = 9\frac{3}{4}\%\right)$

$$\frac{.0975}{1,200 \overline{)\,117.0000}} \text{ Ans.}$$

The example shows that the principal (loan total) changes each month as a payment is made and the cost of buying the construction materials on the installment plan is $117.00 or 9.75%. Before buying, the consumer should not hesitate to get the rates from several sources of installment credit. These costs should be compared.

The consumer must compare the benefits of buying the item on credit and paying additional finance costs. The benefits may relate to increased earning power, emergency situations which might result in a greater loss if the purchase is not made, leaving savings and other investments intact, or better living conditions brought about by purchases of labor-saving devices. The importance of the loan, the ease and time it takes to make the loan, and the collection methods are additional considerations.

Comparison of Typical Credit Charges

EXAMPLE 1: Repayment of a loan where the credit charge is added to the beginning balance and the total is repaid in 12 equal monthly payments.

Specified Rate		Actual Annual Rate
Per $100	% Per Year	
$6.00	6	11.1
8.00	8	14.8
10.00	10	18.5
12.00	12	22.2
14.00	14	25.9

EXAMPLE 2: Repayment of a loan where the credit rate is applied against the total amount of the loan and payments are made monthly.

Rate per Month On Unpaid Balance	Actual Annual Rate
$\frac{1}{2}$ of 1%	6%
$\frac{5}{6}$ of 1%	10%
1%	12%
$1\frac{1}{2}$%	18%
2%	24%
3%	36%

Example 2 shows a high actual annual rate because the rate of the loan is being charged against the total amount of the loan (principal) instead of the true balance. The true balance changes each month as payments on principal and other charges are made.

RULE FOR CALCULATING THE ACTUAL CREDIT RATE

- Use the following formula to determine how much a consumer pays on installment purchases in terms of the *annual percent rate* for credit.

$$R = \frac{2(I \times F)}{U(n + 1)}$$

R denotes installment credit as a percent.
I is the number of installment payments in one year.
F represents the amount of the finance charges.
U is the unpaid balance.
n equals the number of payments in the contract.

EXAMPLE: The unpaid cost of tools and equipment is $600. This amount is to be paid in 12 equal installments of $58 each. Determine the rate for the installment credit at an annual percent.

Step 1   Substitute given or computed values in the formula.

$I = 12$

$F = (12 \times \$58) - \$600 = \$96$

$$R = \frac{2 (I \times F)}{U (n + 1)}$$

$U = \$600$

$n = 12$

$$R = \frac{2 (12 \times 96)}{600 (13)} = .295$$

Step 2   Carry out the indicated mathematical processes.

Step 3   Convert the decimal value (.295) to a percent.

$.295 = 29.5\%$ Ans.

Comparison of Financing Patterns

As the number and value of installment purchases increase, merchants incur greater losses due to unpaid bills, added administrative and bookkeeping expenses, and the loss of interest on substantial amounts of money that are outstanding. Installment costs must reflect these conditions. The extra charge for credit may be called a "service, carrying, or interest finance charge." Different finance patterns and charges on an $800 credit purchase are illustrated in the following table.

Finance Items	Financing Patterns				
	Dealer	Mail Order Company	Bank	Loan Company	Credit Union
Amount of Credit	$800.00	$ 800.00	$800.00	$800.00	$800.00
Monthly Payments	76.57	83.33	78.13	84.17	⊿ 73.17
Total Payments (12 months)	918.80	1,000.00	937.60	1,010.00	878.00
Service Charge	118.80	200.00	137.60	210.00	78.00
Actual Annual % Rate of Interest (AAPR)	14.85% (AAPR)	25.00% (AAPR)	17.20% (AAPR)	26.25% (AAPR) (30% on first $300; 20% on all over $300)	9.75% (AAPR) ⊿ (1.5% per month on unpaid balance)

ASSIGNMENT UNIT 42 REVIEW AND SELF TEST

PRETEST *The Review and Self-Test items that follow may be used as a pretest. Pretests are designed to measure a student's beginning level of mathematical skills competency and to determine the starting point of instruction.*

POSTTEST *The Review and Self-Test items may also be applied as a post test. Post tests are planned to establish the student's level of mathematical skills competency after instruction.*

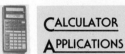

CALCULATOR APPLICATIONS *The problems in this Unit may be solved and/or checked either by conventional mathematical processes or by using a calculator. Each answer displayed on the calculator register is rounded off to the required number of decimal places.*

A. Mathematics Applied to the Purchasing of Goods
Determining Manufacturing Costs (Practical Problems)

1. Determine the manufacturing costs according to the charges given for jobs A to E.

	Material Costs	Labor Costs	Overhead
A	$ 75.00	$232.00	$ 74.00
B	96.00	384.50	142.50
C	84.46	295.23	216.24
D	165.29	347.63	210.98
E	220.06	400.97	$309.58

2. Compute the amounts spent for materials, labor, and overhead for parts A to E. Use the manufacturing costs and percents shown in the table.

	Mfg. Costs	Materials	Labor	Overhead
A	$1,000	50%	30%	20%
B	2,428	25%	60%	15%
C	1,378	12%	70%	18%
D	2,532	18.5%	72.5%	9%
E	1,864	22%	$57\frac{1}{2}$%	$20\frac{1}{2}$%

3. The cost of manufacturing 1,000 parts is $792.00. Of this, 35% is spent for materials, 62% for labor, and the remainder for overhead.
   a. Determine the cost for materials.
   b. Find the cost for labor.
   c. Find the overhead costs.
   d. During the next quarterly period, the materials increased 10% in price, the labor costs rose 5%, and the overhead increased 5%. Determine the new cost for materials, labor, and overhead.
   e. What is the new cost for 1,000 parts?

Application of Single and Multiple Discounts (Practical Problems)

1. High speed steel twist drills of a certain size list at $58.50 a dozen less a trade discount of 22%. Determine the cost of each drill to the nearest cent.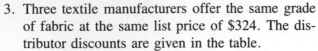

2. The materials cost of making a machine guard is 18%, the labor cost is 75%, and the overhead is 7%. Fifty guards are sold for $3,050 less a discount of 10%, 8%, and 2% for cash. Determine
   a. The cost of each guard.
   b. The materials, labor, and overhead costs for the lot, based on discounted price.

3. Three textile manufacturers offer the same grade of fabric at the same list price of $324. The distributor discounts are given in the table.
   a. Establish which distributor offers the best total discount.
   b. Compute the net cost of the fabric from the distributor that provides the best discount.

Distributor	Multiple Discounts
A	18% and 17%
B	20% and 15%
C	12%, 18%, and 5%

4. Compute (a) the total discount and (b) the net cost for equipment items A through D, using (1) the single discount and (2) the successive discount rates given.

Equipment	List Price	Single Discount			Successive Discounts				
		Discount %	Total Discount	Net Cost	First	Second	Third	Total Discount	Net Cost
A	$ 750	18%			12%	6%	X		
B	$ 975	$16\frac{2}{3}\%$			10%	7%	X		
C	$1,440	$37\frac{1}{2}\%$			20%	10%	$7\frac{1}{2}\%$		
D	$3,945	$\frac{1}{3}$ off			15%	15%	5%		
		Grand Total					Grand Total		

5. Add the separate discounts and costs in problem 4 and determine (a) the grand total of the discounts and (b) the net cost of the equipment, applying the (1) single discount and (2) successive discounts.

6. Compare the single discount and the total successive discount rates for each equipment item with the net costs in problems 4 and 5. State the importance of computing and comparing the net costs of equipment items A through D.

B. MATHEMATICS APPLIED TO CREDIT FINANCING

The cost of lumber, electrical and plumbing fixtures, and other supplies is $3,600.00. A down payment of 15% is required.

The unpaid balance is payable in 12 installments. Each installment consists of uniform reductions of the initial unpaid balance plus interest. Interest is payable at the rate of $1\frac{1}{2}\%$ per month on the unpaid balance.

1. List the unpaid initial balance (column A). Compute the uniform monthly payment on the initial balance. Record each successive unpaid monthly balance (column A).
2. Compute each monthly interest payment on the unpaid balance. Record in column B. Round off interest payments to two decimal places.
3. Add each monthly payment on the balance plus the monthly interest. Record the amount in column C.
4. Total the monthly interest payments in column B and record.
5. Total the monthly loan and interest payments (column C) and record.
6. Use the formula for calculating the annual percent rate for credit and find the AAPR for the loan.

Refer to the following table for data relating to problems 7 through 10.

7. Find the total of all payments on the $4,200.00 loan for each institution (line 3).
8. Determine and record the total carrying charges (item 4) for institutions A, B, C, and D.
9. Calculate and record the actual annual percent rate of interest in each case (item 5).
10. Identify the least expensive of the four institutions.

	A	B	C
	Unpaid Balance	Monthly Interest	Monthly Payments
1			
2			
3			
4			
5			
6			
7			
8			
9			
10			
11			
12			
Total			

Items of Financing	Financial Institution Loan Patterns			
	A	B	C	D
	Dealer	Bank/Credit Card	Mail Order	Credit Union
1 Credit Amount	$4,200	$4,200	$4,200	$4,200
2 Monthly Payments (18)	$271.37	$296.33	$320.83	$233.33 + Interest∠[1]
3 Total of All Payments				
4 Total Carrying Charges				
5 Actual Annual % Rate of Interest (AAPR)				

∠[1] $1\frac{1}{2}\%$ per month on unpaid balance

Unit 43 Mathematics Applied to Banking, Home Ownership, and Investments

OBJECTIVES OF THE UNIT

After satisfactorily studying this unit, the student/trainee will be able to
- *Compute simple and compound interest problems.*
- *Balance a bank statement.*
- *Use mortgage amortization tables to calculate monthly, annual, and total mortgage payments.*
- *Apply tax rates to establish property taxes.*
- *Read and interpret stock market quotations and financial tables for stocks, mutual funds, and bonds.*
- *Compute the yield on stocks and bonds.*

PRETEST   *Use the Review and Self-Test items provided in the Unit Assignment to establish the level of mathematical skills competency and to determine the starting point of instruction.*

A. MATHEMATICS APPLIED TO BANKING SERVICES

A bank may be classified as a *savings bank,* or a *commercial bank,* or a *trust company* which administers trust funds and estates. National banks are corporations that are chartered by the federal government. They must conform to federal banking laws and are regulated by the Federal Reserve System.

The sign in a state or national bank ⎣Member of FDIC⎦ means that the account of an individual depositor is protected against loss up to the amount specified. All banks pay for the use of the depositor's savings. The return on money is often referred to as *interest* or *dividend.* The interest rates governing the use of money vary. Some banks pay interest from the day-of-deposit to the day-of-withdrawal; others, from the day-of-deposit to the end of a quarterly period. Money placed in a savings account on a long-term plan (two, three, four, or more years) yields the highest dividend (interest) rate. However, there is a financial penalty on the amount of a "term deposit" that must be withdrawn before the term expires.

Simple and Compound Interest

Interest may be simple or compounded. *Compounded* means that the principal increases continuously because the rate of interest at each interest posting date is applied to the previously accumulated interest and principal.

RULE FOR SIMPLE INTEREST

- Simple interest is equal to the product of the principal (P), yearly interest rate (R), and the length of deposit (T) in years.
- $I = (P) \times (R) \times (T)$

EXAMPLE: Compute the simple interest on $800 at 8.8% interest for six months.

$$I = (P) \times (R) \times (T)$$

$$I = (\$800) \times (.088) \times \left(\tfrac{1}{2}\right) = \$35.20 \quad \textbf{Ans.}$$

Note: The formula may be used to determine the principal, rate, or time when any of the three other factors are known and substituted in the formula.

RULE FOR COMPUTING THE ACCUMULATED PRINCIPAL

- Use the following formula to find the accumulated principal for loans where compound interest applies.
- $A = (P) \times (1 + R)^n$

              A = Accumulated principal
              P = Principal
              R = Annual rate of interest
              n = Number of (year) periods

Note: If interest is compounded semiannually, the (R) becomes (R/2) and (n) becomes (n) (2). Similarly, for interest that is compounded quarterly, (R) becomes (R/4) and (n), (n) (4).

EXAMPLE: Compute the accumulated principal on $500 at 8.88% interest compounded annually for four years.

$$A = (P) \times (1 + R)^n$$

$$= (\$500) \times (1 + .0888)^4$$

$$= \$702.69 \text{ Accumulated principal} \quad \textbf{Ans.}$$

Banking Services: Mortgages and Property Taxes

Real estate investments are another source of income. Unlike bonds, income from property may not follow a uniform yield on the money invested. Income may result from purchasing at one amount and selling at a higher price, or through renting. The cost of the property is the sum of the purchase price and the closing fees. When property is to be sold, consideration must be given to maintenance and other construction improvement costs, assessment payments (for sewers and roads), amount of taxes and insurance premiums, and the income that the dollars invested in property would have produced from savings or bonds. The real profit from the sale or rental of property should be based on all of these factors.

Property ownership resides with the *mortgagee* (money lender) until all repayment conditions are fulfilled. The home loan is secured by a legal document known as a *mortgage,* a bond and schedule of payments, and a home property insurance policy as a guarantee against possible loss and (sometimes) the payment of all taxes by the *mortgagor*.

Some banks and home loan companies include a property package. A single monthly payment is made which includes the interest, payment on principal, and advance tax and insurance payments which are accumulated to pay such bills when they are due. These institutions charge a *fixed* interest rate on the unpaid principal, subject to a periodic review when the rate may be increased or decreased depending on economic and banking conditions.

Property taxes provide one source of revenue for local and state government expenditures. The state sets a ceiling on the property tax rate. Within this limitation, the local community prepares its budget and determines the amount of property taxes that must be collected. The tax rate is applied on the *assessed valuation,* which is the amount a selected group of qualified people *(assessors)* estimate the property to be worth in relation to other property in the community.

Mathematics problems in relation to property involve the interpretation of graphs on sources of income and expenditures; budgeting for fixed and variable expenses for purchasing property, taxes, insurance, maintenance, and other household items; and many simple computations for comparing the effect of different interest rates and payment schedules on income, expenditures, and investments.

RULE FOR COMPUTING MONTHLY MORTGAGE PAYMENTS

- Review the mortgage to establish the amount of the principal, the down payment, the balance on the principal (loan), the rate of interest, the number of years of the loan, and whether the mortgage is fixed rate or variable rate.
- Secure an *amortization table* similar to the one illustrated. Amortization tables list the monthly payments to be made to repay the principal and interest for each $1,000.00 of a mortgage.

(A) Duration of Mortgage (Years)	(B) Percent of Interest Rate									
	11	$11\frac{1}{2}$	12	$12\frac{1}{2}$	13	$13\frac{1}{2}$	14	$14\frac{1}{2}$	15	$15\frac{1}{2}$
	Schedule of Monthly Payments for Each $1,000 of Mortgage									
15	$11.34	11.67	12.00	12.33	12.66	12.99	13.32	13.66	14.00	14.34
20	10.28	10.64	11.00	11.36	11.72	12.08	12.44	12.80	13.17	13.54
25	9.76	10.14	10.52	10.90	11.28	11.66	12.04	12.43	12.81	13.20
30	9.51	9.90	10.29	10.68	11.07	11.46	11.85	12.25	12.65	13.05

(Partial) Amortization Table

- Determine the number of thousands represented by the mortgage (loan).
- Check column A of the amortization table for the loan period and the appropriate rate of interest (under column B) that applies.
- Multiply the monthly payment per $1,000 by the number of thousands (units) in the loan. The product represents the actual monthly payment.

EXAMPLE: A cottage is purchased for $40,000. A 15% down payment is required. The mortgage is to be paid off at a fixed rate of interest of 14% over 20 years. Determine the amount of each monthly payment.

Step 1   Establish the amount of the mortgage and the number of $1,000 units.

$$\$40,000 \times 0.15 = \$6,000 \text{ Down payment}$$

$$\$40,000 - \$6,000 = \$34,000 \text{ Mortgage} = 34(\$1,000) \text{ units}$$

Step 2   Find the monthly payment in the amortization table for each $1,000 for 20 years under the 14% loan rate.

$$20 \text{ years @ } 14\% = \$12.44 \text{ Monthly payment per } \$1,000$$

$$\$12.44 \times 34 = \$422.96 \text{ Monthly payment on loan}$$

The Tax Rate and Property Taxes

School district and city/county property taxes are generally expressed in terms of: mills per dollar, dollars per hundred dollars, or dollars per $1,000 of assessed valuation. When a mills rate is used, the mills are converted to the equivalent value in terms of the dollar (1,000 mills = one dollar).

RULE FOR COMPUTING PROPERTY TAXES

- Check the tax bill to establish whether the tax rate is in units of $100 or $1,000 of assessed value.
- Divide the assessed valuation by the tax unit to establish the number of units.
- Multiply the number of tax units by the tax rate per unit. The product is the tax on the assessed valuation (property).

EXAMPLE: A city school tax is levied at the rate of $70.71 per $1,000 of assessed valuation. Find the school taxes on a parcel of property assessed at $12,500.

Step 1   Divide the assessed valuation ($12,500) by the tax unit ($1,000).

Step 2   Multiply the tax units (12.5) by the tax rate ($70.71). The product ($883.88) = Amount to be paid as school tax. **Ans.**

Banking and Checking Account Services

Money is deposited in a checking account by completing a *deposit slip*. This slip is prepared in the name of the depositor, using a special form on which the account number is usually recorded. The deposits are posted in a checking account passbook.

Withdrawals or payments are made with checks prepared in the name of the individual or organization (or even the depositor) to whom money is to be paid. All withdrawals are posted automatically at the bank. The depositor keeps a control on the checking account by noting and adding the amount of every deposit and subtracting the value of each withdrawal. These transactions are recorded on the check *stub*.

A bank statement is sent to the depositor at regular intervals. These include *cancelled checks* that have been *cleared* and returned to the bank by the last date recorded on the statement. The bank statement shows the account number and date and has an entry for each check and other debits (like a service charge) and deposits. The balance shown for the previous period $(B)_p$ and the deposits (D) should equal the sum of the closing balance $(B)_c$ and all withdrawals (W).

RULE FOR CHECKING THE BALANCE OF A BANK STATEMENT

$$\bullet \ (B)_p + (D) = (B)_c + (W)$$

EXAMPLE: The closing balance reported on the bank statement of a checking account is $324.22. Deposits total $217.64 and cancelled checks and other debits total $284.89. The previous balance was $391.47. The total of uncashed checks recorded on the checkbook stubs is $57.29; deposits, $46.65. Determine the actual balance at a specified date.

Step 1   Examine all cancelled checks and debit entries to establish that all are chargeable against the account.

Step 2   Check all items against entries on the bank statement and in the checkbook ledger (stub). List any checks that were not cashed.

Step 3   Check the deposits noted in the checkbook with those posted on the statement. List any unposted deposits.

Step 4   Check the balance as shown on the previous statement against a similar notation in the checkbook ledger.

Step 5   Use the formula to check the bank statement.

$$(B)_p + (D) = (B)_c + (W)$$
$$(\$391.47) + (217.64) = (324.22) + (284.89)$$
$$\$609.11 = \$609.11$$

Step 6   Add the total of the unposted deposits ($46.65) to the reported closing balance ($324.22).

$324.22 $(B)_C$
+ 46.65 (D)
$370.87 New balance

Step 7 Subtract from the new balance ($370.87) the total of the unposted withdrawals ($57.29). This difference represents the actual balance on the date the checkbook is *balanced*.

$370.87 **New balance**
− 57.29 **(W)**

$313.58 **Actual balance** **Ans.**

B. MATHEMATICS APPLIED TO INVESTMENTS

Investments in Stocks: The Stock Market

The income earned by working at an occupation may be supplemented by income from investments. Large corporations, governments (local, state, and federal), and other businesses depend on the availability of outside money to operate or to expand. These organizations issue stocks in exchange for money. Through the ownership of stocks, an individual becomes a stockholder and a part owner of a corporation.

A stock issuance is no guarantee that the corporation will make a profit or that it will continue in business. Thus, money diverted to stocks carries an element of gamble, with possible loss in part or in total of the money invested. Usually, the corporation holds an annual business meeting. Stockholders attend these meetings. Voting is carried on in person or by proxy. Specific policy matters and other items brought up at the business meeting are acted upon. The assets and liabilities, management, conditions of growth or decline in the organization, and its products and services are reviewed. A decision is made about profits and how much is to be distributed among the stockholders. The stocks are classified as *preferred* or *common*. The holders of *preferred stocks* receive a *fixed earning (dividend)* if there is a profit and a dividend is declared. The holders of shares of common stock may receive a larger or smaller dividend per share only after preferred stock dividends are paid.

There is a brokerage charge for buying and selling stocks. Therefore, the *original cost* of stocks equals the purchase price plus the commission charge. The *net return* from the sale of stocks equals the original cost minus the sum of the selling price and sales commission.

Stock transactions are summarized throughout each day that the Exchange is open. The transactions are printed in newspapers in at least three different versions: at the opening of the business day, midday, and as final or closing market prices. A portion of a typical closing stock table is illustrated to show *quotations* (prices) and other important information.

RULE FOR READING STOCK MARKET INFORMATION

- Use the latest published stock market listing of each particular stock for which quotations or other information are needed.
- Study the description that is provided with the stock market report listings of what each abbreviated column head means.

- Locate the required stock and read the information that is provided on the horizontal stock line under each column heading.
  Note. When a quotation ends in a fraction (in multiples of $\frac{1}{8}$), the fraction relates to that part of \$1.00. For example, $71\frac{1}{4}$ means \$71.25.

EXAMPLE: Read the stock market report and provide current information about Exxon Corporation common stock.

	(B)		(A)		(C)		(D)	(E)	(F)	(G)	(H)	(I)
	52 Weeks Hi	Lo	Stock	Sym	Div	Yld %	PE	Vol 100s	Hi	Lo	Close	Net Chg
	12 1/8	6 7/8	EuroWtFd	EWF		...	...	132	7 1/4	7 1/8	7 1/8	− 1/4
	16 3/8	14 7/8	Excelsior	EIS	1.41e	9.2	...	4	15 1/4	15 1/4	15 1/4	...
(A)	55 1/8	43 1/4	Exxon	XON	2.40	4.8	16	7745	50 3/8	49 1/2	49 5/8	+ 1/8
			-F-F-F-									
	12 1/2	5 1/4	FAI Insur	FAI	.32e	4.7	7	4	6 3/4	6 3/4	6 3/4	...
▼	47 1/4	27 3/4	FMC Cp	FMC		...	6	359	28 1/4	27 3/8	27 3/8	− 3/4
	14 1/4	9 3/8	FMC Gold	FGL	.05e	.5	16	132	10 5/8	10 3/8	10 1/2	...
	36 3/4	26 1/8	FPL Gp	FPL	2.36	8.3	10	967	29	28 3/8	28 3/8	− 1/2
	23 1/2	12 1/4	FabriCtrs	FCA		...	10	17	16 7/8	16 5/8	16 7/8	+ 1/8
	40 1/8	32 3/4	Fairchild pf		3.60	10.8	...	3	33 3/8	33 1/4	33 1/4	− 1/8
▼	**6 1/4**	**9/16**	**Fairfield**	**FCI**		...	...	**549**	5/8	1/2	1/2	− 1/16
	15 1/4	9 5/8	FamDollr	FDO	.40	3.4	12	487	12	11 3/4	11 3/4	− 1/8
	13 1/2	6 3/4	Fansteel	FNL	.60a	4.8	13	20	12 5/8	12 1/2	12 1/2	− 1/4
	11 1/4	1/2	FarWestFnl	FWF		...	...	126	5/8	9/16	5/8	...
	8	2 5/8	Farah Inc	FRA		...	...	35	3 1/8	3 1/8	3 1/8	− 1/8
	13 5/8	7 1/2	FaysInc	FAY	.20	2.5	11	82	8 1/8	7 7/8	8	+ 1/8
	17 1/8	5 5/8	Fedders	FJQ	.48	7.2	6	505	6 7/8	6 3/8	6 5/8	− 1/8
	58	33 5/8	FedlExp	FDX		...	15	785	37 1/8	36 3/4	36 3/4	− 1/2
	104 3/4	50	FedlHmLoan	FRE	1.60	3.1	7	2553	53 1/4	51 1/2	51 7/8	−1 5/8
	24 7/8	13 5/8	FedlMogul	FMO	.92	6.6	21	311	14	13 3/4	14	+ 1/4
	46 3/8	24 7/8	FedlNMtg	FNM	.72	2.5	7	9546	29 3/8	28 1/2	28 1/2	− 3/4
	32 5/8	11	FedlNMtg wt					4059	14 7/8	14 1/8	14 1/2	− 3/8

Step 1 Locate the Exxon Corporation common stock listing (A).

Step 2 Start at the two left columns (B). Note the *high-low* range of prices during the year of $55\frac{1}{8}$ High; $43\frac{1}{4}$ Low.

Step 3 Move to the next column on the right (C). The *current dividend* shows the expected *stock yield annual dividend* to be \$2.40 per share; yield, 4.8%.

Step 4 Read the *P/E Index* in column (D). This shows that the index of earning capability of the stock is 16.

Step 5 Translate the value shown in column (E) (vol 100 s) by multiplying this number by 100. In this case, 774,500 (7,745 × 100) shares of Exxon common stock were sold on this day.

Step 6 Read the values under columns (F), (G), (H), and (I). Throughout the day, the *high price* paid for the stock was $50\frac{3}{8}$ (\$50.375); the *low price* $49\frac{1}{2}$ (\$49.50); and at the close of the business day, the *last price* (H) was $49\frac{5}{8}$ (\$49.625).

Step 7 Use the final right column (I) to find the *net change*. The (+) or (−) signs reflect the difference in price between the last sale of the previous day and the last sale of the day. The $(+\frac{1}{8})$ posted for the Exxon stock means that it gained $\frac{1}{8}$ or \$0.125 per share over the previous day.

RULE FOR COMPUTING THE YIELD ON STOCKS

- Use the formula

$$Y = \frac{D}{P}$$

Y = Yield or interest rate for one year
D = Yearly dividend
P = Price of one share of stock

EXAMPLE: One share of stock in Corporation X sells for $60 a share. An annual dividend of $4.20 is paid. Find the yield.

$$Y = \frac{D}{P} \qquad Y = \frac{4.20}{60} = 7\% \quad \textbf{Ans.}$$

Investments in Mutual Funds

Securities consisting of stocks, bonds, and other investments are also handled by *mutual fund investment companies,* which represent the shareholders. Monies received from the shareholders are entered into a common *pool,* which is used by the investing agency to purchase stocks and other securities. Shares in the mutual fund are issued to each shareholder at the market price at the day of investment in the mutual fund.

Reading Mutual Fund Quotations

The form in which mutual fund price ranges are quoted by the National Association of Securities Dealers in newspapers each business day is illustrated by the accompanying excerpt.

Note that in some cases (for example, ABT funds) a whole series of mutual funds are handled by one overall investment company.

The price ranges in the table show the *net asset value per share (NAV).* The *offering* price in the next column includes the net asset value plus a maximum sales charge (if applicable). The term *NL* indicates there is no brokerage charge.

The last column on the right *(NAV Chg.)* shows daily changes in the value of each mutual fund. Dividends are declared on the portfolio of investments in which the shareholders funds are pooled. Dividends are distributed according to the proportionate number of shares held by each shareholder in relation to the total number of shares in the fund. Mutual fund investment houses claim the advantages of safety, diversification, and convenience to the small investor.

	NAV	Offer NAV Price Chg.			NAV	Offer NAV Price Chg.
AAL Mutual:				CalTrst	11.24	NL ....
CaGr p	10.36	10.88 .09		CalUS	9.39	NL.....
Inco p	9.48	9.95+ .01		**Calvert Group:**		
MuBd p	9.75	10.24.....		Ariel	21.46	22.47− .21
AARP Invst:				ArielA	13.26	13.88− .19
CaGr	24.13	NL− .09		Capitl p	19.24	20.15− .11
GiniM	15.01	NL+ .01		GvLtd	14.55	14.85+ .01
Gthinc	22.77	NL− .07		Inco	15.65	16.39− .02
HQ Bd	14.73	NL+ .01		Social p	26.21	27.45− .12
TxFBd	16.18	NL.....		SocBd	15.47	16.20+ .01
TxFSh	15.12	NL.....		SocEq	16.22	16.98− .14
ABT Funds:				TxF Lt	10.58	10.80.....
Emrg p	7.76	8.15− .14		TxF Lg	14.96	15.66− .01
FL TF	9.99	10.49.....		US Gov	14.58	15.27+ .01
Gthin p	8.07	8.47− .07		WshA p	11.12	11.64− .14
Secin p	9.18	9.64− .03		**Capstone Group:**		
Utiln p	11.65	12.23− .02		EqGrd	6.76	7.10.....
AHA Bal	9.76	NL− .01		Fd SW	12.15	12.76− .15
AdsnCa p	15.92	16.41− .15		Incom	4.65	4.88 ....
ADTEK	8.76	8.76− .11		MedRs	14.31	15.02+ .07
AFA NAv	9.34	9.81− .13		PBHG	8.86	9.30− .21
AFA Tele	13.15	13.81− .31		Trend	12.23	12.84− .15
AIM Funds:				CarilCa	10.07	10.60−- .06
Chart p	6.66	7.05− .04		**Carneg Cappielo:**		
Const p	6.95	7.35− .13		EmGr p	7.71	8.07− .12
CvYld p	9.22	9.68− .08		Grow p	14.85	15.55− .24
HiYld p	5.65	5.93+ .01		TRetn p	10.20	10.68− .11
LimM p	9.80	9.97.....		**Carnegie Funds:**		
Sumit	7.34.....	− .09		Govt p	9.09	9.52.....
Weing fp	11.48	12.15− .16		TEOhG	8.95	9.37.....
A M A Family:				TENHi	9.45	9.90.....
ClaGt p	7.79	NL− .04		Cardnl	9.45	10.33− .06
GlbGt p	19.74	NL− .23		CrdnlGv	8.74	9.18+ .01
Glbin p	19.13	NL+ .04		Cnt Shs	15.37	NL− .31
GIST p	9.99	NL+ .02		ChnHY p	10.44	10.96− .01

Investments in Bonds

A *bond* is a document by which federal, state, or local governments, and business, industrial, agricultural, or other establishments promise to repay money by a stated date and to pay a fixed amount of interest according to a schedule of payments. Three common terms are used with bonds. (1) The *face value* is the amount stated on the bond, usually as a multiple of $1,000. (2) The *market value* is the price at which the bond is sold. (3) *Par value* is used when the market and face values of a bond are the same. Corporation bonds are also purchased through brokers who charge a commission. The interest on a bond is paid on the face value.

Municipality bonds and other special bonds have tax-exempt provisions. These permit local governments and other privileged organizations to obtain money for public use and projects at low interest rates. In return, the holders of such bonds (which yield a lower interest) receive a tax break because the yield is tax free.

RULE FOR READING BOND MARKET INFORMATION

* Secure and read the latest daily bond market quotations published in newspapers and business journals.
* Study the descriptions provided for each vertical column designation.
* Locate the required bond listing and read the specific information that is provided horizontally according to each vertical column heading.

EXAMPLE: Provide the latest stock exchange bond market information for the performance of Eastman Kodak (EKod) bonds (using the abstracted portion of Bond Market Information that follows).

Step 1 Locate and read the identifying information about EKod in the Ⓐ columns; *company name, guaranteed income payments* on par value (of $1,000) at $8\frac{5}{8}\%$, and a *maturity date* of 16 (2,016).

Step 2 Read the values in columns Ⓑ through Ⓔ
* The *current yield* Ⓑ is 10.3%.
* The *volume traded* that day Ⓒ was 20,800 (208 × 100) shares.
* The *closing price* of each bond Ⓓ was $838.75 ($83\frac{7}{8}\%$ of $1,000).
* The net change Ⓔ (...) shows that the last sale of the day was at the same price as the previous day.

Bonds	Cur Yld	Vol	Close	Net Chg.	Bonds	Cur Yld	Vol	Close	Net Chg.
DetEd 9.15s00	9.3	2	98	+ 1½	GaPw 10½09	10.4	27	101⅛	+ ⅜
DetEd 8.15s00	9.1	2	89⅜	+ 1	GaPw 11s09	10.9	20	101	− ½
DetEd 7⅜01	8.9	11	83¼	+ 1¼	GaPw 13½12	12.6	21	104⅛	...
DetEd 9⅞04	9.9	23	100	+ ½	GaPw 10s16A	10.2	18	98¼	− ¾
CitEd 11⅞00	11.4	8	104½	...	GdNgF 13¼95	32.7	1692	40½	− ½
DetEd 10⅝06	10.6	29	100		Grumn 9¼09	cv	34	76	− ½
Disney zr05	...	105	37⅞	+ ⅛	GlfRes 10⅞97	14.3	5	76	...
Dow 8.92000	9.1	5	97½	+ 3⅛	HalwdGp 13½09	...	11	70½	
Dow 8⅝08	9.6	7	90	...	HarDav 7¼15	cv	5	80½	+ ½
duPnt 8.45s04	9.1	12	93¼	+ ⅜	HeclMn zr04	...	15	30¾	− ¼
duPnt 8½06	9.2	15	92¾	+ ⅛	HmeDep 6s97	cv	22	91½	− 1½
duPnt dc6s01	7.8	15	76⅝	− ⅛	HomFSD 6½11	cv	40	42	+ 1
duPnt 8½16	9.6	64	88½	− 1⅛	HmGrp 14⅞99	29.2	106	51	− 6
duPnt 7½93	7.7	30	97⅛	...	HousF 7⅛95	8.3	15	90¼	− 2¾
duPnt zr10	cv	540	21⅛	+ ⅛	HudFd 8s06	cv	1	65	...
DukeP 10⅛09	9.9	50	102¼	− ⅛	HudFd 14s08	cv	10	94½	+ 1
EKod 8⅝16	10.3	208	83⅞		Huffy 7¼14	cv	33	86	− 7½
Ens 10s01	cv	21	106½	− ¼	ICN 12⅞98	28.1	40	45⅞	+ ⅞
EnvSys 6¾11	cv	30	50⅛	− 2⅜	IIIBel 7⅝06	9.0	5	84½	− 1⅜
viEqutc 10s04	cv	10	4½	+ ½	IIIBel 8¼16	9.5	47	87¼	...
Exxon 6s97	7.1	73	84½	− ⅜	IIIPw 10⅛16	10.1	20	99⅞	+ ¾
Exxon 6½98	7.6	20	85½	...	IIIPw 9⅜16	10.4	5	90	− 2
Fairfd 13¼92	56.1	193	23⅝	− 1	Inco 6.85s93	7.4	2	93	...

RULE FOR COMPUTING THE YIELD ON A BOND

- Divide the interest on the face value by the cost of the bond.

$$Y = \frac{\text{Interest on Face Value}}{\text{Cost of Bond}}$$

EXAMPLE: • A $100.00 face value bond is purchased for $93.00. The annual interest is $7\frac{3}{4}\%$. Determine the yield.

$$Y = \frac{7.75 \, (7\frac{3}{4}\% \text{ on } \$100 \text{ face value})}{93.00 \text{ (cost of bond)}} = 8.33\% \quad \textbf{Ans.}$$

United States Government Savings Bonds

Two common forms of U.S. Government Savings Bonds are the Series EE and the Series HH bonds. The *Series EE bonds* provide a system of investing by which the interest accumulates during the life of the bond and the face value may be automatically reinvested at maturity. The income may be declared when the bond is *cashed*. This is an important feature for persons who want to reduce income taxes during high earning years by cashing the bonds during periods of low income. Series EE bonds are popular; employees and others invest regularly through a payroll savings and other plans. Series EE bonds are sold in denominations of $50, $75, $100, $200, $500, $1,000, $5,000, and $10,000 at issue (cost) prices of $25.00, $37.50, $50.00, $100, $250, $500, $2,500, and $5,000.

The value of series EE bonds is printed on each bond. There is a penalty (no interest) for redeeming a bond before it is held for six months. Interest is accrued at the beginning of each successive half-year period with the rate increasing gradually until at maturity the yield reaches the percent declared from the issue date to maturity. The yield on U.S. Government Savings Bonds has steadily increased over the years to remain competitive with other investments. Increases have been applied automatically on these bond holdings and are reflected in the increased value which is received when such a bond is cashed.

Series HH bonds are issued in denominations of $500, $1,000, $5,000, and $10,000. Semiannual interest is mailed regularly to bondholders. This income must be included as part of the ongoing income. These bonds have an expiration date at which time they may be reinvested.

ASSIGNMENT UNIT 43 REVIEW AND SELF TEST

PRETEST
The Review and Self-Test items that follow may be used as a pretest. Pretests are designed to measure a student's beginning level of mathematical skills competency and to determine the starting point of instruction.

POSTTEST
The Review and Self-Test items may also be applied as a post test. Post tests are planned to establish the student's level of mathematical skills competency after instruction.

CALCULATOR APPLICATIONS
The problems in this Unit may be solved and/or checked either by conventional mathematical processes or by using a calculator. Each answer displayed on the calculator register is rounded off to the required number of decimal places.

A. MATHEMATICS APPLIED TO BANKING SERVICES
Simple and Compound Interest

1. Write the formulas for computing (a) simple interest and (b) compound interest on principal.
2. Compute the accumulated principal (column c) for accounts A through D at the simple rates of interest given in the table.
3. Compute the accumulated principal (column d) for the same accounts at the interest rates and compounding schedule indicated in the table.

(a) Formula:				(b) Formula:	
Account Information				**Accumulated Principal**	
Account	Original Principal	Rate of Interest	Number of Periods	Simple Interest (c)	Compound Interest (d)
A	$ 500	5.15%/year compounded annually	2 years		
B	$ 750	7.96%/year compounded semiannually	$1\frac{1}{2}$ years		
C	$2,400	10.71%/year compounded quarterly	15 years		
D	$1,500	8%/year compounded quarterly	4 years		

Mortgages and Property Taxes

1. A building is purchased for $78,500. A down payment is made of $18,500. A 25-year mortgage is granted by a bank at a rate of 12.5%. Use the amortization table and determine (a) the monthly payments and (b) the yearly payments throughout the life of the mortgage (property loan).

(A) Duration of Mortgage (Years)	(B) Percent of Interest Rate									
	11	$11\frac{1}{2}$	12	$12\frac{1}{2}$	13	$13\frac{1}{2}$	14	$14\frac{1}{2}$	15	$15\frac{1}{2}$
	Schedule of Monthly Payments for Each $1,000 of Mortgage									
15	$11.34	11.67	12.00	12.33	12.66	12.99	13.32	13.66	14.00	14.34
20	10.28	10.64	11.00	11.36	11.72	12.08	12.44	12.80	13.17	13.54
25	9.76	10.14	10.52	10.90	11.28	11.66	12.04	12.43	12.81	13.20
30	9.51	9.90	10.29	10.68	11.07	11.46	11.85	12.25	12.65	13.05

(Partial) Amortization Table

2. Find the amount of the (a) monthly and (b) yearly payments on mortgages A, B, and C. Use the data from an amortization table in computing the payments.

Property Loan (Mortgage)		Interest Rate (%)	Time (Years)	Payments	
				Monthly	Yearly
A	$38,000	11	20		
B	52,600	12.5	25		
C	76,480	13.5	30		

Banking and Checking Account Services

1. Compute each amount that is not recorded on the checkbook stubs for accounts A, B, C, and D.

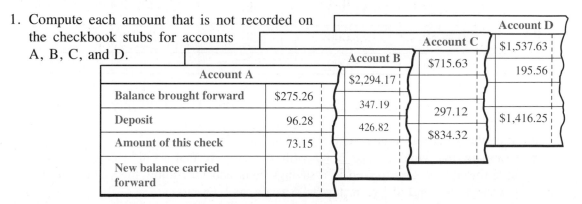

Account A	
Balance brought forward	$275.26
Deposit	96.28
Amount of this check	73.15
New balance carried forward	

Account B	
$2,294.17	
347.19	
426.82	

Account C	
$715.63	
297.12	
$834.32	

Account D	
$1,537.63	
195.56	
$1,416.25	

2. Determine the amount of the uncashed checks for the accounts in Problem 1, using the following closing balances from the bank statements of the four accounts.

Account	A	B	C	D
Closing balance	$426.72	$1,923.61	$1,022.28	$637.64
Uncashed checks				

B. MATHEMATICS APPLIED TO INVESTMENTS

Investments in Stocks, Corporation Bonds, and U.S. Government Savings Bonds

1. Secure the financial section of a newspaper for information about common stocks and broker commissions.
2. Select and name a common stock and read the stock market quotation for particulars on the selected stock.
3. Find the purchase price of 250 shares of the selected stock, including broker commissions. Assume the stock is purchased at the *high* quotation of the day.
4. Compute the profit or loss on the sale of the 250 shares under the following stock market conditions.
   Note. Broker commissions for selling are to be deducted to obtain the actual return for the shares.
   a. Selling at the *low* of the day.
   b. Selling at the *high* of the year.
   c. Selling at the *low* of the year.
5. Determine the (a) interest at par value and (b) annual yield for bonds A, B, and C. Use the formula: Current Yield = Interest at Par Value ÷ Market Value. Round off the yield to two decimal places.

Bond Values Purchase Price (Market Value)		Face (Par) Value	Annual Interest Rate (%)	Interest (at Par Value)	Current Yield (%)
A	$ 871.25	$1,000	$8\frac{1}{4}$		
B	1,075.50	1,000	$8\frac{3}{4}$		
C	5,750.00	5,000	9.75		

6. Review the payment schedule of a current issue of a U.S. Government Savings Bond, Series EE.
   a. State the value of holding such a bond to maturity.
   b. Compare the yield on the bond with the interest that would accrue on the same investment if it were placed in a savings bank account. Show the yield in dollars and percent at the end of $\frac{1}{2}$ year, 1 year, 5 years, and 7 years.

Amount Invested	Source	Condition of Investment		Comparison of Income			
		Rate	Period	$\frac{1}{2}$ year	1 year	5 years	7 years
	Savings Bank						
	U.S. Gov't. Bond						

Unit 44 Mathematics Applied to Payrolls, Income, and Taxes

OBJECTIVES OF THE UNIT

After satisfactorily studying this unit, the student/trainee will be able to
- *Understand different taxes on wages and personal benefit deductions that affect take-home pay and other transactions.*
- *Apply the four basic mathematical processes to compute Social Security and withholding payments, income taxes, and take-home pay.*
- *Prepare income tax forms, wage and tax estimates, and back up income tax schedules.*

PRETEST *Use the Review and Self-Test items provided in the Unit Assignment to establish the level of mathematical skills competency and to determine the starting point of instruction.*

A. MATHEMATICS APPLIED TO PAYROLLS AND WAGES

Basic Payroll Classifications of Workers

The four basic classifications under which workers are paid include: *salaried, hourly, piece-work,* and *commission.*

- **Salaried Employees** are paid a fixed annual salary, regardless of the hours worked. Companies with a bonus-sharing plan for increased productivity, profit level, or other incentive, distribute an additional annual payment. The bi-monthly or monthly salary (before taxes and other benefit payments are deducted) is determined by dividing the annual salary by the number of payroll periods.
- **Hourly employees** are paid for the numbers of hours worked, plus additional *overtime* in excess of the standard work week. Three terms are used in relation to the hours worked per week: *straight time* that is credited to a worker for the standard work week; *overtime* for hours worked in excess of the standard number of hours; and *double time* for hours worked on Sundays, legal holidays, and all time over a stated number of overtime hours. The overtime rate is usually $1\frac{1}{2}$ times the standard hourly rate; double time, twice the hourly rate.
- **Piece-work employees** receive wages based on productivity (quantity and quality). A worker is paid a specified amount of money for each piece or unit produced. Total wages are found by adding the number of pieces (units) produced weekly and multiplying this quantity by the *piece-rate.*

- **Commission employees** are usually salespersons who are paid a wage based totally on commission (a percent of the total dollar sales volume). Salary is determined by multiplying the total dollar sales by the rate of commission. In other instances, salary is based upon a minimum salary plus commission.

RULE FOR COMPUTING WAGES (BASED ON HOURLY, OVERTIME, AND DOUBLE TIME RATES)

- Multiply the straight time each day by 1.
- Multiply the overtime by 1.5.
- Multiply the double time by 2.

EXAMPLE: Compute the total working time according to the daily hours entered in the table, using a standard work week of 40 hours.

Day	Mon.	Tues.	Wed.	Thur.	Fri.	Sat.	Sun.
Time in Hours	8	9	9	9	9	4	8

Step 1   Multiply the straight time each day by 1 and add to get the total.

M $8 \times 1 = 8$
T $8 \times 1 = 8$
W $8 \times 1 = 8$
Th $8 \times 1 = 8$
F $8 \times 1 = 8$
Straight time = 40 hr

Step 2   Multiply the overtime hours by 1.5 and add to find the total.

T $1 \times 1.5 = 1.5$
W $1 \times 1.5 = 1.5$
Th $1 \times 1.5 = 1.5$
F $1 \times 1.5 = 1.5$
S $4 \times 1.5 = 6$
Overtime = 12 hr

Step 3   Multiply the double time hours by 2.

Sun. $8 \times 2 = 16$ hr

Step 4   Add the straight time (40), overtime (12), and double time (16).

$40 + 12 + 16 = 68$ hr   **Ans**

Step 5   Multiply the total number of hours (68) by the hourly rate to get wages due.

An added incentive to improve quality and to get greater production is to increase the hourly rate by a bonus after a quota is reached. Bonus plans are given to individuals, to groups, and to departments. The bonus for additional production or quota is added either to the weekly wage or given over longer periods of time.

B. MATHEMATICS RELATED TO FEDERAL AND STATE INCOME TAXES

Any individual with a prescribed minimum gross income must file an income tax return annually. This return, when accurately completed and accepted, shows the conditions and qualification for each entry and how the final tax computations were determined. If the total withholding payments made by one or more companies for income tax and Social Security tax were larger than the one computed on the income tax return, the difference will either be returned to the wage earner or credited on the next year's tax, depending on the block the wage earner checks on the income tax form.

Employee's Withholding Exemption Certificate for Income Taxes

Since taxes are withheld by an employer each payroll period, a worker who is employed for the first time by a company, or whose family status is changed, completes an *Employee's Withholding Exemption Certificate* (Form W-4). The employee is identified by full name and Social Security number. The employee indicates the number of exemptions (theoretically the number of persons who are dependent on the wage earner). The exemptions and allowances, and any additional withholding payments per pay period the employee wants to be deducted from wages, are totaled on this form (W-4). The amounts withheld are credited to the employees federal account with the Internal Revenue Service as income tax payments. An adjustment is made before April 15 for taxable income earned to December 31 of the previous year. This is done by filing a U.S. Individual Federal Income Tax Return.

Federal Withholding Tax Schedule

Printed *Federal Withholding Tax Schedules* (one for single persons; one for married persons) are used to establish the amount of federal income taxes that are to be paid each payroll period. The tables are simple to read. The vertical column displays a $10.00 range for weekly wages. Horizontal columns show the number of dependents and the tax to be paid on a specific weekly wage for a given number of dependents.

Withholding taxes on weekly wages that exceed the range provided in the tables are calculated. The formula given with each table requires that the excess amount be multiplied by a specified percent. The product is added to the maximum amount given in the table.

RULE FOR COMPUTING WITHHOLDING TAX FOR FEDERAL INCOME TAXES

- Round off employee's earnings to the nearest dollar.
- Determine the exemptions claimed (Form W-4).
- Select the appropriate Schedule (single person or married person).
- Locate the weekly bracket for the wage in the left column of the Withholding Tax Schedule.
- Read the tax to be withheld that is given in the box under the number of dependents.

EXAMPLE: Determine the income withholding tax on a single person earning $348.41 weekly and claiming two dependents.

Step 1   Round off the earned wage from $348.41 to $348.00.

Step 2   Select the single person weekly payroll schedule.

Step 3   Locate the wage within the $340 to $350 range.

Step 4   Read the amount of income tax to be withheld under the (2) withholding allowances claim column for the weekly withholding amount.

Wages		Dependents			
		0	1	2	3
330	340				
340	350	56.70	51.90	47.10	42.30
350	360				

Ans

The Federal Income Tax "Package"

Annually, each person who previously filed a federal income tax receives a package called *Federal Income Tax Forms and Instructions*. Actually, the mailing includes duplicates of blank U.S. Individual Income Tax Returns, schedules for further description, reporting and computations, specific directions to clarify every item on the forms and schedules, various tax tables, and other processing information.

The number 1040 identifies the Federal Income Tax Return form. There are simplified (1040 EZ) and other adaptations of this form. Form 1040 is complemented with *Schedules* that provide additional information about financial conditions and adjustments to income.

- *Schedules A and B* (1040) are used to record *Itemized Deductions and Dividend and Interest Income*.

- *Schedule C* (1040) relates to *Business Profit and Loss.*
- *Schedule D* (1040) is used to report *Capital Gains and Losses.*
- *Schedule E* (1040) deals with *Supplemental Income and Loss.*
- *Schedule SE* (1040) (short and long forms) covers *Social Security Self-Employment Tax.*
- *Form 2441* relates to *Child and Dependent Care Expenses.*
- *Form 4562* is used to report *Depreciation and Amortization* expenses.

Wage and Tax Statement (Form W-2)

Each employer is required to withhold Social Security tax (F.I.C.A.—Federal Insurance Contributions Act) payments on wages up to a stated amount. The employer must also contribute an equal amount as a tax. The federal income tax withheld, the total wages, F.I.C.A. tax, and state and city taxes that are withheld are all summarized by the employer on a *Wage and Tax Statement (W-2)*. Multiple copies of this form are given to the employee, who attaches them to federal and state income tax returns and retains the employee's copy.

State Income Forms and Taxes

Many states provide taxpayers with a packet of materials that is similar to the U.S. Internal Revenue Service mailing. The state forms are comparatively simple to prepare provided the U.S. Individual Income Tax Return Form 1040 and supplemental schedules are prepared first. Much of the information on the federal form can be transferred. There are differences between the two forms, however. The allowances for taxes, state pensions, exemption amounts, and the tax rate schedule are different.

C. MATHEMATICS APPLIED TO SOCIAL SECURITY AND INSURANCE

Social Security is associated with the personal protection of the worker, the family, and heirs under certain conditions. This plan provides either financial payments at retirement, benefits to survivors in the event of death, or disability benefits. In addition, health insurance and medical care are available. This protection package is administered by the Social Security Administration, U.S. Department of Health and Human Services. Also, other medical and health insurance plans are provided through bargaining agreements with employers, unions, and members, or through other arrangements. In some instances, the employee or employer pays the total cost; sometimes both employee and employer contribute in prearranged proportions.

Many workers also contribute to pension plans that are adminis-tered for a local government, state government, or a business organiza-tion. Such participation provides a pension at retirement under qualify-ing conditions of years of service and payments into the system. This earned pension may be in addition to Social Security payments.

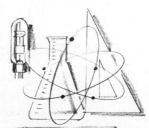

The premiums the worker pays into the Social Security Insurance Fund are deducted from wages and are considered federal government taxes. These amounts, together with the contributions of the employer, are paid regularly to the federal government. The contributions are recorded as covered periods of employment (specific amounts of time each year) in a Social Security account for the employee. All transactions with a state or the federal government are identified by each individual's Social Security identification number.

The Social Security (F.I.C.A.) tax rate for employee and employer; the maximum amount of combined wages and earnings subject to F.I.C.A. taxes; and the self-employment tax rate are subject to annual congressional action. Up to this point, the F.I.C.A. tax rate and level of earning against which the rate is applied have increased consistently during the lifetime of the system.

RULE FOR CALCULATING SOCIAL SECURITY TAXES

* Multiply the weekly earnings by the prevailing tax rate. The product represents the weekly Social Security deduction.
  Note. Social Security deductions withheld by an employer (or for self-employment) stop each year when the prescribed base yearly income is reached.
* Add the employer's contribution.
  Note. The total (either self-employment or individual and employer contributions) represents the amount credited to the employee's Social Security account.

EXAMPLE 1: A worker earns $348 a week. Using a 7.65% tax rate, determine the amount withheld from the employee earnings for Social Security.

Step 1 Check the earned income for the year to determine whether it is within the prescribed base amount. Social Security taxes are deductible on this amount only.

Step 2 Multiply the weekly earning ($348) by the Social Security rate. The decimal equivalent .0765 (for 7.65%) is used.

$348
× .0765

Step 3 Point off two decimal places. The result ($26.62) is the Social Security tax withheld.

$ 26.62   Ans

EXAMPLE 2: A self-employed worker earned $24,500 during the year. Find the amount of Social Security (F.I.C.A.) tax that is payable on the earnings. Use a 15.30% F.I.C.A. tax rate.

Step 1 Multiply the total earnings by the tax rate.

($24,500) × (0.1530) =

Step 2 Identify the product as the self-employment F.I.C.A. tax.

F.I.C.A. tax: $3,748.50   Ans

D. COMPUTING TAKE-HOME PAY

While only samples of Social Security and income taxes are covered in this unit, all other deductions are computed in a similar manner. The total is subtracted from the wages earned to determine the take-home pay.

RULE FOR COMPUTING TAKE-HOME PAY

- Determine the total number of hours worked in the payroll period.
- Multiply hours worked by the hourly rate and add any amount due from a bonus plan to get the wages due.
- Compute the Social Security, income tax, and any other taxes or deductions.
- Total any payment deductions for health, accident, or expenses taken directly out of wages.
- Add all deductions to be paid by the employee. Subtract the total of deductions from wages due. The difference is the take-home pay.

ASSIGNMENT UNIT 44   REVIEW AND SELF TEST

PRETEST   *The Review and Self-Test items that follow may be used as a pretest. Pretests are designed to measure a student's beginning level of mathematical skills competency and to determine the starting point of instruction.*

POSTTEST   *The Review and Self-Test items may also be applied as a post test. Post tests are planned to establish the student's level of mathematical skills competency after instruction.*

CALCULATOR APPLICATIONS   *The problems in this Unit may be solved and/or checked either by conventional mathematical processes or by using a calculator. Each answer displayed on the calculator register is rounded off to the required number of decimal places.*

A. Computing Payrolls and Wages (Practical Problems)

1. Determine from the time sheet the total hours worked and the amount due each worker. The overtime rate applies over 40 hours.

Worker	Hours Worked						Hourly Rate	Worker	Hours Worked						Hourly Rate
	M	T	W	Th	F	S			M	T	W	Th	F	S	
A	9	9	9	9	9	4	$18.60	D	8	0	8	8	8	4	8.30
B	8	8	8	9	9	4	14.30	E	9	9	9	9	9	6	4.75
C	8	8	8	8	8	4	11.60	F	8	8	8	8	8	0	6.35

2. Find the gross wage earned by piece-work employees A, B, and C in manufacturing electronic components according to the payroll information table. A weekly productivity bonus rate of 95 cents additional per unit is given for all quantities over 100 units.

Employee	Production (in units)					Rate per Unit
	M	T	W	Th	F	
A	16	18	24	18	20	$2.05
B	24	23	23	27	26	$1.91
C	31	26	29	28	24	$1.78

B. Mathematics Applied to Social Security (F.I.C.A.) and Federal Income Taxes

1. Find the amount of weekly F.I.C.A. taxes to deduct from (a) weekly wages A, B, and C and (b) annual salaries of D, E, and F. Use a F.I.C.A. rate of 7.65% and round off each tax deduction to the nearest dollar.

(a) Weekly Wage		F.I.C.A. Tax Deduction	(b) Annual Salary		F.I.C.A. Tax Deduction
A	$252.00		D	$8,775.00	
B	$475.50		E	$16,087.50	
C	$721.25		F	$32,760.00	

2. a. Secure a current Federal Income Tax Withholding Schedule (Weekly Payroll Periods) for single persons and married persons.

   b. Use the Schedule to determine the amount to be withheld weekly for federal income tax for employees A, B, and C, and D and E, according to the following weekly earnings and number of claimed allowances.

Employee	Single Persons			Married Persons	
	A	B	C	D	E
Amount Earned	$700.00	$575.50	$397.28	$810.00	$422.75
Claimed Allowances	2	2	2	3	3

C. Computing Take-Home Pay (Practical Problems)

1. Determine the weekly earnings (f) of workers A, B, C, and D based on the total of regular and overtime hours worked (c) and (d). Use a standard work week of 40 hours; overtime equal to time and a half; and double time for Sunday work.

(a) Worker	Hours Worked (b)							Regular Hours (c) Worked	Overtime (d)		(e) Hourly Rate	(f) Weekly Earnings
	M	T	W	Th	F	S	Su		Hours Worked	Hours Pay		
A	8	8	8	8	8	6	6				$12.50	
B	9	9	9	9	9	4	0				8.70	
C	8	8	0	9	9	0	0				9.80	
D	9	9	9	9	9	8	4				5.24	

2. Use the weekly earnings (f) in the preceding problem and the following information to find the taxes (h), (i) and (j) and take-home pay (k). Workers A and B are single persons; C and D, married persons. Round off F.I.C.A. taxes to the nearest dollar.

Worker	Claimed Allowances (W-4) (g)	Withholding Tax (h)	F.I.C.A. Tax (i) (7.65%)	State, Local, Other Taxes and Benefits Package (18.6%) (j)	Take-Home Pay (k)
A	1				
B	2				
C	3				
D	2				

Unit 45 Mathematics Essential to Careers and Career Planning

OBJECTIVES OF THE UNIT

After satisfactorily studying this unit, the student/trainee will be able to

- *Appreciate the importance of work ethics in career planning and the preparation of each worker for a career in the work force.*
- *Identify the characteristics of the work force in terms of: economic sectors, education and training patterns for developing occupationally competent workers at all levels of employment; retraining and upgrading; and job mobility.*
- *Interpret mathematical requirements and applications at successive stages of career development.*

PRETEST   *Use the Review and Self-Test items provided in the Unit Assignment to establish the level of mathematical skills competency and to determine the starting point of instruction.*

A. WORK ETHICS FOUNDATIONS FOR CAREERS AND CAREER PLANNING

The production of goods and services is recognized as the foundation for individual and national security and well being. Productive work produces wealth . . . human, social, economic, cultural, spiritual, and moral. The following concepts of work are cornerstones in career planning and development.

- Work is essential to human relationships whereby one person helps another to reach a state of effective

citizenship. Goods are produced and services are performed that are necessary to individuals and to society.

- Work provides self-fulfillment so that each person can find meaning and purpose, ideals and values, and happiness in responsible citizenship and wholesome living.
- Work is recognized as useful of human resources; unemployment, underemployment, idleness, and separation from work, as wasteful.
- Work requires preparation for entry into the work force, retraining for career changes, and occupational upgrading throughout a worker's life span to remain occupationally competent and continuously employable.

B. CHARACTERISTICS OF THE WORK FORCE

The work force consists of people who have developed skills that are essential to employment in an occupation that falls within one of the *economic sectors*. The most common economic sectors in which jobs in the work force are grouped include:

- Agriculture and Agri-Business
- Business and Banking
- Distribution and Marketing
- Home and Institutional Services
- Health and Allied Services
- Public Services
- Trade and Industrial-Technical

Occupations that relate to each of these (and other) economic sectors may be grouped in *occupational constellations*. For instance, all the job titles in the health and allied services economic sector are identified within this occupational constellation.

Job Families and Major Occupational Fields

Within each occupational constellation there are *job families*. For instance, in the trade and industrial-technical sector constellation there are well over 100 major *occupational clusters* or *job families*. Some sample job families include: building industry occupations, automotive industries occupations, electronic technology, to name a few.

Each occupational cluster includes other *job clusters*. Continuing with an analysis of the building industries occupations cluster reveals that it consists of such trades and technical clusters as: architecture and design, masonry and trowel trades, electrical installation and servicing trades, cabinetmaking and millwork trades, and others.

Job Levels and Career Planning

There are many different levels of skill requirements within each trade or technical occupation. Some workers perform within a narrow range of job skills that are classified as *unskilled*. Other jobs beyond this basic level require higher degree manipulative (performance) skills and technology, and the ability to solve job-related problems requiring mathematical skills, science skills, and communication skills in dealing with drawings, prints, and other graphic information. Clusters of jobs that have these skill requirements range from *semiskilled* to *skilled craftspersons* and *technicians,* including middle-management and supervisory personnel.

The next job levels in each occupational field reach up to *semiprofessional* and *professional. Job mobility* laterally from one occupational field to another field or from one level to the next higher level often requires the mastery of advanced occupational skills and technology for the job level accompanied by the capability to solve mathematical problems and communicate accurately. Experience is another major factor in job mobility.

C. MATHEMATICS APPLIED TO CAREERS AND CAREER PLANNING

The terms: *awareness, orientation, exploration, occupational entry preparation, retraining,* and *upgrading* training are all part of a career development ladder. The first step begins in elementary school. Second, and continuing steps, are taken through the middle school and high school into post-secondary institutions and on into special (less formalized) training programs. Thus, there is a continuum to career development throughout the occupational work life of each worker. Other important, and often parallel, career development programs are provided by business, industry, the military (through occupational specialty training), labor organizations, and others.

Successive Stages of Occupational Awareness, Orientation, Exploration, Technology Education, and Vocational-Technical Education

Educational systems provide a broad-based general education curriculum and other guidance, counseling, testing, and human resource services that impact on career planning and career development. The earlier mentioned needs for each individual to develop functional communication skills, computational skills, science skills, general education technical skills, and social/citizenship skills are fulfilled within the learning areas that comprise the curriculum.

Beginning in the elementary grades, the closely integrated learning areas have as an objective the development within each person of an *awareness* of the work world; the technology that is encompassed throughout the different economic sectors.

Orientation and *exploratory* experiences are essential parts of the middle/junior high school curriculum. More advanced exploratory experiences are provided in general education curriculums in senior high schools. The three major stages of instruction from early awareness, to orientation, to exploration, are part of the common core or foundational skills and technology of all youth and adults. Technology at each stage is based upon a broad understanding and sampling of job families that constitute the world of work. In other words, technology instruction

constitutes a wide spectrum of activities that are related to the economic sectors. Computational skills that are foundational to technology education are tightly interlocked in the curriculum.

Further enrichment in career planning is provided through work-related, practical, manipulative skill and related technology experiences. Activities are provided within shop, laboratory, and clinical facilities wherein hands-on experiences, representative of different economic sectors, are sampled.

Industrial arts (often identified as industrial technology) provides learning experiences relating to the trade and industrial-technical sector. Agri-business relates to the important sector of agriculture. Similarly, business, market and merchandising, home and family related services, health and allied medical services, and other programs represent particular work force areas within the economic sector.

As a result of these experiences, many students begin preparation to enter into the work force by electing a vocational-technical curriculum at the high school level. Others pursue a job objective in postsecondary institution programs. Apprenticeship training; military occupational specialty training; adult preemployment, retraining, and occupational extension programs provide other opportunities for continuous career development.

Mathematics Competencies Based on Analyses of Occupational Needs

Mathematics for each different vocational-technical curriculum must be based on the computational skill needs within a job family. Mathematics content differs for each specific occupation and for each employment level within the occupation. For example, mathematics content for such curriculums as: licensed practical nurse, dietary technician, computer programmer, ornamental horticulture, and machine technology show the range and differences in *applications* of basic vocational-technical mathematics. While there is a *common core of mathematical principles, concepts,* and *rules,* each occupational family has specific applications. Also, the range and depth of mathematical skills vary with the occupational level.

Some jobs require the ability to apply just the four basic arithmetic processes. At the next stage of development and level of employment, other fundamental principles relating to common fractions, decimal fractions, direct and indirect measurements, and Customary units and SI metric units of measurement, are mastered.

At still higher levels of employment, the worker must be able to deal with statistics and graphs; percentages, ratios and proportions, and finances. The individual must use fundamentals of algebra, geometry, and trigonometry that are directly related to and applied to an occupational job family. Occupational skills depend on a worker's ability to accurately interpret formulas, geometric lines, shapes, and constructions in terms of plane and solid parts and mechanisms. Practical solutions are needed that relate to right, acute, and oblique triangles.

Applications: Basic Function, Scientific, Engineering, and Graphics Calculators

Electronic calculators are used in applications of mathematical principles, concepts, and rules at each major level of employment. Calculators eliminate laborious computations, simplify problem solving, and speed up computations. Calculators are recommended as an alternate method of checking solutions after the long-hand problem-solving method is mastered.

Four-process arithmetic calculators have all the capacity needed for most unskilled level job applications. Engineering and scientific calculators provide added problem-solving capability, especially in related algebra, geometry, and trigonometry. Engineering and scientific calculators have the power required to solve mathematics problems at intermediate and terminal career levels for skilled craftspersons, technicians, semi-professional and engineering support personnel.

The graphics calculator has add-on features for displaying equations, trigonometric and other functions, statistics and data representation, geometric forms, and complex mathematical processes. Additional keyboard capacity permits manipulation of variables, tracing, and instant display of results.

ASSIGNMENT UNIT 45   REVIEW AND SELF TEST

PRETEST   *The Review and Self-Test items that follow may be used as a pretest. Pretests are designed to measure a student's beginning level of mathematical skills competency and to determine the starting point of instruction.*

POSTTEST   *The Review and Self-Test items may also be applied as a post test. Post tests are planned to establish the student's level of mathematical skills competency after instruction.*

A. Work Ethics, Characteristics of the Work Force, and Employment Demands

1. State two work ethics (concepts) that are foundational to career planning.
2. Secure labor market information for the geographic area of your residence from such publications as State Labor Market Demand reports, the U.S. Government Occupational Outlook Handbook, Dictionary of Occupational Titles, and other documents.
3. Name four economic sectors in the immediate labor market area.
4. Give the predominant economic sector.
5. Identify one occupational cluster within an occupational constellation.
6. Select and name one occupational field in the occupational cluster.
7. List four different job levels in the selected occupational field.
8. Give an example of a job title for each of the four different job levels.

B. Mathematics Applied to Careers and Career Planning

1. Describe briefly how the scope, depth, and content is determined for vocational-technical mathematics that is appropriate to occupational preparation and a career.
2. Compare the mathematical skill competency needs of an unskilled worker with those of a skilled technician.
3. Describe briefly how awareness, orientation, and exploratory experiences in education programs are valuable in career planning.
4. Study a job title in a cluster at the skilled technician level and another at the unskilled level.
   a. List the unskilled job title in one column; the skilled technician in another.
   b. Take the table of contents of this book. Make a simple analysis of the mathematics needed by the unskilled worker and the technician. Then, name three *Parts* in which the mathematics units are grouped for the unskilled worker and four additional *Parts* for mathematics competencies needed by the technician.
5. Make a simple line graph to illustrate the impact of years of experience at successively higher levels of employment and increased competency in related computational skills to wages. *Note*. Use a published job level and wage rate schedule.

Section 9
Unit 46 Achievement Review on Mathematics Applied to Consumer and Career Needs

OBJECTIVES OF THE UNIT

This achievement review serves as an overall test for Section 9. The unit is designed to measure the student's/trainee's ability to apply mathematical concepts and principles in solving problems faced by the consumer as related to

- *Money Management and Budgeting.*
- *Purchasing Goods and Credit Financing.*
- *Payrolls, Income, and Taxes.*
- *Loans, Banking, and Insurance.*
- *Career Planning and Development.*

SECTION PRETEST/ POSTTEST

The Review and Self-Test items that follow relate to each Unit within this Section. The test items may be used as a Unit-by-Unit pretest and/or Section post test.

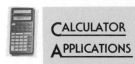

UNIT 41. MATHEMATICS APPLIED TO MONEY MANAGEMENT AND BUDGETING

A. Mathematical Processes Applied to Money

Refer to the accompanying table of budget items and expense percents.

Monthly Income	$1,200			$2,400		
Exemptions	2	3	4	2	3	4
Budget Items	Percent of Income					
Food	22	24	28	14	15	16
Rent, Light, Fuel	35	35	35	34	34	34
Clothing	10	11	12	8	10	12
Medical, recreation	6	7	7	7	7	8
Auto (transportation)	10	10	8	10	10	10
Social Security, Insurance, Taxes	12	8	5	20	16	12
Education, Church	5	5	5	7	8	8

1. Compute the amount spent monthly for each budget item for an income of $1,200 and three dependents.
2. Find the difference in the amounts paid at the $1,200 and $2,400 income levels for three dependents, respectively.

B. Mathematics Applied to Money Management and Budgeting

Problems 1 through 6 relate to (a) the preparation of a trial budget, (b) an actual or simulated working budget, and (c) a system for controlling and managing the budget.

1. Prepare a worksheet for listing major categories of (a) income and (b) fixed and variable expenditure items that are to be included in an actual or a simulated trial budget.
2. Add columns for both income and expenses. Then, record (a) the amounts for exact or anticipated income and expenses and (b) prepare a schedule showing when each item is due and payable.
3. Add still other columns to show monthly and annual income for each item. (a) Compute and record monthly income and expenditures. (b) Add the monthly totals and post the yearly amounts. (c) Add and record the gross annual totals.
4. Develop a balanced budget by establishing priorities and making adjustments until the income and expenditures are equal. Then, prepare the working budget.
5. Develop a system for the control and management of the one-year budget.
6. Compute the percent of the total annual income that each major category of (a) income and (b) expenditures constitutes. Add another column on the budget forms and record these percents.

UNIT 42. MATHEMATICS APPLIED TO PURCHASING GOODS AND CREDIT FINANCING

A. Mathematics Applications to Costs, Discounts, and Purchasing Goods

1. Use the information for articles A through E and determine the cost of: (a) materials, (b) labor, and (c) overhead. Round off each cost to two decimal places.

	Quantity Produced	Total Mfg. Cost	Materials Cost (%) (a)	Labor Cost (%) (b)	Overhead (%) (c)	Mark-up for List Price (%)	Discounts (%)
A	1,000	$2,400	30	65	5	20	10
B	750	3,250	25	68	7	25	6
C	10 gross	2,750	$16\frac{1}{2}$	$77\frac{1}{2}$	6	$37\frac{1}{2}$	15
D	15 doz	675	12.5	82.7	4.8	45	12 and 8
E	78 bbl	1,982	$8\frac{1}{2}$	$79\frac{1}{4}$	$12\frac{1}{4}$	$62\frac{1}{2}$	20, 10, 2

2. Compute (a) the list price and (b) selling price of articles A through E.

B. Mathematics Applied in Credit Financing

1. Secure and read a loan payment schedule and installment purchase plan from three sources: a bank, a retailer, and another loan organization.

   a. Prepare a table for materials A, B, and C. The following items may be stated and compared: (1) the conditions under which an installment credit loan is made, (2) the rates for credit charges, (3) number of payments, (4) the amount of each payment, (5) the total amounts of payments, and (6) the total cost of installment buying.

   b. Determine which financial organization to use. State the reasons for the decision.

Materials	Principal of Loan ($ Cost)	Period of Financing (Months)	Rates for Credit (2)	Payment Number (3)	Amount Each Payment (4)	Total of Payments (5)	Total Cost of Service (b)
			LOAN ORGANIZATIONS				
A	1,200	18	(B) (R) (O)	(B) (R) (O)	(B) (R) (O)	(B) (R) (O)	(B) (R) (O)
B	1,650	24	(B) (R) (O)	(B) (R) (O)	(B) (R) (O)	(B) (R) (O)	(B) (R) (O)
C	3,480	30	(B) (R) (O)	(B) (R) (O)	(B) (R) (O)	(B) (R) (O)	(B) (R) (O)

Conditions of Installment Loan (1)
Bank (B)_____
Retailer (R)_____
Other Service (O)_____

2. Establish the credit rate for equipment items A through D. Use the formula:

$$R = \frac{2(I \times F)}{U(n + 1)}$$ State the credit rate as a percent, rounded to two decimal places.

Equipment	Principal (Unpaid Balance) (U)	Installment Payments in One Year (I)	Number of Payments in Contract (n)	Finance Charges (F)	Credit Rate (%) (R)
A	$ 1,200	6	18	$ 288	
B	$ 2,400	12	18	$ 288	
C	$ 3,050	6	24	$1,704	
D	$10,400	4	16	$5,056	

UNIT 43. MATHEMATICS APPLIED TO BANKING, HOME OWNERSHIP, AND INVESTMENTS

A. Mathematics Applications to Banking Services and Real Estate

1. State the formula for computing simple interest. Then, compute the missing values for accounts A through F.

	Formula			
Account	Simple Interest	Principal	Rate of Interest	Time
A		$2,500	8.7%	1 year
B	$ 31.05		$5\frac{3}{4}\%$	1 year
C	$118.40	$ 750		2 years
D		$1,200	$5\frac{1}{4}\%$	6 months
E	$175.72		9.76%	8 months
F	$318.00	$3,600		10 months

2. Calculate the actual cash balance in the following checking account. Bank statement balance ($1,672.57); outstanding (uncashed) checks of $171.25, $417.49, and $27.78, and a mortgage payment of $572.50. Since the date of the bank statement, deposits were made in the amounts of $122.16 and $308.15.

3. The sum of $550 per month has been budgeted for the reduction of principal and the payment of interest on real estate. The purchase price of the property (including all closing and other expenses) is $55,000. A 20% down payment is made. The balance is covered by

a binder and a schedule of payments. The interest rate is 13.5% per year on the unpaid balance. A schedule of payments is to be prepared for the first six months.

a. Determine the amounts paid monthly for (1) interest and (2) the reduction of principal, and (3) the monthly unpaid balance on the principal.

b. Find the totals of the six months of (1) the monthly payments, (2) interest, (3) amount paid on the principal, and (4) the reduction of principal.

	Month	1	2	3	4	5	6	Totals (b)
a.(3)	Balance of Principal							
	Monthly Payments							
a.(1)	Monthly Interest							
a.(2)	Reduction of Principal							

B. Mathematics Applied to Investments

Refer to the abstracted portion of common stock market quotations.

1. Indicate the price per share for stocks A, B, and C according to the condition of the market at the time of purchase.
2. Find the net price of 150 shares of stock in each company, exclusive of stock brokers' fees and other transaction expenses.

Net Chg	52 Weeks Hi	52 Weeks Lo	Stock	Sym	Div	Yld %	PE	Vol 100s	Hi	Lo	Close	Net Chg
5/8	6 7/8	3	Winnebago	WGO	.10j	...	...	4	3 1/8	3 1/8	3 1/8	...
7/8	32 1/8	26 5/8	WiscEngy	WEC	1.76	6.1	10	846	29 1/4	28 5/8	28 7/8	± 1/4
3/4	24 1/2	19 3/4	WiscPS	WPS	1.66	7.8	11	469	21 3/8	20 3/4	21 1/4	+ 1/4
1/2	42 3/4	24 7/8	WitcoCp	WIT	1.72	6.6	19	149	26 3/8	25 7/8	25 7/8	− 1/8
1/8	▼ 13 1/8	7 7/8	WolverineWW	WWW	.16	2.1	8	88	8	7 3/4	7 3/4	− 1/4
1/8	s 36 5/8	22 7/8	Woolworth	Z	1.04	4.1	10	2375	25 5/8	24 1/2	25 1/4	+ 1/8
	205	138	Woolworth pf		2.20	1.5	...	2	142	142	142	+4
1/4	15 1/2	3	WorldCp	WOA		...	...	70	4 5/8	4 1/2	4 5/8	+ 1/8
.	19 3/8	12 3/4	WorldValFd	VLU	1.94e	14.2	...	19	13 5/8	13 1/2	13 5/8	+ 1/8
/8	59 1/4	38 3/4	Wrigley	WWY	.88	1.9	16	150	47 7/8	46 3/8	46 3/8	−1
..	15 1/8	8 1/4	WyleLabs	WYL	.28	2.6	10	868	10 5/8	10 3/8	10 5/8	+ 3/8
...	25 7/8	14 3/4	WynnsInt	WN	.60	4.0	10	5	15	15	15	− 1/4
...			**-X-Y-Z-**									
32												
	66 5/8	32 1/4	Xerox	XRX	3.00	8.6	6	2726	36 3/8	34 5/8	34 7/8	−1 3/8
3	32 3/8	16 1/2	XTRA	XTR	.72	4.3	...	17	16 7/8	16 5/8	16 5/8	− 1/4
	25 1/4	16 1/8	XTRA pf		1.94	11.9	...	4	16 1/4	16 1/4	16 1/4	...

	Company		Market Condition When Bought	Price per Share	Net Price (150 Shares)
A	Woolworth	Z	High		
B	Wrigley	WWY	Low		
C	Xerox	XRX	Last		

3. Refer to the financial section of a newspaper for bond market information. Also, secure a schedule of broker commissions on corporation bonds.
4. Read the bond market quotations for a selected corporation bond. Record (a) the name of the bonds, (b) face value, (c) interest, (d) and yield.
5. Find the broker's commission on 375 shares of the selected bonds.
6. Determine the cost of the bonds, including brokerage fees. Assume the bonds are purchased at the *low* price for the *day*.

7. Compute the profit or loss from the sale of the 375 bonds under the following market conditions. Broker's commissions for selling (if applicable) are to be deducted to establish the actual selling price.
   a. Selling at the *high* for the *year*.
   b. Selling at the *low* for the *year*.
   c. Selling at the *high* for the *day*.

UNIT 44. MATHEMATICS APPLIED TO PAYROLLS, INCOME, AND TAXES

A. Mathematics Applied to Payrolls, Wages, and Take-Home Pay

1. Given the annual wages (a) and additional taxable income (b) for single employees A and B and married employee C (who is filing a joint return), according to the information provided in the table, find the total income (c) and the F.I.C.A. tax (d) using a tax rate of 7.65%.

		F.I.C.A. Tax			Federal Income Tax					
	Annual Wages (a)	Additional Taxable Income (b)	Total Income (c)	F.I.C.A. Tax (7.65%) (d)	Standard Deduction (e)	Exemptions No. (f)	Exemptions Amount (g)	Total Taxable Income (h)	Income Tax (i)	Annual Take-Home Pay (j)
A	$16,200	$ 800				1				
B	28,750	1,275				3				
C	41,680	2,670				5				

2. Find and record the amount for the standard deductions (e) for the three employees and the exemption amount (g) for the number of exemptions claimed (f).
3. Calculate the total taxable income (h) and find the income tax (i) in the Tax Table.
4. Start with the annual wages (a) and determine the annual take-home pay (j).

B. Mathematics Applied to Income Taxes

1. Secure a *U.S. Individual Income Tax Return (Form 1040)* packet which includes directions, 1040 Schedules, and Tax Tables.
2. Prepare the U.S. Individual Federal Income Tax return as a simulation exercise. Use the following information to compute the federal tax and to record any overpayment of tax or balance that may be payable to the Internal Revenue Service.

   Given: Single person return; one claimed exemption

Income from wages and other taxable compensation	$14,560
Dividends from stocks	420
Interest on savings account	480
Federal income tax withheld from wages	1,950
Credit on overpayment from previous year	150
Estimated payments made during the year	200

UNIT 45. MATHEMATICS ESSENTIAL TO CAREERS AND CAREER PLANNING

A. Characteristics of the Work Force and Employment Demands

1. Secure U.S. Government Printing Office, Department of Labor, Labor Statistics Tables and/or Annual Reports of the President on Manpower/Womanpower for data relating to employment by occupational levels for the nation for the 1980s and 1990s.
2. Go back to the 1980 reports and give the *projected total employment* for the 1980s.
3. Review the data for three selected levels of employment: (a) professional and technical workers, (b) craftspersons and forepersons, and (c) operatives.
   a. Give the projected employment requirements at each of these levels for the 1980s.
   b. Determine the percent distribution of the total employed work force that is represented for each of the three levels for the 1980s.
4. Provide the same information as required in problem 3 for the three levels of employment for the 1990s.
5. Find the percent rate of change for the employment of (a) professional and technical workers, (b) craftspersons and forepersons, and (c) operatives from the 1980s to the 1990s.

B. Mathematics Applications to Careers and Career Planning

1. State the objectives of instruction in mathematics for each of the following education and training programs.
   a. Early childhood education.
   b. Middle/junior high school education.
   c. High school: general education curriculum.
   d. High school: vocational-technical education curriculum.
   e. Adult business/industry cooperative training.
   f. Apprenticeship training: related and supplemental mathematics instruction.
2. Make a simple graphic of a career plan. Start with preparation for an entry job through instruction in a vocational-technical high school curriculum. Assume a series of progressively higher job levels over a period of years.
   a. Represent three job levels on the graphic at different career stages.
   b. Indicate, also, that at each level a greater degree of mathematical skills in dealing with occupational problems is required.

APPENDIX

TABLE 1 Standard Tables of Customary Units of Measure

Linear Measure		
12 inches (in.)	=	1 foot (ft).
3 ft.	=	1 yard (yd.)
16 1/2 ft.	=	1 rod (rd.)
5 1/2 yd.	=	1 rd.
320 rd.	=	1 mile
1760 yd.	=	1 mile
5280 ft.	=	1 mile

Surface Measure		
144 sq. in.	=	1 sq. ft.
9 sq. ft.	=	1 sq. yd.
30 1/4 sq. yd.	=	1 sq. rd.
160 sq. rd.	=	1 acre
640 acres	=	1 sq. mile
43,560 sq. ft.	=	1 acre

Cubic Measure (Volume)		
1728 cu. in.	=	1 cu. ft.
27 cu. ft.	=	1 cu. yd.
128 cu. ft.	=	1 cord

Angular (Circular) Measure		
60 sec. (")	=	1 min. (')
60'	=	1 degree (°)
90°	=	1 quadrant
360°	=	1 circle

Time Measure		
60 seconds (sec.)	=	1 minute (min.)
60 min.	=	1 hour (hr.)
24 hr.	=	1 day
7 days	=	1 week
52 weeks	=	1 year
365 days	=	1 year
10 years	=	1 decade

Liquid Measure		
4 gills	=	1 pint (pt.)
2 pt.	=	1 quart (qt.)
4 qt.	=	1 gallon (gal.)
231 cu. in.	=	1 gal.
31.5 gal.	=	1 barrel (bbl.)
42 gal.	=	1 bbl. of oil
8 1/2 lb.	=	1 gal. water
7 1/2 gal.	=	1 cu. ft.

Weights of Materials		
0.096 lb.	=	1 cu. in. aluminum
0.260 lb.	=	1 cu. in. cast iron
0.283 lb.	=	1 cu. in. mild steel
0.321 lb.	=	1 cu. in. copper
0.41 lb.	=	1 cu. in. lead
112 lb.	=	1 cu. ft. Dowmetal
167 lb.	=	1 cu. ft. aluminum
464 lb.	=	1 cu. ft. cast iron
490 lb.	=	1 cu. ft. mild steel
555.6 lb.	=	1 cu. ft. copper
710 lb.	=	1 cu. ft. lead

Avoirdupois Weight		
16 ounces (oz.)	=	1 pound (lb.)
100 lb.	=	1 hundredweight (cwt.)
20 cwt.	=	1 ton
2000 lb.	=	1 ton
8 1/2 lb.	=	1 gal. of water
62.4 lb.	=	1 cu. ft. of water
112 lb.	=	1 long cwt.
2240 lb.	=	1 long ton

Dry Measure		
2 cups	=	1 pt.
2 pt.	=	1 qt.
4 qt.	=	1 gal.
8 qt.	=	1 peck (pk.)
4 pk.	=	1 bushel (bu.)

Miscellaneous		
12 units	=	1 dozen (doz.)
12 doz.	=	1 gross
144 units	=	1 gross
24 sheets	=	1 quire
20 quires	=	1 ream
20 units	=	1 score
6 ft.	=	1 fathom

TABLE 2 Standard Tables of Metric Units of Measure

Linear Measure

Unit	Value in Meters	Symbol or Abbreviation
micron	0.000 001	μ
millimeter	0.001	mm
centimeter	0.01	cm
decimeter	0.1	dm
meter (unit)	1.0	m
dekameter	10.0	dam
hectometer	100.0	hm
kilometer	1 000.00	km
myriameter	10 000.00	Mm
megameter	1 000 000.00	

Surface Measure

Unit	Value in Square Meters	Symbol or Abbreviation
square millimeter	0.000 001	mm^2
square centimeter	0.000 1	cm^2
square decimeter	0.01	dm^2
square meter (centiare)	1.0	m^2
square dekameter (are)	100.0	a^2
hectare	10 000.0	ha^2
square kilometer	1 000 000.0	km^2

Volume

Unit	Value in Liters	Symbol or Abbreviation
milliliter	0.001	mL
centiliter	0.01	cL
deciliter	0.1	dL
liter (unit)	1.0	L
dekaliter	10.0	daL
hectoliter	100.0	hL
kiloliter	1 000.0	kL

Mass

Unit	Value in Grams	Symbol or Abbreviation
microgram	0.000 001	μg
milligram	0.001	mg
centigram	0.01	cg
decigram	0.1	dg
gram (unit)	1.0	g
dekagram	10.0	dag
hectogram	100.0	hg
kilogram	1 000.0	kg
myriagram	10 000.0	Mg
quintal	100 000.0	q
ton	1 000 000.0	t

Cubic Measure

Unit	Value in Cubic Meters	Symbol or Abbreviation
cubic micron	10^{-18}	μ^3
cubic millimeter	10^{-9}	mm^3
cubic centimeter	10^{-6}	cm^3
cubic decimeter	10^{-3}	dm^3
cubic meter	1	m^3
cubic dekameter	10^3	dam^3
cubic hectometer	10^6	hm^3
cubic kilometer	10^9	km^3

TABLE 3 Metric and Customary Decimal Equivalents for Fractional Parts of an Inch

DECIMAL EQUIVALENTS					
Fraction	Decimal Equivalent		Fraction	Decimal Equivalent	
	Customary (in.)	Metric (mm)		Customary (in.)	Metric (mm)
1/64	.015625	0.3969	33/64	.515625	13.0969
1/32	.03125	0.7938	17/32	.53125	13.4938
3/64	.046875	1.1906	35/64	.546875	13.8906
1/16	.0625	1.5875	9/16	.5625	14.2875
5/64	.078125	1.9844	37/64	.578125	14.6844
3/32	.09375	2.3813	19/32	.59375	15.0813
7/64	.109375	2.7781	39/64	.609375	15.4781
1/8	.1250	3.1750	5/8	.6250	15.8750
9/64	.140625	3.5719	41/64	.640625	16.2719
5/32	.15625	3.9688	21/32	.65625	16.6688
11/64	.171875	4.3656	43/64	.671875	17.0656
3/16	.1875	4.7625	11/16	.6875	17.4625
13/64	.203125	5.1594	45/64	.703125	17.8594
7/32	.21875	5.5563	23/32	.71875	18.2563
15/64	.234375	5.9531	47/64	.734375	18.6531
1/4	.250	6.3500	3/4	.750	19.0500
17/64	.265625	6.7469	49/64	.765625	19.4469
9/32	.28125	7.1438	25/32	.78125	19.8438
19/64	.296875	7.5406	51/64	.796875	20.2406
5/16	.3125	7.9375	13/16	.8125	20.6375
21/64	.328125	8.3384	53/64	.828125	21.0344
11/32	.34375	8.7313	27/32	.84375	21.4313
23/64	.359375	9.1281	55/64	.859375	21.8281
3/8	.3750	9.5250	7/8	.8750	22.2250
25/64	.390625	9.9219	57/64	.890625	22.6219
13/32	.40625	10.3188	29/32	.90625	23.0188
27/64	.421875	10.7156	59/64	.921875	23.4156
7/16	.4375	11.1125	15/16	.9375	23.8125
29/64	.453125	11.5094	61/64	.953125	24.2094
15/32	.46875	11.9063	31/32	.96875	24.6063
31/64	.484375	12.3031	63/64	.984375	25.0031
1/2	.500	12.7000	1	1.000	25.4000

TABLE 4 Decimal and Millimeter Equivalents

Millimeter Equivalents of Decimals (0.01″ to 0.99″)										
Dec.	0	1	2	3	4	5	6	7	8	9
0.0	. . . .	0.254	0.508	0.762	1.016	1.270	1.524	1.778	2.032	2.286
0.1	2.540	2.794	3.048	3.302	3.556	3.810	4.064	4.318	4.572	4.826
0.2	5.080	5.334	5.588	5.842	6.096	6.350	6.604	6.858	7.112	7.366
0.3	7.620	7.874	8.128	8.392	8.636	8.890	9.144	9.398	9.652	9.906
0.4	10.160	10.414	10.688	10.922	11.176	11.430	11.684	11.938	12.192	12.446
0.5	12.700	12.954	13.208	13.462	13.716	13.970	14.224	14.478	14.732	14.986
0.6	15.240	15.494	15.748	16.022	16.256	16.510	16.764	17.018	17.272	17.526
0.7	17.780	18.034	18.288	18.542	18.796	19.050	19.304	19.558	19.812	20.066
0.8	20.320	20.574	20.828	21.082	21.336	21.590	21.844	22.098	22.352	22.606
0.9	22.860	23.114	23.368	23.622	23.876	24.130	24.384	24.638	24.892	25.146

Example   0.1″ = 2.540 mm, 0.75″ = 19.050 mm

Decimal Equivalents of Millimeters (1 mm to 99 mm)										
mm	0	1	2	3	4	5	6	7	8	9
0	. . . .	0.0394	0.0787	0.1181	0.1575	0.1968	0.2362	0.2756	0.3150	0.3543
1	0.3937	0.4331	0.4724	0.5118	0.5512	0.5906	0.6299	0.6693	0.7087	0.7480
2	0.7874	0.8268	0.8661	0.9055	0.9449	0.9842	1.0236	1.0630	1.1024	1.1417
3	1.1811	1.2205	1.2598	1.2992	1.3386	1.3780	1.4173	1.4567	1.4961	1.5354
4	1.5748	1.6142	1.6535	1.6929	1.7323	1.7716	1.8110	1.8504	1.8898	1.9291
5	1.9685	2.0079	2.0472	2.0866	2.1260	2.1654	2.2047	2.2441	2.2835	2.3228
6	2.3622	2.4016	2.4409	2.4803	2.5197	2.5590	2.5984	2.6378	2.6772	2.7165
7	2.7559	2.7953	2.8346	2.8740	2.9134	2.9528	2.9921	3.0315	3.0709	3.1102
8	3.1496	3.1890	3.2283	3.2677	3.3071	3.3464	3.3858	3.4252	3.4646	3.5039
9	3.5433	3.5827	3.6220	3.6614	3.7008	3.7402	3.7795	3.8189	3.8583	3.8976

Example   10 mm = 0.3937″, 57 mm = 2.2441″

TABLE 5 Conversion of English and Metric Units of Measure

Linear Measure								
Unit	Inches to milli- metres	Milli- metres to inches	Feet to metres	Metres to feet	Yards to metres	Metres to yards	Miles to kilo- metres	Kilo- metres to miles
1	25.40	0.03937	0.3048	3.281	0.9144	1.094	1.609	0.6214
2	50.80	0.07874	0.6096	6.562	1.829	2.187	3.219	1.243
3	76.20	0.1181	0.9144	9.842	2.743	3.281	4.828	1.864
4	101.60	0.1575	1.219	13.12	3.658	4.374	6.437	2.485
5	127.00	0.1968	1.524	16.40	4.572	5.468	8.047	3.107
6	152.40	0.2362	1.829	19.68	5.486	6.562	9.656	3.728
7	177.80	0.2756	2.134	22.97	6.401	7.655	11.27	4.350
8	203.20	0.3150	2.438	26.25	7.315	8.749	12.87	4.971
9	228.60	0.3543	2.743	29.53	8.230	9.842	14.48	5.592

Example 1 in. = 25.40 mm, 1 m = 3.281 ft., 1 km = 0.6214 mi.

Surface Measure										
Unit	Square inches to square centi- metres	Square centi- metres to square inches	Square feet to square metres	Square metres to square feet	Square yards to square metres	Square metres to square yards	Acres to hec- tares	Hec- tares to acres	Square miles to square kilo- metres	Square kilo- metres to square miles
1	6.452	0.1550	0.0929	10.76	0.8361	1.196	0.4047	2.471	2.59	0.3861
2	12.90	0.31	0.1859	21.53	1.672	2.392	0.8094	4.942	5.18	0.7722
3	19.356	0.465	0.2787	32.29	2.508	3.588	1.214	7.413	7.77	1.158
4	25.81	0.62	0.3716	43.06	3.345	4.784	1.619	9.884	10.36	1.544
5	32.26	0.775	0.4645	53.82	4.181	5.98	2.023	12.355	12.95	1.931
6	38.71	0.93	0.5574	64.58	5.017	7.176	2.428	14.826	15.54	2.317
7	45.16	1.085	0.6503	75.35	5.853	8.372	2.833	17.297	18.13	2.703
8	51.61	1.24	0.7432	86.11	6.689	9.568	3.237	19.768	20.72	3.089
9	58.08	1.395	0.8361	96.87	7.525	10.764	3.642	22.239	23.31	3.475

Example 1 sq. in. = 6.452 cm^2 1 m^2 = 1.196 sq. yd., 1 sq. mi. = 2.59 km^2

Cubic Measure								
Unit	Cubic inches to cubic centi- metres	Cubic centi- metres to cubic inches	Cubic feet to cubic metres	Cubic metres to cubic feet	Cubic yards to cubic metres	Cubic metres to cubic yards	Gallons to cubic feet	Cubic feet to gallons
1	16.39	0.06102	0.02832	35.31	0.7646	1.308	0.1337	7.481
2	32.77	0.1220	0.05663	70.63	1.529	2.616	0.2674	14.96
3	49.16	0.1831	0.08495	105.9	2.294	3.924	0.4010	22.44
4	65.55	0.2441	0.1133	141.3	3.058	5.232	0.5347	29.92
5	81.94	0.3051	0.1416	176.6	3.823	6.540	0.6684	37.40
6	98.32	0.3661	0.1699	211.9	4.587	7.848	0.8021	44.88
7	114.7	0.4272	0.1982	247.2	5.352	9.156	0.9358	52.36
8	131.1	0.4882	0.2265	282.5	6.116	10.46	1.069	59.84
9	147.5	0.5492	0.2549	371.8	6.881	11.77	1.203	67.32

Example 1 cm^3 = 0.06102 cu. in., 1 gal. = 0.1337 cu. ft.

Volume or Capacity Measure									
Unit	Liquid ounces to cubic centi- metres	Cubic centi- metres to liquid ounces	Pints to litres	Litres to pints	Quarts to litres	Litres to quarts	Gallons to litres	Litres to gallons	Bushels to hecto- litres	Hecto- litres to bushels
1	29.57	0.03381	0.4732	2.113	0.9463	1.057	3.785	0.2642	0.3524	2.838
2	59.15	0.06763	0.9463	4.227	1.893	2.113	7.571	0.5284	0.7048	5.676
3	88.72	0.1014	1.420	6.340	2.839	3.785	11.36	0.7925	1.057	8.513
4	118.3	0.1353	1.893	8.454	3.170	4.227	15.14	1.057	1.410	11.35
5	147.9	0.1691	2.366	10.57	4.732	5.284	18.93	1.321	1.762	14.19
6	177.4	0.2029	2.839	12.68	5.678	6.340	22.71	1.585	2.114	17.03
7	207.0	0.2367	3.312	14.79	6.624	7.397	26.50	1.849	2.467	19.86
8	236.6	0.2705	3.785	16.91	7.571	8.454	30.28	2.113	2.819	22.70
9	266.2	0.3043	4.259	19.02	8.517	9.510	34.07	2.378	3.171	25.54

Example 1 L = 2.113 pt., 1 gal. = 3.785 L

TABLE 6 Conversion Factors for SI, Conventional Metric, and Customary Units Used in Physical Science Applications

Category	Conversion of Customary and Conventional Metric Units to SI Metrics			Conversion of SI Metrics to Customary and Conventional Metric Units		
	From Customary or Conventional Metric Unit	To SI Metric	Factor (A) (Multiply by)	From SI Metric	To Customary or Conventional Metric Unit	Factor (Multiply by the reciprocal of the multiplier (Factor A) which is used in conversion to the SI metric unit)
acceleration	ft/s^2	m/s^2	0.304 8*	m/s^2	ft/s^2	
	in/s^2		2.540 0 $\times 10^{-2}$ *		in/s^2	
area	ft^2	m^2	9.290 3 $\times 10^{-2}$	m^2	ft^2	
	in^2		6.451 6 $\times 10^{-4}$*		in^2	
density	g/cm^3	kg/m^3	1.000 0 $\times 10^{-3}$*	kg/m^3	g/cm^3	
	lb (mass)/ft^3		16.018 5		lb (mass)/ft^3	
	lb (mass)/in^3		2.768 0 $\times 10^4$		lb (mass)/in^3	
energy	Btu (thermochemical)		1.054 3 $\times 10^3$		Btu (thermochemical)	
	cal (thermochemical)		4.184 0*		cal (thermochemical)	
	eV	J	1.602 1 $\times 10^{-19}$	J	eV	
	erg		1.000 0 $\times 10^{-7}$*		erg	
	ft · lb (force)		1.355 8		ft · lb (force)	
	kW·h		3.600 0 $\times 10^6$*		kW·h	
	Wh		3.600 0 $\times 10^3$*		Wh	
flow (liquid and solid)	ft^3/min		4.719 5 $\times 10^{-4}$		ft^3/min	
	ft^3/s	m^3/s	2.831 7 $\times 10^{-2}$	m^3/s	ft^3/s	
	in^3/min		2.731 2 $\times 10^{-7}$		in^3/min	
	lb (mass)/s		0.453 6		lb (mass)/s	
	lb (mass)/min	kg/s	7.559 9 $\times 10^{-3}$	kg/s	lb (mass)/min	
	tons (short, mass)/h		0.252 0		tons (short, mass)/h	
force	dyne		1.000 0 $\times 10^{-5}$ *		dyne	
	kg (force)	N	9.806 6	N	kg (force)	
	lb (force)		4.448 2		lb (force)	
heat	Btu (thermochemical)/ft^2	J/m^2	1.134 9 $\times 10^4$	J/m^2	Btu (thermochemical)/ft^2	
	cal (thermochemical) /cm^2		4.184 0 $\times 10^4$*		cal (thermochemical)/cm^2	
	ft^2/h	m^2/s	2.580 6 $\times 10^{-5}$	m^2/s	ft^2/h	
length	yd		0.914 4		yd	
	ft		0.304 8*		ft	
	in	m	2.540 0 $\times 10^{-2}$ *	m	in	
	μ (micron)		1.000 0 $\times 10^{-6}$*		μ (micron)	
	mil		2.540 0 $\times 10^{-5}$*		mil	

(continued)

* Denotes exact value.

TABLE 6 Conversion Factors for SI, Conventional Metric, and Customary Units Used in Physical Science Applications (continued)

Category	Conversion of Customary and Conventional Metric Units to SI Metrics			Conversion of SI Metrics to Customary and Conventional Metric Units		
	From Customary or Conventional Metric Unit	To SI Metric	Factor (A) (Multiply by)	From SI Metric	To Customary or Conventional Metric Unit	Factor
mass	lb (mass avoirdupois)	kg	0.453 6	kg	lb (mass, avoirdupois)	Multiply by the reciprocal of the multiplier (Factor A) which is used in conversion to the SI metric unit
	oz (mass, avoirdupois)		$2.835\ 0 \times 10^{-2}$		oz (mass, avoirdupois)	
	ton, long = 2 240 lb (mass)		$1.016\ 0 \times 10^{3}$		ton, long = 2240 lb (mass)	
	ton, metric		$1.000\ 0 \times 10^{3}$ *		ton, metric	
	ton, short = 2 000 lb (mass)		$0.907\ 2 \times 10^{3}$		ton, short = 2 000 lb (mass)	
power	Btu (thermochemical)/min	W	17.572 5	W	Btu (thermochemical)/min	
	cal (thermochemical)/min		$6.973\ 3 \times 10^{-2}$		cal (thermochemical)/min	
	erg/s		$1.000\ 0 \times 10^{-7}$ *		erg/s	
	ft · lb (force)/min		$2.259\ 7 \times 10^{-2}$		ft · lb (force)/min	
	hp (550 ft · lb/s)		$7.457\ 0 \times 10^{2}$		hp (550 ft · lb/s)	
pressure (stress)	atm (760 torr)	N/m^2	$1.013\ 2 \times 10^{5}$	N/m^2	atm (760 torr)	
	dyne/cm^2		0.100 0 *		dyne/cm^2	
	g (force)/cm^2		98.066 5 *		g (force)/cm^2	
	kg (force)/cm^2		$9.806\ 6 \times 10^{4}$		kg (force)/cm^2	
	lb (force)/in^2 (or psi)		$6.894\ 8 \times 10^{3}$		lb (force)/in^2 (or psi)	
	lb (force)/in^2 (or psi)	kg (force)/mm^2	$7.030\ 7 \times 10^{-4}$	kg (force)/mm^2		
	torr (mm mercury at 0°C)	N/m^2	$1.333\ 2 \times 10^{2}$	N/m^2	torr (mm mercury at 0°C)	
velocity	ft/min	m/s	$5.080\ 0 \times 10^{-3}$ *	m/s	ft/min	
	in/s		$2.540\ 0 \times 10^{-2}$ *		in/s	
	mph		0.447 0			
	mph	km/h	1.609 3	km/h	mph	
volume	ft^3	m^3	$2.831\ 7 \times 10^{-2}$	m^3	ft^3	
	in^3		$1.638\ 7 \times 10^{-5}$		in^3	
	liter		$1.000\ 0 \times 10^{-3}$ *		liter	
temperature	deg C	K	$t_K = t_C + 273.15$	K	deg C	

*Denotes exact value

TABLE 7 Numerical Values of Trigonometric Functions for Decimal-Degree (Metric) Angles
(Partial tables 0.0° to 15.0° and 74.0° to 90.0°)

0.0° to 8.0

Dec/Deg	Sin	Cos	Tan	Cot	Dec/Deg
0.0	0.000 00	1.000 0	0.000 00	Infinite	90.0
.1	0.001 75	1.000 0	0.001 75	572.957	.9
.2	0.003 49	1.000 0	0.003 49	286.477	.8
.3	0.005 24	1.000 0	0.005 24	190.984	.7
.4	0.006 98	1.000 0	0.006 98	143.237	.6
.5	0.008 73	1.000 0	0.008 73	114.589	.5
.6	0.010 47	0.999 9	0.010 47	95.489	.4
.7	0.012 22	0.999 9	0.012 22	81.847	.3
.8	0.013 96	0.999 9	0.013 96	71.615	.2
.9	0.015 71	0.999 9	0.015 71	63.657	.1
1.0	0.017 45	0.999 8	0.017 46	57.290	89.0
.1	0.019 20	0.999 8	0.019 20	52.081	.9
.2	0.020 94	0.999 8	0.020 95	47.740	.8
.3	0.022 69	0.999 7	0.022 69	44.066	.7
.4	0.024 43	0.999 7	0.024 44	40.917	.6
.5	0.026 18	0.999 7	0.026 19	38.188	.5
.6	0.027 92	0.999 6	0.027 93	35.801	.4
.7	0.029 67	0.999 6	0.029 68	33.694	.3
.8	0.031 41	0.999 5	0.031 43	31.821	.2
.9	0.033 16	0.999 5	0.033 17	30.143	.1
2.0	0.034 90	0.999 4	0.034 92	28.636	88.0
.1	0.036 64	0.999 3	0.036 67	27.271	.9
.2	0.038 39	0.999 3	0.038 42	26.031	.8
.3	0.040 13	0.999 2	0.040 16	24.898	.7
.4	0.041 88	0.999 1	0.041 91	23.859	.6
.5	0.043 62	0.999 0	0.043 66	22.904	.5
.6	0.045 36	0.999 0	0.045 41	22.022	.4
.7	0.047 11	0.998 9	0.047 16	21.205	.3
.8	0.048 85	0.998 8	0.048 91	20.446	.2
.9	0.050 59	0.998 7	0.050 66	19.740	.1
3.0	0.052 34	0.998 6	0.052 41	19.081	87.0
.1	0.054 08	0.998 5	0.054 16	18.464	.9
.2	0.055 82	0.998 4	0.055 91	17.886	.8
.3	0.057 56	0.998 3	0.057 66	17.343	.7
.4	0.059 31	0.998 2	0.059 41	16.832	.6
.5	0.061 05	0.998 1	0.061 16	16.350	.5
.6	0.062 79	0.998 0	0.062 91	15.895	.4
.7	0.064 53	0.997 9	0.064 67	15.464	.3
.8	0.066 27	0.997 8	0.066 42	15.056	.2
.9	0.068 02	0.997 7	0.068 17	14.669	.1
4.0	0.069 76	0.997 6	0.069 93	14.301	86.0
.1	0.071 50	0.997 4	0.071 68	13.951	.9
.2	0.073 24	0.997 3	0.073 44	13.617	.8
.3	0.074 98	0.997 2	0.075 19	13.300	.7
.4	0.076 27	0.997 1	0.076 95	12.996	.6
.5	0.078 46	0.996 9	0.078 70	12.706	.5
.6	0.080 20	0.996 8	0.080 46	12.429	.4
.7	0.081 94	0.996 6	0.082 21	12.163	.3
.8	0.083 68	0.996 5	0.083 97	11.909	.2
.9	0.085 42	0.996 3	0.085 73	11.664	.1
5.0	0.087 16	0.996 2	0.087 49	11.430	85.0
.1	0.088 89	0.996 0	0.089 25	11.205	.9
.2	0.090 63	0.995 9	0.091 01	10.988	.8
.3	0.092 37	0.995 7	0.092 77	10.780	.7
.4	0.094 11	0.995 6	0.094 53	10.579	.6
.5	0.095 85	0.995 4	0.096 29	10.385	.5
.6	0.097 58	0.995 2	0.098 05	10.199	.4
.7	0.099 32	0.995 1	0.099 81	10.019	.3
.8	0.101 06	0.994 9	0.101 58	9.845	.2
.9	0.102 79	0.994 7	0.103 34	9.677	.1
6.0	0.104 53	0.994 5	0.105 10	9.514	84.0
.1	0.106 26	0.994 3	0.106 87	9.357	.9
.2	0.108 00	0.994 2	0.108 63	9.205	.8
.3	0.109 73	0.994 0	0.110 40	9.058	.7
.4	0.111 47	0.993 8	0.112 17	8.915	.6
.5	0.113 20	0.993 6	0.113 94	8.777	.5
.6	0.114 94	0.993 4	0.115 70	8.643	.4
.7	0.116 67	0.993 2	0.117 47	8.513	.3
.8	0.118 40	0.993 0	0.119 24	8.386	.2
.9	0.120 14	0.992 8	0.121 01	8.264	.1
7.0	0.121 87	0.992 5	0.122 78	8.144	83.0
.1	0.123 60	0.992 3	0.124 56	8.028	.9
.2	0.125 33	0.992 1	0.126 33	7.916	.8
.3	0.127 06	0.991 9	0.128 10	7.806	.7
.4	0.128 80	0.991 7	0.129 88	7.700	.6
.5	0.130 53	0.991 4	0.131 65	7.596	.5
.6	0.132 26	0.991 2	0.133 43	7.495	.4
.7	0.133 99	0.991 0	0.135 21	7.396	.3
.8	0.135 72	0.990 7	0.136 98	7.300	.2
.9	0.137 44	0.990 5	0.138 76	7.207	.1
8.0	0.139 17	0.990 3	0.140 54	7.1154	82.0

Dec/Deg	Cos	Sin	Cot	Tan	Dec/Deg

82.0 to 90.0

8.0 to 15.0

Dec/Deg	Sin	Cos	Tan	Cot	Dec/Deg
8.0	0.139 17	0.990 3	0.140 54	7.1154	82.0
.1	0.140 90	0.990 0	0.142 32	7.0264	.9
.2	0.142 63	0.989 8	0.144 10	6.9395	.8
.3	0.144 36	0.989 5	0.145 88	6.8547	.7
.4	0.146 08	0.989 3	0.147 67	6.7720	.6
.5	0.147 81	0.989 0	0.149 45	6.6912	.5
.6	0.149 54	0.988 8	0.151 24	6.6122	.4
.7	0.151 26	0.988 5	0.153 02	6.5350	.3
.8	0.152 99	0.988 2	0.154 81	6.4596	.2
.9	0.154 71	0.988 0	0.156 60	6.3859	.1
9.0	0.156 43	0.987 7	0.158 38	6.3138	81.0
.1	0.158 16	0.987 4	0.160 17	6.2432	.9
.2	0.159 88	0.987 1	0.161 96	6.1742	.8
.3	0.161 60	0.986 9	0.163 76	6.1066	.7
.4	0.163 33	0.986 6	0.165 55	6.0405	.6
.5	0.165 05	0.986 3	0.167 34	5.9758	.5
.6	0.166 77	0.986 0	0.169 14	5.9124	.4
.7	0.168 49	0.985 7	0.170 93	5.8502	.3
.8	0.170 21	0.985 4	0.172 73	5.7894	.2
.9	0.171 93	0.985 1	0.174 53	5.7297	.1
10.0	0.173 6	0.984 8	0.176 3	5.6713	80.0
.1	0.175 4	0.984 5	0.178 1	5.6140	.9
.2	0.177 1	0.984 2	0.179 9	5.5578	.8
.3	0.178 8	0.983 9	0.181 7	5.5026	.7
.4	0.180 5	0.983 6	0.183 5	5.4486	.6
.5	0.182 2	0.983 3	0.185 3	5.3955	.5
.6	0.184 0	0.982 9	0.187 1	5.3435	.4
.7	0.185 7	0.982 6	0.189 0	5.2924	.3
.8	0.187 4	0.982 3	0.190 8	5.2422	.2
.9	0.189 1	0.982 0	0.192 6	5.1929	.1
11.0	0.190 8	0.981 6	0.194 4	5.1446	79.0
.1	0.192 5	0.981 3	0.196 2	5.0970	.9
.2	0.194 2	0.981 0	0.198 0	5.0504	.8
.3	0.195 9	0.980 6	0.199 8	5.0045	.7
.4	0.197 7	0.980 3	0.201 6	4.9594	.6
.5	0.199 4	0.979 9	0.203 5	4.9152	.5
.6	0.201 1	0.979 6	0.205 3	4.8716	.4
.7	0.202 8	0.979 2	0.207 1	4.8288	.3
.8	0.204 5	0.978 9	0.208 9	4.7867	.2
.9	0.206 2	0.978 5	0.210 7	4.7453	.1
12.0	0.207 9	0.978 1	0.212 6	4.7046	78.0
.1	0.209 6	0.977 8	0.214 4	4.6646	.9
.2	0.211 3	0.977 4	0.216 2	4.6252	.8
.3	0.213 0	0.977 0	0.218 0	4.5864	.7
.4	0.214 7	0.976 7	0.219 9	4.5483	.6
.5	0.216 4	0.976 3	0.221 7	4.5107	.5
.6	0.218 1	0.975 9	0.223 5	4.4737	.4
.7	0.219 8	0.975 5	0.225 4	4.4373	.3
.8	0.221 5	0.975 1	0.227 2	4.4015	.2
.9	0.223 3	0.974 8	0.229 0	4.3662	.1
13.0	0.225 0	0.974 4	0.230 9	4.3315	77.0
.1	0.226 7	0.974 0	0.232 7	4.2972	.9
.2	0.228 4	0.973 6	0.234 5	4.2635	.8
.3	0.230 0	0.973 2	0.236 4	4.2303	.7
.4	0.231 7	0.972 8	0.238 2	4.1976	.6
.5	0.233 4	0.972 4	0.240 1	4.1653	.5
.6	0.235 1	0.972 0	0.241 9	4.1335	.4
.7	0.236 8	0.971 5	0.243 8	4.1022	.3
.8	0.238 5	0.971 1	0.245 6	4.0713	.2
.9	0.240 2	0.970 7	0.247 5	4.0408	.1
14.0	0.241 9	0.970 3	0.249 3	4.0108	76.0
.1	0.243 6	0.969 9	0.251 2	3.9812	.9
.2	0.245 3	0.969 4	0.253 0	3.9520	.8
.3	0.247 0	0.969 0	0.254 9	3.9232	.7
.4	0.248 7	0.968 6	0.256 8	3.8947	.6
.5	0.250 4	0.968 1	0.258 6	3.8667	.5
.6	0.252 1	0.967 7	0.260 5	3.8391	.4
.7	0.253 8	0.967 3	0.262 3	3.8118	.3
.8	0.255 4	0.966 8	0.264 2	3.7843	.2
.9	0.257 1	0.966 4	0.266 1	3.7583	.1
15.0	0.258 8	0.965 9	0.267 9	3.7321	75.0
.1	0.260 5	0.965 5	0.269 8	3.7062	.9
.2	0.262 2	0.965 0	0.271 7	3.6806	.8
.3	0.263 9	0.964 6	0.273 6	3.6554	.7
.4	0.265 6	0.964 1	0.275 4	3.6305	.6
.5	0.267 2	0.963 6	0.277 3	3.6059	.5
.6	0.268 9	0.963 2	0.279 2	3.5816	.4
.7	0.270 6	0.962 7	0.281 1	3.5576	.3
.8	0.272 3	0.962 2	0.283 0	3.5339	.2
.9	0.274 0	0.961 7	0.284 9	3.5105	.1
16.0	0.275 6	0.961 3	0.286 7	3.4874	74.0

Dec/Deg	Cos	Sin	Cot	Tan	Dec/Deg

74.0 to 82.0

Note. Numerical values for secant and cosecant functions may be determined by calculator as reciprocals for cosine and sine functions, respectively.

TABLE 8 Powers and Roots of Numbers (1 through 100)

Num-ber	Powers Square	Cube	Roots Square	Cube	Num-ber	Powers Square	Cube	Roots Square	Cube
1	1	1	1.000	1.000	51	2,601	132,651	7.141	3.708
2	4	8	1.414	1.260	52	2,704	140,608	7.211	3.733
3	9	27	1.732	1.442	53	2,809	148,877	7.280	3.756
4	16	64	2.000	1.587	54	2,916	157,464	7.348	3.780
5	25	125	2.236	1.710	55	3,025	166,375	7.416	3.803
6	36	216	2.449	1.817	56	3,136	175,616	7.483	3.826
7	49	343	2.646	1.913	57	3,249	185,193	7.550	3.849
8	64	512	2.828	2.000	58	3,364	195,112	7.616	3.871
9	81	729	3.000	2.080	59	3,481	205,379	7.681	3.893
10	100	1,000	3.162	2.154	60	3,600	216,000	7.746	3.915
11	121	1,331	3.317	2.224	61	3,721	226,981	7.810	3.936
12	144	1,728	3.464	2.289	62	3,844	238,328	7.874	3.958
13	169	2,197	3.606	2.351	63	3,969	250,047	7.937	3.979
14	196	2,744	3.742	2.410	64	4,096	262,144	8.000	4.000
15	225	3,375	3.873	2.466	65	4,225	274,625	8.062	4.021
16	256	4,096	4.000	2.520	66	4,356	287,496	8.124	4.041
17	289	4,913	4.123	2.571	67	4,489	300,763	8.185	4.062
18	324	5,832	4.243	2.621	68	4,624	314,432	8.246	4.082
19	361	6,859	4.359	2.668	69	4,761	328,509	8.307	4.102
20	400	8,000	4.472	2.714	70	4,900	343,000	8.367	4.121
21	441	9,261	4.583	2.759	71	5,041	357,911	8.426	4.141
22	484	10,648	4.690	2.802	72	5,184	373,248	8.485	4.160
23	529	12,167	4.796	2.844	73	5,329	389,017	8.544	4.179
24	576	13,824	4.899	2.884	74	5,476	405,224	8.602	4.198
25	625	15,625	5.000	2.924	75	5,625	421,875	8.660	4.217
26	676	17,576	5.099	2.962	76	5,776	438,976	8.718	4.236
27	729	19,683	5.196	3.000	77	5,929	456,533	8.775	4.254
28	784	21,952	5.292	3.037	78	6,084	474,552	8.832	4.273
29	841	24,389	5.385	3.072	79	6,241	493,039	8.888	4.291
30	900	27,000	5.477	3.107	80	6,400	512,000	8.944	4.309
31	961	29,791	5.568	3.141	81	6,561	531,441	9.000	4.327
32	1,024	32,798	5.657	3.175	82	6,724	551,368	9.055	4.344
33	1,089	35,937	5.745	3.208	83	6,889	571,787	9.110	4.362
34	1,156	39,304	5.831	3.240	84	7,056	592,704	9.165	4.380
35	1,225	42,875	5.916	3.271	85	7,225	614,125	9.220	4.397
36	1,296	46,656	6.000	3.302	86	7,396	636,056	9.274	4.414
37	1,369	50,653	6.083	3.332	87	7,569	658,503	9.327	4.418
38	1,444	54,872	6.164	3.362	88	7,744	681,472	9.381	4.448
39	1,521	59,319	6.245	3.391	89	7,921	704,969	9.434	4.465
40	1,600	64,000	6.325	3.420	90	8,100	729,000	9.487	4.481
41	1,681	68,921	6.403	3.448	91	8,281	753,571	9.539	4.498
42	1,764	74,088	6.481	3.476	92	8,464	778,688	9.592	4.514
43	1,849	79,507	6.557	3.503	93	8,649	804,357	9.644	4.531
44	1,936	85,184	6.633	3.530	94	8,836	830,584	9.695	4.547
45	2,025	91,125	6.708	3.557	95	9,025	857,375	9.747	4.563
46	2,116	97,336	6.782	3.583	96	9,216	884,736	9.798	4.579
47	2,209	103,823	6.856	3.609	97	9,409	912,673	9.849	4.595
48	2,304	110,592	6.928	3.634	98	9,604	941,192	9.900	4.610
49	2,401	117,649	7.000	3.659	99	9,801	970,299	9.950	4.626
50	2,500	125,000	7.071	3.684	100	10,000	1,000,000	10.000	4.642

TABLE 9 Calculator Processes for Common Algebraic, Geometric, and Trigonometric Applications

Geometric Calculation		Formula	Calculator Processes	Measurement Unit Value
Areas (A)	Square	$A = s^2$	(s) × (s)	square units
	Rectangle	$A = (\ell)(w)$	(ℓ) × (w)	" "
	Circle	$A = \pi r^2$	π to required accuracy (π) × (r) × (r)	" "
	Triangle	$A = \dfrac{(b \times h)}{2}$	(b × h) ÷ 2	" "
Perimeter (P)	Square	$P = 4(s)$	(s) × (4)	base unit
	Rectangle	$P = 2(\ell \times w)$	(ℓ × w) × 2	" "
	Triangle	$P = s_1 + s_2 + s_3$	s_1 + (s_2) + (s_3)	" "
	Circumference	$C = \pi d$	(π) × (d)	" "
Degrees (°) and Radians (r)	Angle (as a radian value)	$\angle A = \dfrac{n^\circ}{57.295\,78^\circ}$	n° ÷ 57.295 78	radian
	Radian angle (r) (in degrees)	$d = r \times 57.295\,78$	(r) × (57.295 78)	degrees
Volume (V)	Cube	$V = s^3$	(s) × (s) × (s)	cubic units
	Rectangular solid	$V = (\ell) \times (w) \times (h)$	(ℓ) × (w) × (h)	" "
	Cylindrical solid	$V = \pi r^2 h$	(π) × (r) × (r) × (h) (π to required accuracy)	" "
	Sphere	$V = \dfrac{4\pi r^3}{3}$	(4) × (π) × (r) × (r) × (r) ÷ 3	" "
	Cone	$V = 0.2618h\,(D^2 + Dd + d^2)$	(D × D) + (D × d) + (d × d) × (h) × (0.2618)	" "

ANSWERS TO ODD-NUMBERED PROBLEMS

PART ONE FUNDAMENTALS OF BASIC MATHEMATICS

Section 1 Whole Numbers

Unit 1 Addition of Whole Numbers

A. 1. a. 59   b. 120   c. 1,008   d. 12,987

    3. a. 57   b. 99   c. 105   d. 457   e. 1,650   f. 8,057

B. 1. A. 37   C. 91   E. 965   G. 1,021   I. 219   K. 880

       B. 70   D. 144   F. 823   H. 1,362   J. 229   L. 24,134

C. 1. 12 inches   5. A. 84″   7. 67″   13. 94,468 kW

    3. 65 yards   B. 120″   9. 504,861 motors   15. 699 sq ft

                C. 156″   11. 437′

Unit 2 Subtraction of Whole Numbers

A. 1. A. 44   C. 16   E. 144   G. 528   I. 184   K. 1,186

       B. 61   D. 39   F. 206   H. 778   J. 509   L. 878

    3. a. 8,108   b. 1,778   c. 1,018 feet   d. 6,638 miles

B. 1. 2,875 board feet

    3. Week 1: 780 miles   Week 3: 1,988 miles   Week 5: 1,936 miles

        Week 2: 1,089 miles   Week 4: 1,089 miles

    5. a. 149 milliamperes   b. 213 watts   c. 24 750 ohms

7.

	Production for Processes				
	A	**B**	**C**	**D**	**E**
a. Fastest	2	2	2	3	1
b. Second Fastest	14	42	1,133	889	436
c. Slowest	21	126	1,222	2,199	878

    9. a. 1. 182   b. A. 19,557   c. 2,136 calories

           2. 598   B. 17,421

           3. 1,888

           4. 168

   11. 3875 Ω

Unit 3 Multiplication of Whole Numbers

A. 1. A. 84   B. 154   C. 702   D. 1,920   E. 847

B. 1. 20,808 work hours   5. Ⓐ = 36″   Ⓑ = 58″   7. 94,500 shingles

    3. $8,370.00   Ⓒ = 25 ″   Ⓓ = 36″   9. $162.16

                                              11. 15,042 miles

13.

Part A			
Machine	(a)	(b)	(c)
1	32	288	864
2	48	432	1,296
3	72	648	1,944

Part B			
Machine	(a)	(b)	(c)
1	80	720	3,600
2	88	792	3,960
3	136	1,224	6,120

Part C			
Machine	(a)	(b)	(c)
1	800	7,200	86,400
2	960	8,640	103,680
3	1,056	9,504	114,048

Part D			
Machine	(a)	(b)	(c)
1	1,872	16,848	454,896
2	1,976	17,784	480,168
3	3,032	27,288	736,776

15. 222 in.^{3} (222 cu in.)

Unit 4 Division of Whole Numbers

A. 1. A. 21   C. 21   E. 26   G. 11   I. 14
      B. 30   D. 45   F. 10   H. 30   J. 22

B. 1. 85 square yards   7. 8 inches   13. 1,475 calories
     3. 3 tubes   9. 1,292 watts   15. 182 milliamperes
     5. 21 gallons   11. $6.00

Section 2 Common Fractions

Unit 6 The Concept of Common Fractions

A. 1.
a. $\frac{1}{8}$ f. $\frac{5}{6}$ k. $\frac{11}{16}$ p. $\frac{20}{32}$ u. $\frac{31}{64}$
b. $\frac{3}{8}$ g. $\frac{3}{16}$ l. $\frac{15}{16}$ q. $\frac{25}{32}$ v. $\frac{57}{64}$
c. $\frac{5}{8}$ h. $\frac{1}{4}$ m. $\frac{5}{32}$ r. $\frac{30}{32}$ w. $\frac{1}{2}$
d. $\frac{1}{6}$ i. $\frac{5}{16}$ n. $\frac{10}{32}$ s. $\frac{11}{64}$ x. $\frac{3}{4}$
e. $\frac{1}{2}$ j. $\frac{7}{16}$ o. $\frac{15}{32}$ t. $\frac{21}{64}$ y. $\frac{5}{6}$
                                                z. $\frac{7}{8}$

B. 1.
a. $\frac{1}{64}$ c. $\frac{3}{32}$ e. $\frac{1}{4}$ g. $\frac{9}{32}$ i. $\frac{1}{2}$ k. $\frac{3}{4}$
b. $\frac{1}{16}$ d. $\frac{7}{32}$ f. $\frac{17}{64}$ h. $\frac{5}{16}$ j. $\frac{5}{8}$ l. $\frac{7}{8}$

3. a. $\frac{1''}{8}$ b. $\frac{3''}{8}$ c. $\frac{5''}{8}$ d. $\frac{7''}{8}$

C. 1. a. $\frac{14''}{32} = \frac{7''}{16}$ b. $\frac{48''}{64} = \frac{3''}{4}$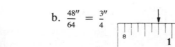

c. $\frac{10''}{16} = \frac{5''}{8}$

e. $\frac{10''}{64} = \frac{5''}{32}$

d. $\frac{44''}{64} = \frac{11''}{16}$

f. $\frac{18''}{128} = \frac{9''}{64}$

3. e c d f b a

5. Check lines drawn to the lengths indicated.

7. A. $\frac{1''}{2}$ B. $1\frac{5''}{8}$ C. $1\frac{5''}{16}$ D. $2\frac{1''}{8}$ E. $6\frac{19''}{32}$ F. $1\frac{3''}{4}$ G. $1\frac{5''}{8}$ H. $\frac{11''}{16}$ I. $1\frac{1''}{4}$ J. $2\frac{9''}{16}$

Unit 7 Addition of Fractions

A. 1. 1   3. $\frac{3}{4}$ 5. $\frac{7}{8}$ 7. $1\frac{3}{4}$ 9. $2\frac{3}{4}$

B. 1. $128\frac{5}{12}$ 3. $34\frac{1}{4}$ 5. $15\frac{59}{64}$

C. 1. A. $5\frac{5''}{8}$ B. $7\frac{9''}{32}$ C. $7\frac{21''}{64}$ D. $7\frac{21''}{32}$ E. $10\frac{15''}{32}$

3. A. $33\frac{1''}{4}$ B. $32\frac{5''}{8}$ C. $33\frac{7''}{32}$ D. $36\frac{13''}{32}$

5. $102\frac{3''}{32}$

Unit 8 Subtraction of Fractions

A. 1. $\frac{1}{8}$ 3. $\frac{1}{2}$ 5. $\frac{1}{4}$ 7. $\frac{13}{32}$ 9. $\frac{1}{16}$

B. 1. $3\frac{1}{4}$ 3. $31\frac{19}{32}$ 5. $71\frac{3}{64}$

C. 1. $\frac{2}{3}$ 3. $25\frac{11}{16}$ 5. $350\frac{59}{64}$ 7. $\frac{7}{64}$ 9. $\frac{29}{32}$

D. 1. $\frac{2}{5}$ 3. $9\frac{1}{2}$ 5. $150\frac{11}{64}$ 7. $2\frac{1}{2}$ 9. $2\frac{7}{64}$

E. 1. $\frac{7''}{8}$ B. $1\frac{3''}{4}$ C. $2\frac{3''}{16}$ D. $2\frac{5''}{64}$ E. $\frac{37''}{64}$ 3. 7 inches   5. $\frac{5''}{8}$

7. $1'-11\frac{7''}{8}$ 9. A. $11\frac{5''}{16}$ B. $6\frac{14''}{16}$ or $6\frac{7''}{8}$ C. $16\frac{25''}{32}$ D. $8\frac{57''}{64}$ 11. $\frac{13''}{16}$ 13. $9'-6\frac{3''}{4}$

Unit 9 Multiplication of Fractions

A. 1. $\frac{1}{8}$ 3. $\frac{7}{72}$ 5. $\frac{25}{72}$ 7. $\frac{13}{32}$ 9. $\frac{195}{512}$

B. 1. $\frac{7}{36}$ 3. $\frac{19}{32}$ 5. $2\frac{71}{72}$ 7. $\frac{255}{256}$ 9. $1\frac{169}{256}$

(Note: All multiplication and division steps precede all addition and subtraction steps unless otherwise shown by parentheses.)

C. 1. $2\frac{8}{9}$ 3. $14\frac{179}{256}$ 5. $14\frac{257}{512}$

D. 1. $\frac{1}{8}$ 3. $1\frac{3}{64}$

E. 1. A. $5\frac{9}{16}''$ B. $14\frac{29}{32}''$ C. $156\frac{7}{8}''$ D. $120\frac{13}{64}''$ E. $94\frac{51}{64}''$

3. A. $3\frac{9}{32}''$ B. $4\frac{3}{8}''$ C. $2\frac{3}{16}''$ D. $4\frac{3}{8}''$

5.

Method	Part A		Part B	
	Daily Production	Unit Cost	Daily Production	Unit Cost
1	75	150	$157\frac{1}{2}$	441
2	105	$262\frac{1}{2}$	$202\frac{1}{2}$	$632\frac{13}{16}$
3	$112\frac{1}{2}$	$309\frac{3}{8}$	$221\frac{1}{4}$	$912\frac{21}{32}$

7. 211′ 3″

Unit 10 Division of Fractions

A. 1. 3   3. $\frac{2}{3}$ 5. $\frac{3}{4}$ 7. $\frac{7}{10}$ 9. $\frac{39}{424}$

B. 1. 9   3. 8   5. $\frac{1}{64}$ 7. $26\frac{2}{3}$ 9. 3

C. 1. 1   3. $1\frac{17}{26}$ 5. $1\frac{95}{101}$ 7. $\frac{47}{102}$

D. 1. $\frac{1}{4}$ 3. $87\frac{3}{7}$ 5. $4\frac{49}{152}$ 7. $\frac{225}{1552}$

E. 1. 8 parts   3. 13 pieces   5. 386 pieces   7. 7.5 lbs/ft
9. a. A. 5 minutes 20 seconds B. 8 minutes C. 4 minutes D. 3 minutes 37 seconds E. 45 seconds
b. 21 minutes 42 seconds   11. $1\frac{1}{2}$ amperes (rounded to nearest $\frac{1}{2}$ ampere)

Section 3 Decimal Fractions

Unit 12 The Concept of Decimal Fractions

A. 1. A. .5   B. .3   C. .6   D. .3   E. .9   F. .1   G. .5   H. .77   I. .62
3. a. .7   b. .16   c. .015   d. .0011   e. .2152   f. 3.1875
5. a. 1.1   b. 3.09   c. 25.91   d. 272.067   e. 2525.0021   f. 362.2007
B. 1. Dimensions located on the tenth scale.

3. A. 1.007   B. 2.03   C. 1.15   D. .803   E. .90
5. a. .76   b. 1.95   c. 7.32   d. 29.41   e. 2.56   f. 18.27   g. 221.76   h. .90   i. 21.00
7. a. 9.094 ohms   b. 6.375 inches   c. 4.333 hours   d. 3.938 watts

Unit 13 Addition of Decimals

A. 1. a. 9.4   c. 220.72   e. 1.5   g. 31.52   i. 354.4096
b. 99.4   d. 287.726   f. 470.4   h. 727.505   j. 3226.08606
B. 1. A. 3.4375   B. 3.4688   C. 5.0001   D. 6.5313   E. 10.0001
3. A. 5.125   5. 1.7875″ thick   7. 9.39 amperes

Unit 14 Subtraction of Decimals

A. 1. A. .250   B. .2565   C. .8442   D. 1.8438   E. 1.2969
3. A. 2.972   B. 2.097   C. 1.847   D. .500   E. 5.127
5. A. .906   B. 1.625   C. 4.406   D. .370   E. .344
7. 0.00125″ shim required   9. $I_C = 1.062$ A (amperes)

Unit 15 Multiplication of Decimals

A. 1. a. 7.2   c. 8.64   e. 127.4   g. 21.294   i. 680.58144   k. 267.50428
    b. 24   d. 37   f. 7.22   h. 69.4434   j. 3.0289281   l. 20.42789
B. 1. A. 64.395 pounds   B. 5.1675 pounds   C. 5.76375 pounds   D. 4.17375 pounds
C. 1. A. $6.56   B. $10.63   C. $9.01   D. $107.21   E. $0.56 Total Cost. $133.97
3. A. 6.8 inches   B. 7.6 inches   C. 37.3 inches   D. 119.8 inches   E. 200.4 inches   Total = 371.9 inches
5. A. 41.48 hr   B. 46.75 hr   C. 60.76 hr
7. a. $61.35   b. $2.56   c. $0.57   d. $59.36   9. Power = 3.49 watts

Unit 16 Division of Decimals

A. 1. a. .118   b. .0237   c. 4   d. 4.01   e. .333   f. 2.540
B. 1. a. .750   b. .625   c. .563   d. .167   e. .097   f. .406   g. .453
C. 1. a. .500   b. .750   c. .375   d. .3125   e. .21875   f. .296875
D. 1. A. .010   B. .003   C. .0017   D. .0021   E. .0022
3. A. 16   B. 37   C. 49   D. 30   E. 17
5. A. $14\frac{7}{16}$ oz   B. $25\frac{2}{5}$ gr   C. $175\frac{3}{8}$ qt   D. $9\frac{9}{16}$ lb

Section 4 Measurement: Directed and Computed (Customary Units)

Unit 18 Principles of Linear Measure

A. 1. a. A. $\frac{1}{2}''$ B. $\frac{1}{4}''$ C. $41\frac{3}{4}''$ D. $42\frac{5}{8}''$ E. $63\frac{7}{8}''$

    b.
A. $\frac{1}{8}$ C. $\frac{3}{4}$ E. $2\frac{5}{8}$ G. $\frac{5}{8}$ I. $1\frac{15}{16}$ K. $\frac{21}{32}$ M. $1\frac{13}{32}$ O. $2\frac{29}{32}$ Q. $\frac{47}{64}$ S. $1\frac{19}{32}$

B. $\frac{1}{2}$ D. $1\frac{7}{8}$ F. $\frac{5}{16}$ H. $1\frac{1}{2}$ J. $2\frac{7}{16}$ L. $1\frac{1}{4}$ N. $2\frac{5}{32}$ P. $\frac{15}{64}$ R. $1\frac{1}{16}$ T. $2\frac{37}{64}$

3. a. $3\frac{3}{4}$ b. $1\frac{3}{4}$ c. $3\frac{1}{4}$ d. $3\frac{5}{8}$ e. $3\frac{11}{16}$ f. $2\frac{5}{16}$ g. $2\frac{31}{32}$ h. $3\frac{11}{16}$ i. $1\frac{3}{64}$ j. $3\frac{7}{16}$

5. A. $\frac{15}{32}$ B. $\frac{1}{2}$ C. $\frac{27}{32}$ D. $\frac{1}{2}$ E. $2\frac{1}{16}$ F. $\frac{11}{32}$ G. 1   H. $5\frac{23}{32}$

7. a. $3\frac{7}{10}$ b. 1   c. $3\frac{17}{50}$ d. $2\frac{33}{50}$

B. 1. A. 1.5002   B. 2.7502   C. 1.3333   D. 4.0835   E. 5.4168   F. 6.7501
3. A. 1.660   C. 1.073   E. .391   G. 1.266   I. .953
    B. 2.385   D. 1.510   F. .500   H. 2.609   J. 5.328
C. 1. A. .012   B. .075   C. .200   D. .250   E. .562   F. .9375
D. 1. A. .5003   B. .3902   C. .2553
E. 1. A. .500   B. .525   C. 2.463   D. 9.906
F. 1. A. .500 (one block) or .300 + .200 or .350 + .150 or .400 + .100
    B. .750 (one block) or .500 + .250 or .400 + .350 or .450 + .300
    C. .750 + .125
    D. .150 + .115 or .130 + .135 or .140 + .125 or .120 + .145
    E. .1001 + .500
    F. .1007 + .650
    G. .1003 + .149 + 1.000

H. .1008 + .100 + 2.000

I. .1009 + .104 + .800 + 4.000 + 3.000 + 2.000 + 1.000

J. .135 + .750 + .600 + .400 + 4.000 + 3.000 + 2.000 + 1.000

(*Note.* Many other combinations are possible and acceptable.)

Unit 19 *Principles of Angular and Circular Measure*

A. 1. a. 60°   b. 80°   c. 135°   d. 50°   e. 105°

3. a. $\frac{1°}{2}$ b. $\frac{3°}{4}$ c. $1\frac{1°}{4}$ d. 300′   e. 450′   f. 84′   g. 300″   h. 330″

B. 1. a. 30°   b. 255°   c. 100°   d. 1°16′30″   e. 82°48′   f. 93°5′20″

C. 1. A. 80°   B. 150°   C. 32°   D. 170°   E. 105°   F. 160°   G. 85°   H. 20°

3. A semicircular protractor is to be used to lay out the following angles: A. 10°   B. 25°   C. 100°   D. 120°

5. A. 237°   B. 134°   C. 48°   D. 217°

D. 1. A. 2.094   B. 7.854   C. 20.617   D. 26.495

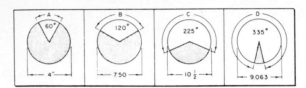

A. $\frac{60}{360}$ × 4 × 3.1416 = 2.0944″   B. $\frac{120}{360}$ × 7.5 × 3.1416 = 7.854″

C. $\frac{225}{360}$ × 10.5 × 3.1416 = 20.61675 or 20.617″   D. $\frac{335}{360}$ × 9.063 × 3.1416 = 26.4950763 or 26.495″

3. A. 4′–9″   B. 6′–10″   C. 8′–3″

Unit 20 *Principles of Surface Measure*

A. 1. A. $(8)^2$ = 64 sq in.   D. 20.6 × 12.4 = 255.44 sq in.

B. $(10.50)^2$ = 110.25 sq in.   E. 60 × 16 = 960 sq in. or 6.667 sq ft

C. $(76)^2$ = 5776 sq in. or 40.111 sq ft   F. 4.75 × 3.10 = 14.725 sq ft

3. 1,904 sq ft

B. 1. A. 10.25 × 8.50 = 87.125 sq in. or $87\frac{1}{8}$ sq in.

B. (3.4 × 1.2) − .4(3.4 − 2.2) = 3.6 sq in.

C. (30.25 × 16.75) − (14.5 × 8.5) = 383.4375 or $383\frac{7}{16}$ sq in.

C. 1. A. .5 × .6 × (1.4 + .9) = .69 sq in.

B. .5 × 9.5 × (12.5 + 10.25) = 108.0625 or $108\frac{1}{16}$ sq in.

C. .5 × 5.8 × (13.4 + 17.6) − $(2.4)^2$ = 84.14 sq in.

3. Lot A is 12,348 sq ft   Lot B is 12,299 sq ft

D. 1. A. .5 × 12 × 8.5 = 51 sq in.   C. .5 × 1.02 × 15.8 = 8.058 sq in.

B. .5 × 3.2 × 5.8 = 9.28 sq in.   D. .8 × 2.6 + .5 × .8 × 2.6 − $(4)^2$ = 2.96 sq in.

E. 1. A. .7854 × $(.8)^2$ = .50 sq in.   D. 3.1416 × $(.5)^2$ = .79 sq in.

B. .7854 $(6)^2$ = 28.27 sq in.   E. 3.1416 × $(2)^2$ = 12.57 sq in.

C. .7854 × $(5.25)^2$ = 21.65 sq in.   F. 3.1416 × $(3.8)^2$ = 45.36 sq in.

3. 79.97 sq in.

F. 1. A. $\frac{1}{3}$ × $(12)^2$ × .7854 = 37.699 sq in.

B. $\frac{1}{5}$ × $(2.125)^2$ × 3.1416 = 2.837 sq in.

C. $\frac{1}{12}$ × $(4.8^2 − 1.92^2)$ × .7854 = 1.267 sq in.

G. 1. A. $7.75 \times 3.1416 \times 8.5 = 206.95$ sq in. lateral surface

  B. $5 \times 3.1416 \times 7.5 + 2 \times (5)^2 \times .7854 = 157.08$ sq in. total surface

Unit 21 Principles of Volume Measure

A. 1. a. $2 \times 1728 = 3456$ cu in.

  b. $1.5 \times 1728 = 2592$ cu in.

  c. 3.625 or $\frac{29}{9} \times 1728 = 6264$ cu in.

  d. $10 \times 1728 + 19 = 17,299$ cu in.

  e. $3456 \div 1728 = 2$ cu ft

  f. $18.144 \div 1728 = .0105$ cu ft

  g. $8640 \div 1728 = 5$ cu ft

  h. $1944 \div 1728 = 1$ cu ft 216 cu in.

  i. $3 \times 27 = 81$ cu ft

  j. $\frac{13}{3} \times 27 = 117$ cu ft

  k. $5 \times 27 + 7 = 142$ cu ft

  l. $7 \times 27 + 19 = 208$ cu ft

B. 1. A. $6 \times 6 \times 6 = 216$ cu in.   B. $(8.5)^3 = 614.125$ cu in.   C. $(1.5)^3 = 3.375$ cu ft

  3. A. $\frac{1}{2}$ B. $\frac{1}{4}$ C. $2\frac{1}{4}$

C. 1. 512 cu yd

  3. A. $(6 \times 4 - 2 \times 3) \times 12 = 216$ cu in.

  B. $(20.2 \times 6.8 - 4.2 \times 4.2) \times 17 + 7.6 \times 12.6 \times 3.4 = 2360.8$ cu in.

D. 1. A. $(4)^2 \times .7854 \times 10 = 125.66$ cu in.

  B. $(12.5)^2 \times .7854 \times 24.5 = 3006.61$ cu in.

  C. $(1.6)^2 \times 3.1416 \times 6.4 = 51.47$ cu in.

  3. A. $(5^2 - 2^2) \times .7854 \times 10 = 164.9$ cu in.   B. $(4.25^2 - 1.5^2) \times .7854 \times 12 = 149$ cu in.

  5. $(3.625 \times 250 + 107) \times .75^2 \times .7854 \times .28 \times 1.26 = \$157.92$

E. 1. $\{12 \times 6 - (3 \times 2 \times 2 \times .7854)\} \times 16 \times 20 \times .26 = 5206.3$ lb

F. 1. a. 16 qt   c. 15 qt   e. $10\frac{1}{2}$ g. 7 pt 3 gills   i. 9 gal 1 qt   k. 3 bbl 2 gal   m. 2 gal 2 qt

  b. 25 qt   d. 13 pt   f. 17 pt 2 gills   h. 4 gal 1qt   j. 2 bbl   l. 3 gal   n. 1,155 cu in.

  o. 1,097.25 cu in.

Section 5 Percentage and Averages

Unit 23 The Concepts of Percent and Percentage

A. 1. 25%   51% $37\frac{1}{2}\%$

  3. A. 65%   B. $6\frac{1}{2}\%$ C. .6%   D. $6\frac{2}{3}\%$ E. $12\frac{1}{4}\%$

  5. A. 150%   B. 75%   C. $12\frac{1}{2}\%$ D. 205%   E. $2\frac{1}{2}\%$ F. .4%

B. 1. A. 200   B. 325   C. 62.5′   D. 1.25 tons   E. .563 lb

C. 1. A. \$20.00   B. 9.6 sheets   C. 12.6 acres   D. .65 cu yd   E. 26 kW   F. 364.5 lb

  3. A. 106 lb tin   B. 43.2 lb tin   C. 75 lb tin

  106 lb lead   4.8 lb lead   50 lb lead

  5. 53.592 hp

Unit 24 Application of Percentage, Base, and Rate

A. 1. A. 1920 tons   B. 843.75 gal   C. 5.4188 in.   D. 147.0825 sq ft   E. 18.531 sheets

  3. (a) The amount of profit is \$456.   (b) The percent of profit is 12%.

  5. a. $1290 \times (100 - 12.5) = 1,128$ (whole) castings   b. $1290 \times 1.95 \div 1,128 = \$2.23$

B. 1. A. $120 \div .90 = 133\frac{1}{3}$ hp   C. $137.8 \div .072 = 1913.89$ lb   E. $126.5 \div .0325 = 3892.3$ in.

  B. $2016 \div 54 = 3733\frac{1}{3}$ cables   D. $78.5 \div .0625 = 1207.69$ bars

  3. $\frac{2.4}{120} = .02 = 2\%$

C. 1. $\frac{1}{5} = .20 = 20\%$ acid $100 - 20 = 80\%$ water   3. $346 \times .82 = 283.7$ rpm

5.

	Copper	Tin	Zinc	Phosphorus	Lead	Iron	
A	320	44	32.8	1.6	1.2	.4	pounds
B	435.7	45.7	38.8	1.8	2.7	.3	pounds

Unit 25 Averages and Estimates
A. 1. 11.675″   3. 1.2498″
B. 1. 1235 − (212 + 224 + 232 + 275) = 292 units

Section 6 Graphs and Statistical Measurements

Unit 27 Development and Interpretation of Bar Graphs
A. 1. The vertical scale of the graph shows percents from 0% to 50%. The vertical bars represent: fuel (15%); products (9%); miscellaneous (15%).
B. 1. a. 1984   b. 1979   c. Using 610 in 1981 and 720 in 1984, percent of increase is 18%.

Unit 28 Development and Interpretation of Line Graphs
A. 1.

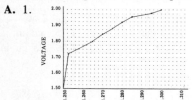

B. 1.

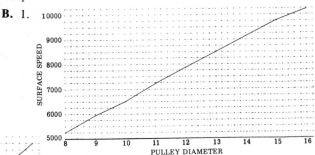

3.

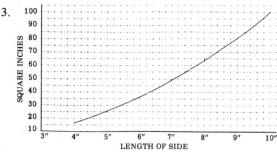

5. a. (1) 275′/min,   (2) 550′/min,   (3) 840′/min surface speed
   b. (1) 6″   (2) $9\frac{1}{2}″$ (3) 11″

C. 1. a. 42.5 to 60.0 hours of production time
     b. System A. 600 units   System B. 500 units   System C. 740 units
     c. System A. 600 to 880 units   System B. 500 to 1030 units   System C. 740 to 1300 units
     d. System A has the greatest variation in hourly production.
     e. System A production gradually increases from 600 to 630 to 675 units.
     System B drastically increases from 500 to 660 units and then at a constant rate to 720 units at 47.5 hours.
     System C has a constant rate rise from 740 to 820 to 900 units at 47.5 hours.

Unit 29 Development and Application of Circle Graphs
A. 1. The circle is divided into two areas. The larger area $(81\frac{1}{4}\%)$ represents the 1,300 men. The smaller area $(18\frac{3}{4}\%)$ represents the 300 women.

B. 1. a. 35,280   c. Men: 18–35   10,080   Women: 18–35   3,024
        b. 71% men   29% women   36–55   12,600   36–55   6,250

Unit 30 Statistical Measurements
A. 1. a. (1) 34.13% in the + range; 34.13% in the − range
          (2) 3,413 bearings in the + range; 3,413 bearings in the − range.
       b. 2,718 second quality bearings are acceptable   c. 456 total number of rejects
B. 1. a. Single sample size of 75 is required
       b. The lot is accepted if no more than four parts in the sampling are defective. The lot is rejected when five or more parts are defective.
       c. 42 parts are required in the first sample in the double sampling plan.
       d. If there are three or four rejects in the first batch of 42 parts, a second sampling of 84 parts is required, totaling 126 parts.
       e. (1) Eight sampling sequences are used.   (2) Total number of parts in the eight samples is 128.
C. 1. a. The tensile strength range is from 20,000 to 90,000 lb/sq in.
       b. (1) The average (mean) tensile strength is 39,429 lb/sq in. correct to the nearest pound.
          (2) The median tensile strength is 26,000 lb/sq in.
          (3) The tensile strength mode is 20,000 lb/sq in

PART TWO FUNDAMENTALS OF SI METRIC MEASUREMENTS

Section 7 Metrication: Systems, Instruments, and Measurement Conversions

Unit 32 SI Metric Units of Measurement
A. 1.1. (f)   2. (a)   3. (i)   4. (h)   5. (d)   6. (e)
B. 1. A. 10 mm   C. 232 mm   E. 288 mm   G. 3.5 cm   I. 27.6 cm
        B. 55 mm   D. 261 mm   F. 2 cm   H. 5.6 cm   J. 29.4 cm
      3. A. 31 mm   B. 15 mm   C. 38 mm   D. 30 mm   E. 126 mm or 12.6 cm
      5. (a) A. 4.1 cm   B. 2.3 cm   C. 3.8 cm   D. 2.6 cm
        (b) E. 12.8 cm
C. 1. A. 3.96 cm   B. 7.16 cm   C. 3.4 cm   D. 10.56 cm   E. 13.96 cm   F. 17.36 cm
      3. A. 4.24 cm   C. 2.62 cm   E. 1.116 cm   G. 3.216 cm   I. 2.316 cm
        B. 6.14 cm   D. 3.84 cm   F. 1.316 cm   H. 6.716 cm   J. 13.516 cm
      (*Note.* In industrial practice, metric dimensions E through I are rounded to two decimal places for machining as indicated.)
D. 1. A. 900 cm^2 B. 225 dm^2
      3. A. 60/360 × (2.6) × 0.785 4 = 88 488 dm^2
        B. 240/360 × (8.4^2 − 2.2^2) × .785 4 = 34.410 992 cm^2
      5. (a) 20.45 square meters   (b) 179 boards
E. 1. A. (6.2)^3 = 238.328 dm^3 B. 12 × 1.8 × .94 = 2.030 4 m^3 (all changed to meters)
      3. A. (8.2)^2 × 0.785 4 × 24.6 = 1 299.13 cm^3
        B. (2.4)^2 × 3.141 6 × 3.2 = 57.91 dm^3
        C. (6.4)^2 × 0.785 4 × 44.8 = 1 441.22 cm^3
F. 1. 7.4 × 3.2 × 2.6 × 1.057 = 61.5 liters
      3. A. 19 305.60 cm^3 B. 2.09 liters   C. 1070.12 qt   D. 827.51 gal

G. 1.

		Larger Units		Smaller Units	
a.	Designation	kilogram	hectogram	milligram	centigram
b.	Unit Symbol	kg	hg	mg	cg
c.	Value in Relation to One Gram	1000 g	100 g	$\frac{1}{1000}$ g or 0.001 g	$\frac{1}{100}$ g or 0.01 g

H. 1. Annealing 1,400°F; tempering 449.6°F.

Unit 33 Precision Measurement: Metric Measuring Instruments

A. 1. Ⓐ. 3.00 mm   Ⓑ. 4.50 mm   Ⓒ. 6.75 mm   Ⓓ. 7.373 mm   Ⓔ. 8.755 mm

B. 1. Ⓐ. 5.00 mm   Ⓑ. 10.52 mm   Ⓒ. 19.83 mm   Ⓓ. 8.504 mm   Ⓔ. 16.874   Ⓕ. 16.925 mm

C. 1. Ⓐ. 0.537″   Ⓑ. 1.884″

D. 1. Hole A, 179.40 mm; hole B, 107.46 mm; hole C, 128.60 mm.

E. 1. Ⓐ. 71.24 mm   Ⓑ. 233.58 mm   Ⓒ. 380.94 mm

F. 1. Ⓐ. 30° 0′   Ⓑ. 5° 15′   Ⓒ. 62° 50′

G. 1.

	Dimension	Series of Gage Blocks (mm)	Sizes of Blocks in Each Series (mm)
A	58.555 (mm)	0.001	1.005
		0.01	1.05
		0.1	1.5
		1.0	5.0
		10.0	50.0
B	155.863 (mm)	0.001	1.003
		0.01	1.06
		0.1	1.8
		1.0	2.0
		25.0	100.0 and 50.0

Unit 34 Base, Supplementary, and Derived SI Metric Units

A. 1. *Examples.* Range of measurement; measurements in two systems; formulas, or conversion factors, or mathematical processes; equivalent values in the second system.

    3. a. Supplementary units: radian, steradian; b. Measurement symbol: rad, sr.

       c. The radian measures plane angles; steradian, solid angles (or the area of a spherical surface).

       d. 1 radian (rad) = 57.295 78°, 1 steradian (sr) = radius (r) of circle squared

B. 1. The addition of nonsignificant zeros helps to identify and control the precise degree of accuracy required for one or more dimensions on a drawing.

    3. **Examples.** SI metric   125.72 mm ± 0.1 mm

             Customary units   7.8125″ ± 0.0001″

C. 1. The controlling dimension identifies the prime system of measurement in which an object and its features are designed.

Unit 35 Conversion: Factors and Processes (SI and Customary Units)

A. 1. *Note.* The fractional inch values are first changed to decimal inch values (column b) and then equivalent mm values correct to three decimal places (column c).

Part	Linear Measurement	(b) Decimal Value (.001")	(c) Equivalent mm Value (.001 mm)	(d) Equivalent (.01 mm) Measurements to (.001") Precision
A	7^7 ˣ ˣ	7.875"	200.025	200.03
B	$4\frac{13''}{64}$	4.203"	106.759	106.76
C	$\frac{31''}{32}$	0.969"	24.606	24.61

3. Example. Hard conversion requires that standards of accuracy and dimensional measurements be in one measurement system without regard to the availability of tooling or the capacity to produce the part to be interchangeable in the second measurement system.

5. a. 2 mm   b. 57.5 dm   c. 13.06 cm   d. 5.94 m   e. 602.36 in.   f. 5 in   g. 2 in.   h. 22.97 ft

7.   a. 0.000 75 m³   c. 5.89 yd³   e. 1,700 ℓ   g. 3 ℓ
     b. 353.14 ft³   d. 0.102 m³   f. 2 ℓ   h. 119.23 ℓ

B. 1. Conversion tables usually contain such information as a. identification of the known unit of measure (given value); b. units in the new required unit of measure; c. conversion factors; or categories under which units are grouped; notation on the precision of the factors.

3. A. $\frac{1}{25.4}$ B. $\frac{0.007\ 560}{1}$ C. $\frac{1}{6\ 894.757}$ D. $\frac{15\ 432.9}{1}$

5.
(a)	(b)
A. 68 947.57	689 450 Pa
B. 627.63	7 657 kg (force)

PART THREE FUNDAMENTALS OF ELECTRONIC CALCULATORS

CALCULATOR APPLICATIONS

The problems in each Unit of this Section may be solved and/or checked either by conventional mathematical processes or by using a calculator. Each answer displayed on the calculator register is rounded-off to the required number of decimal places.

Section 8 Calculators: Basic and Advanced Mathematical Processes

Unit 37 Four-Function Calculators: Basic Mathematical Processes

A. 1. (1) c   (3) a   (5) b

3. (a) The C key may be pressed to reset the contents of the accumulator display and the arithmetic register to zero. (*Note.* On some machines the C key is pressed twice.)

(b) The CE or CD key is pressed to erase an error, followed by the entry of the correct new value.

B. 1.

	Required Significant Digits	Quantity Examples	
		Whole Number Values	Mixed Number Values
A	1	7	0.000 07
B	3	295	0.002 95
C	5	17 346	17.034
D	7	2 528 946	250.096 4

C. 1.

Problem (add)		29 375. Kilometers (km)
		7 645.2 km
		987.96 km
		1 969.7 km

Step (a)	Keyboard Entry (b)	Display Reads (c)
1	CC	0
2	29 375	29 375
3	+	29 375
4	7 645.2	7 645.2
5	+	37 020.2
6	987.96	987.96
7	+	38 008.16
8	1 969.7	1 969.7
9	=	39 977.86 km Ans

3. **Problem**   Example: Divide 27 three times by 2.62 as a constant.

Step	Keyboard Input	Display Readout
1	CC	0
2	27	27
3	÷	27
4	2.62	2.62
5	=	10.305 343
6	=	3.933 337
7	=	1.501 273 6 Ans

Unit 38 Four-Function Calculators: Advanced Mathematical Processes

A. 1. *Overflow.* (a) A condition where the numerical capacity of the calculator is exceeded. Condition may be cleared by dividing by a multiple of ten. **Example.** 10 787 932.873

Underflow. (b) A condition where a significant number of nonsignificant numbers are lost. Again, the condition may be cleared by using submultiples of ten. Example. .000 000 004 98 displays as: .000 000 00

3. 1 161.0 mm

B. 1. The computed values shown in the table are rounded off to four places.

Kind of Measurement	Equivalent Measurements					
	A	B	C	D	E	F
Fractional	$\frac{1''}{16}$	$\frac{1''}{64}$	$\frac{1''}{8}$	$\frac{5''}{16}$	$1\frac{31''}{64}$	$4\frac{2,656''}{10,000}$
Decimal	0.0625″	0.0156″	0.125″	0.3125″	1.4843″	4.2656″
Metric (mm)	1.5875	0.3969	3.1750	7.9375	37.7	108.346

C. 1. 801.018 1 MHz

D. 1. An algorithm is a set of mathematical rules, procedures and logical decisions that are fed into a calculator. The algorithm is used to translate advanced mathematical processes into the four basic processes that are within the capacity of the calculator.

3.

	Quantity	Power Value Display Readout	Power Value Measurement
A	8^6	262144	262 144
B	7.2^3	373.248	373.25
C	$8.4^2 \times 10.6^2$	7928.1216	7928.12
D	$10.04^3 \times 16.2^2$	265 601.87	265 601.87
E	2.2^{12}	12855.001	12 855.00
F	0.46^{15}	0.0000087	0.000 0087
G	π^4 Use $\pi = 3.1416$	97.4100	(4 places) 97.4100
H	α^6 Use $\alpha = 2.7183$	403.44494	(3 places) 403.445

E.

Display Readout (a)	Required Dimension (b)
A. 9.2736	9.273 6
B. 5.6629	5.662 9
C. 7.461457	7.461 5
D. 2.9916291	2.991 6

(*Note.* The algorithm for problem D. is used as an example to show how the dimension was calculated.)

$\sqrt[6]{716.88}$ First estimate is 3.

Substituting this value,

$R = [716.88 \div (3)^5 + (5)(3)] \div 6$

$\quad = 2.99169$ Second estimate $= \sqrt[6]{716.968}$

$R = [716.88 \div (2.9917)^5 + (5)(2.9917)] \div 6$

$\quad = 2.9916291$ Third estimate $= \sqrt[6]{716.880}$

The 6th power of 2.991 6 (rounded to four places) is within the required precision.

Unit 39 Scientific Calculators: Operation, Functions, and Programming
A. 1. **Examples.** · The scientific calculator is internally programmed to perform direct operations in statistical measurement, algebra, geometry, and trigonometry.

· The scientific calculator is designed to store, recall, exchange, and sum data.

3. a. Press the ON/C clear entry/clear key before any function or operation key is pressed to remove an incorrect reading.

b. Press the ON/C key twice.

B. 1.

Numerical Entry (a)	Function Key (b)	Display (c)	Notes (d)
	ON/C	0	Clear display and register
3.1416	× ((	3.1416	3.1416 stored pending evaluation of ( )
8.2	+	8.2	(8.2 + stored)
5.73	)	13.93	(8.2 + 5.73) evaluated
	÷	43.762488	3.1416 × 13.93 evaluated
	((	43.762488	43.762488 stored pending evaluation of next ( ).
9.84	−	9.84	(9.84 − stored)
3.652	)	6.192	(9.84 − 3.652 evaluated)
	=	7.0675852 Ans	6.192 divided into 3.1416 × (8.2 + 5.73)

3. A. 440 lb   B. 428 lb   C. 3,129 lb

C. 1.

Numerical Entry (a)	Key Function and Symbol (b)	Display (c)
25.4	1/x Reciprocal Key	25.4

3. The problem requires the listing of (a) each numerical entry, (b) name and sequence of keys, and (c) display for finding the mass (volume) of a cored bronze casting. The rounded-off final answer to two decimal places is 7 413.33 cm³.

D. 1.

	Circuit A	Circuit B
Total Resistance (R_T) =	87.08 Ω (ohms)	139.72 Ω (ohms)
Mean Resistance =	14.5133 Ω	29.953 Ω
Standard Deviation =	9.069	13.252
Variance =	82.239	175.615

PART FOUR MATHEMATICS APPLIED TO CONSUMER AND CAREER NEEDS (WITH CALCULATOR APPLICATIONS)

CALCULATOR
APPLICATIONS

The problems in each Unit of this Section may be solved and/or checked either by conventional mathematical processes or by using a calculator. Each answer displayed on the calculator register is rounded off to the required number of decimal places.

Section 9 Functional Applications of Mathematics

Unit 41 Mathematics Applied to Money Management and Budgeting
A. 1. A. One dollar and twelve cents   B. Twenty dollars and seven cents

C. Two thousand six hundred ninety-two dollars and seventy-five cents

D. Two cents

E. Thirty-seven cents

F. Two dollars and thirty-seven and one-half cents

G. Forty-six and one-quarter cents

B. 1. $30 \times \dfrac{14}{12} \times 2 \times \$1.732 = \$121.24$

3. $\dfrac{3}{4} \times 24 \times 36 \times 96 \div 231 \times \$5.50 = \$1,481.14$

$\dfrac{3}{4} \times 24 \times 36 \times 96 \div 231 = 269.3$ gal

C. 1. While the examples provided in the unit may be used as a guide, individual worksheets are to be prepared listing major categories of (a) income and (b) fixed and variable expenses for an actual or a trial budget.

3. Monthly income and yearly income projected amounts are to be recorded for each entry in the appropriate column. Gross annual totals are to be recorded.

5. An individual control and management system is to be developed for a one-year budget.

Unit 42 Mathematics Applied to Purchasing Goods and Credit Financing

Determining Manufacturing Costs

1. A. $381.00   B. $623.00   C. $595.93   D. $723.90   E. $930.61
3. a. $277.20   d. $304.92 materials   e. $845.46
   b. $491.04 $515.59 labor
   c. $23.76 $24.95 overhead

Single and Multiple Discounts

1. $19.50 \div 12 \times .78 = \$1.27$
3. a. Distributor B gives the best discount ($103.68). Distributor A gives a $103.49 discount; distributor C, $101.89.
   b. The net cost of the fabric from Distributor B is $220.32.
5.

(a) (1)	(a) (2)	(b) (1)	(b) (2)
$2,152.50	$2,006.74	$4,957.50	$5,103.26

B. MATHEMATICS APPLIED TO CREDIT FINANCING

1. The initial and each successive unpaid monthly balance is recorded in the table (column A, items 1 through 12). The uniform monthly payment on the unpaid balance is $255.90 ($3,060 ÷ 12).

3. The monthly payment on the unpaid balance plus the monthly interest are shown in column C.

5. All monthly loan and interest payments are totaled for column C. The total is $3,358.39.

7. The total of all payments on the $4,200 loan for each of the four financial institutions listed in the accompanying table are recorded as item 3.

9. The calculated actual annual percent rate of interest (AAPR) for each institution is recorded as item 5.

Answers to Odd-Numbered Problems Unit 42

	A Unpaid Balance	B Monthly Interest	C Monthly Payments
1	$3,060	$45.90	$300.90
2	2,805	42.08	297.08
3	2,550	38.26	293.26
4	2,295	34.43	289.43
5	2,040	30.60	285.60
6	1,785	26.78	281.78
7	1,530	22.95	277.95
8	1,275	19.13	274.13
9	1,020	15.30	270.30
10	765	11.48	266.48
11	510	7.65	262.25
12	255	3.83	258.83
	Total	$298.39	$3,358.39

Items of Financing	Financial Institution Loan Patterns			
	A Dealer	B Bank/Credit Card	C Mail Order	D Credit Union
1 Credit Amount	$4,200	$4,200	$4,200	$4,200
2 Monthly Payments (18)	$271.37	$296.33	$320.83	$233.33+ Interest $\angle^1$
3 Total of All Payments	$4,884.66	$5,334.00	$5,775.00	$4,830.00
4 Total Carrying Charges	$684.66	$1,134.00	$1,575.00	$630.00
5 Actual Annual % Rate of Interest (AAPR)	16.30%	18.00%	25.00%	10.00%

$\angle^1$ $1\frac{1}{2}\%$ per month on unpaid balance

Unit 43 Mathematics Applied to Banking, Home Ownership, and Investments

Simple and Compound Interest

A. 1. (a) $I = P \times R \times T$ (b) $A = P (I \times R)^n$
   3. A. $559.15 B. $843.16 C. $4,071.15 D. $2,059.18

Mortgage and Property Taxes

1. (a) Monthly Payments = $645.00 (b) Yearly Payments = $7,848.00

Banking and Checking Account Services

1. A. $298.39 B. $1,178.54 C. $415.81 D. $316.94
B. 1. The financial section of a newspaper is to be used for information about common stocks and broker commissions.
   3. The actual cost of the 250 shares of the selected stock is determined by multiplying the quotation price at the *high* for the day by 250 and adding the brokerage fees.
   5. (a). A. $82.50 B. $87.50 C. $487.50
      (b). A. 9.43% B. 8.14% C. 8.48%

Unit 44 Mathematics Applied to Payrolls, Income, and Taxes

A. 1. A. 49 hours $995.10 C. 46 hours $533.60 E. 51 hours $268.38
      B. 46 hours $700.70 D. 36 hours $315.40 F. 40 hours $254.00
B. 1. Values rounded to the nearest dollar.
      A. $19 B. $36 C. $56 D. $13 E. $24 F. $48
C. 1.

Worker (a)	Regular Hours Worked (c)	Overtime (d)		(f) Weekly Earnings
		Hours Worked	Hours Pay	
A	40	12	21	$762.50
B	40	9	$13\frac{1}{2}$	465.45
C	34			332.20
D	40	17	$27\frac{1}{2}$	353.70

Unit 45 Mathematics Essential to Careers and Career Planning

A. 1. The student/trainee is to state briefly the importance of two work ethic concepts to career planning and a career.

   EXAMPLES: • Work is foundational to the production of all goods and services.

   Society is dependent on adequate food resources produced by workers.

   • Work is essential for personal and national financial security.

   3. Examples of four economic sectors in the immediate labor market area: industry, business, health and medical services, agriculture and agribusiness.

   5. Example of an occupational cluster: machine industries occupations.

   7. Examples of four job levels: (a) operative, (b) semiskilled, (c) skilled, (d) semiprofessional.

B. 1. Analyses are made of the cluster of job titles within an occupational group to establish essential mathematical skills that must be mastered to perform successfully on the job.

   3. The awareness, orientation and exploratory learning experiences provide the student/trainee with realistic workplace information. These experiences are interwoven into all learning to identify foundational occupational and citizenship understandings. Manipulative sampling and broad-based knowledge of occupations is a first step in intelligent career planning.

   5.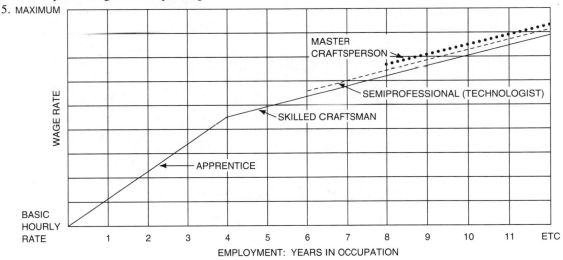

GLOSSARY

Algebra A branch of mathematics requiring symbols, terms, and signed numbers in problems that cannot be solved by the four basic mathematical processes alone.

Algebraic Entry (Calculator) An algebraic logic sequence (order in which a problem is normally solved) for entering values on a calculator.

Algorithm A set of mathematical rules, numerical procedures, or logical decisions that also may be used with an electronic calculator to solve advanced mathematical problems.

Alphanumeric Code Instructions (primarily related to numerical control) that consist of numbers, letters, and other symbols. Letter-number symbols that form the foundation for input-output signals into computerized equipment.

Angular Measure The size of an opening formed by two intersecting lines or surfaces. Angles are measured in radians or steradians and in degrees (°), minutes ('), and seconds ("). Angular measure involves the four basic arithmetical processes with angles and parts of angles.

Area of a Circle $A = \pi$ (radius)^2 or $A = 0.7854$ (diameter)^2

Area of a Triangle The surface area of a triangle equals the base $\times \frac{1}{2}$ altitude.

Axiom A self-evident mathematical truth. For example, the value of an equation is not changed if the same process and quantity are applied to both sides of an equation.

Balanced Equation The values on both sides of an equation are equal.

Bar Graphs Heavy lines or bars of a definite length to represent specific quantities on a grid having vertical and horizontal lines. A scale indicates the value between each line.

Base Units in SI Metrics Seven base units circumscribe all physical science measurements. The base units include meter, kilogram, second, ampere, candela, kelvin, and mole.

Basic Arithmetical Processes The four processes of adding, subtracting, multiplying, and dividing. Included, also, is any combination of the four processes.

Binary Numbers A numbering system comprising only two numeric symbols: 0 and 1. A base-two numerical notation system on which computers depend. Two conditions of operation of electronic circuits: on-off, plus or minus, charge or discharge, and conductor or nonconductor.

Calculator: Four-Function An electronic device used to simplify and obtain consistently accurate answers to mathematical problems. Operations involving addition, subtraction, multiplication, and division are entered through a keyboard. Answers are displayed as whole numbers and decimal values.

Calculator: Scientific/Engineering An electronic computational instrument with capacity to perform algebraic, trigonometric, geometric, statistical, and other higher mathematics operations.

Career Mathematics Relates to the development of computational skills that parallel and complement on-the-job technology and manipulative occupational competency skills. Upgrading mathematical skill competency consistent with vertical job mobility.

Cartesian Coordinate System A graphical system used to describe in mathematical terms any absolute point from any other point in space. Precise positioning of a feature or dimension of a part along three coordinate axes that are mutually perpendicular.

Celsius (°C) A derived unit of temperature measurement in SI metrics, expressed as degrees Celsius (°C). Celsius temperatures replace centigrade temperatures.

Circle A closed curved line on a flat surface. Every point on the closed curved line is the same distance from a fixed given point called a center. *Concentric circles* have a common center. *Eccentric circles* are "off-center" and do not originate from the same center.

Circle Graph The division of a circle into a specific number of parts, each sector denoting a fixed relationship to all other data presented.

Circular Measure The measurement of circles, curved surfaces, cylinders, and other circular shapes.

Circumference The distance around the periphery of a circle, measured either in standard units of linear measure or in degrees.

Clear-Entry Key A key used on a calculator keyboard to correct number-entry errors without interfering with any previous part of a calculating sequence.

Collecting Terms The process of combining letters, symbols, or numbers within grouping symbols.

Complex Equation A formula or equation with two unknown variables. Complex equations are solved by either substitution or addition.

Computer-aided Design (CAD) Generating design features, dimensioning, and other information from data and specifications that are programmed, stored, retrieved, and modified from computer memory and other computer functions.

Computer-aided Manufacturing (CAM) A complete computerized system that combines the capacity of microprocessors, minicomputers, and main-line computers to control multiple machine and instrument functions in manufacturing.

Computer-aided Programming (CAP) Programming machine tools, instruments, and other equipment by computers rather than by manual programming.

Computer Numerical Control (CNC) The automated control of all functions of a machine tool, instrument, or other equipment. Input, output, and storage of command signals that are generated by computer rather than by manual programming.

Congruence A geometric term indicating that all the physical properties of size and shape in one part are identical in another part. Two figures are congruent if they differ only in location. Two line segments of equal length or two circles of equal diameter are congruent.

Constant A fixed mathematical relationship for which a specific symbol is used universally.

Consumer Mathematics The application of mathematics processes, principles, rules, and concepts to meet everyday computational needs as a consumer. Examples of mathematical applications include: money management, budgeting, purchasing goods, credit financing, loans, banking services, payrolls, and taxes.

Conventional Units French and European metric systems using the meter as the basic unit of measure. Unit designations, standards, values, and derivation of base and other units differ from SI metrics.

Conversion of Quantities The changing of a quantity from a unit of measure to an equivalent value in another unit of measure. The process of multiplying or dividing (as may be required) a specified quantity by a conversion factor.

Credit Financing Paying for a product or service over a period of time. Involves the payment of financing costs in addition to the purchase price.

Cubical Solid A figure formed by extending an original square surface so that all corners are square and the edges are the same length.

Cubic Measure (also Volume Measure) Measurement of the space occupied by a body that has three linear dimensions of length, height, and depth. Cubic measure is the product of these three dimensions, expressed in cubic units.

Customary Units United States units of measure based upon the inch, yard, and pound, similar to the British Standard units.

Cylinder of Revolution (Right Cylinder) Two equal and perpendicular bases with a lateral (vertical outside) surface around the bases. The area of the lateral surface of a cylinder of revolution equals the circumference × height.

Decimal-Degree (Trigonometric Function Table) A table of numerical trigonometric function values for metric angle measurements. Metric angle measurements that are expressed in whole and/or decimal parts of a degree.

Decimal Equivalent A conversion of a customary unit of measure to its equivalent value as a decimal.

Decimal Fraction A fraction whose denominator is a multiple of ten: for example, 100; 1,000; and 10,000. Decimal fractions are written on one line with a decimal point in front of the numerical value.

Decimal Places The number of digits to the right of the decimal point. The number of places is usually related to the degree of accuracy required in the answer.

Depreciation The loss in value of a process, material, or product.

Derived Units in SI Metrics Units that are derived from the base units in SI metrics to satisfy varying computational needs. The complementary derived units related to quantities of length, mass, time, electric current, thermodynamic temperature, luminous intensity, and amount of substance.

Discounts A reduction in the price of materials, parts, and mechanisms. Discounts may be single or multiple (which include successive discounts).

Division The process of determining how many times one value is contained in another. Division is a simplified method of subtraction.

Equation A mathematical statement indicating that the quantities or expressions on both sides of an equality sign (=) are equal.

Estimating A shortcut mathematical process of determining a range against which an actual answer may be checked for accuracy.

Exponent A simple mathematical statement that indicates a quantity is to be multiplied by itself a number of times. The exponent is a small number (superscript) written to the right and slightly above a quantity. The exponent indicates the power to which the number is to be *raised* (multiplied by itself).

Expressing Quantities Stating a quantity either in terms of numbers, or a combination of numbers and symbols. *Numerical terms* signifies that numbers are used to express a quantity. *Literal terms* relates to the use of numbers and symbols to express a quantity.

Expression The statement of a problem in abbreviated form using numerical and literal terms.

Extracting the Square Root The process of determining the equal factors that, when multiplied together, give the original value.

Factoring The steps used in determining the series of smaller numbers that, when multiplied by each other, produces the original number.

Formula A shortened method of expressing relationships and a combination of mathematical processes that consistently give the same solution.

Formula Evaluation The process of (1) substituting all known numerical and literal factors in a formula and (2) solving for a particular variable by performing the indicated operations.

Gage Block Measurement A system for establishing a precise measurement by "wringing" together hardened and precisely finished steel blocks against which measurements may be established. Precision blocks used to establish precise linear and angular dimensions.

Geometry A branch of mathematics relating to points, lines, planes, closed flat shapes, and solids. These elements may be used alone or in combination to describe, design, construct, and test every visible object.

Graphics Calculator A mathematical calculator with capability to perform basic mathematical processes, engineering/scientific processes, and graphically display the mathematical results. The graphics calculator permits tracing values on a screen to establish values using variable factors.

Grouping Symbols Symbols that are used to group quantities together to simplify reading and mathematical processes. The four common grouping symbols are parentheses, brackets, braces, and a bar to cover a quantity.

Hertz The frequency of one cycle per second.

Improper Fraction A fraction with a numerator greater than the denominator.

Indirect (Inverse) Proportion Ratios that vary inversely. The two terms in the first or second ratio are reversed.

Interest Payment for the use of money at a simple or compound interest rate. Compound interest involves posting the amount of interest (applying it to the principal) each payment date; accumulating interest on principal and interest.

Interpolation The process of calculating an intermediate value of a function that lies between two known values.

Interpolation (Machining) The method of advancing a workpiece or a tool from one point in a program to the next point. Producing a contoured feature in a workpiece. Continuous path machining controlled by interpolation.

Job Levels The clustering of jobs (titles) within an economic sector according to required levels of skill and technology. Job levels range from unskilled to operative, to semiskilled, to skilled, and on to professional and management levels.

Joule The work done when a force of one newton is moved a distance of one meter in the direction of the force.

Kilogram A base unit of mass in SI metrics. A kilogram is equal to 2.2 pounds or 1,000 grams.

Law of Cosines The square of any side of a triangle is equal to the sum of the squares of the other sides minus twice the product of the two sides and the cosine of their included angle.

Law of Sines An abbreviated method of expressing that the sides of a triangle are proportional to the sines of the opposite angles.

Like Factors (Algebra) A factor that is common to the processes being performed in solving an algebraic equation.

Line Graph A grid of lines at fixed values on which straight, curved, or broken lines are plotted to present information visually.

Line Segment The working length of a line defined by end points.

Literal Equation A specific type of equation in which some or all of the quantities are represented by letters instead of numbers.

Logarithm of a Number The exponent indicating the power to which it is necessary to raise a number to produce a given number.

Lowest (Least) Common Denominator The smallest number into which each number in a set of denominators will divide exactly.

Magnitude A specific quantity, size, or amount of a force.

Mantissa The decimal part of a logarithm.

Members (Equation) The expressions that appear on either side of the equality sign in an equation. The quantity on the left side is the *first member*. The *second member* appears on the right side.

Memory (Calculator) The capability of a calculator to store alphanumeric values and later recall intermediate mathematical calculations.

Meter A base unit of length in SI metrics and conventional metric measurements. For most practical purposes the meter equals 39.37 inches.

Metrication Any policy, act, or process that tends to increase the changeover from Customary (U.S.) and Conventional Metric units of measurement to SI Metrics.

Metricize The process of converting any other unit of measure to its SI metric equivalent.

Micrometer A precision instrument for taking linear measurements in any one of the measurement systems, depending on the calibrations on the micrometer. The distance between the anvil and a spindle is accurately read from a calibrated barrel and thimble. Vernier micrometers have an additional set of graduated lines. These make it possible to read to a more precise degree (0.002 mm and 0.0001″) than regular micrometers.

Money Management Controlling and managing income and expenses by setting up a budget and a system of records and maintaining them.

Multiplication A simplified method of addition.

Multiplying and Dividing Quantities The product of multiplying two like negative terms or any number of positive terms is $(+)$. The product is $(-)$ when an odd number of negative terms is multiplied. These rules for the use of signs apply also to division.

Newton A force that, when applied to a body of one kilogram mass, produces an acceleration of the body of one meter per second.

Normal Frequency Distribution A bell-shaped curve formed by plotting actual production sizes in a random distribution against a design dimension for a part. Distribution of part sizes as contrasted with a required dimension.

Numerical Control (NC) A system of coded instructions that controls processes and movements of machine tools and instruments.

Overflow A calculator condition where the numerical capacity of the instrument is exceeded.

Parallelogram A figure having two pairs of parallel sides. The area = length × height.

Percent A short way of relating a given number of parts of a whole, which is equal to 100 (percent). One percent is $\frac{1}{100}$ of the whole.

Percentage The product of the base times the rate.

Plane Geometry That phase of geometry that relates to objects having two dimensions that lie within a plane.

Plane (Plane Surface) A flat surface on which a straight line connecting two points lies.

Positive and Negative Quantities *Positive numbers* refers to all numbers that are greater than zero. *Negative numbers* relates to values that are less than zero.

Prefixes and Symbols (SI Metrics) Multiple and submultiple values of the base-ten notation stem are identified by a system of prefixes. Each prefix is identified by an SI symbol. SI metric quantities are usually defined by the appropriate prefix followed by the unit of measure.

Probability The number of ways an event can occur divided by the total number of ways the event can occur.

Programmable Calculator A calculator that contains memory to store numbers and instructions for performing a sequence of operations whenever this information and data are required.

Proper Fraction A fraction with a numerator smaller than the denominator.

Proportion Two ratios, equal in value, that are placed on opposite sides of an equality sign.

Protractor A measuring tool for angles. Protractors vary from simple flat tools graduated in degrees to more precise instruments. A movable blade protractor and the universal-bevel protractor are graduated to permit readings in degrees and minutes. The vernier protractor with vernier scales is used for still greater degrees of accuracy.

Pyramid A solid geometric object formed by connecting each corner of a flat shape with a point (*vertex*) outside the base.

Pythagorean Theorem A formula expressing the relationship among the three sides of any right triangle. $A^2 = B^2 + C^2$.

Radical Sign A mathematical shorthand way of indicating that the equal factors of the value under the radical symbol ($\sqrt{\quad}$) are to be determined.

Ratio A comparison of one quantity with another like quantity or value.

Rectangle A figure whose opposite sides are parallel and whose adjacent sides are at right angles to each other. The area = length × height.

Resultant A single vector that represents the sum of a given set of vectors.

Right Prism A solid formed by a flat form as it moves perpendicular to its base to a specified altitude. The shape of the base determines the name of the right prism.

Rounding (Rounding Off) Reducing the number of digits to a lesser number than the total number available. A quantity is reduced to the number of significant digits that produces a desired degree of accuracy in the result.

Scientific Notation System A mathematical system where quantities are written and computed in a simplified form using power of ten multiples.

Sector of a Circle The surface or area between the center and the circumference of a circle that is included within a given angle. The area of a sector equals the area of the circle multiplied by the fractional part that the sector occupies.

Signed or Directed Numbers Positive $(+)$ and negative $(-)$ numbers that indicate either direction from a fixed reference point or that identify the mathematical operation.

Significant Digits Any digit that is needed to define a specific value or quantity. The number of significant digits is based on the implied or required precision associated with the problem.

SI Metrics (SI) An up-to-date International System of Units of measurement established cooperatively among most industrialized nations. SI measurements depend on base, supplementary, and derived units, and on combinations of these units to accurately measure and quantitatively define all measurable objects. SI is the standard abbreviation for *Le Système International d'Unités*.

Soft Conversion Refers to measurements that are computed directly without consideration for product design. Direct conversion of measurement values from either SI metric or customary units of measure.

Solid Angle Represents the enclosed area that is equal to a square whose sides are equal in length to the radius of a sphere. Measured in steradian units of measure.

Solid Geometry That phase of geometry that relates to objects formed and measured by three or more dimensions. Features that lie in two or more planes.

Solids of Revolution The shape taken by outside lines or lines of a flat form when revolved around an axis.

Statistical Measurement Manipulating data and large numbers of measurements by mathematical computations.

Stretchout Shop term applied to a two-dimensional layout that becomes a three-dimensional object when rolled or formed into a closed figure.

Supplementary Units in SI Metrics The radian and steradian constitute the two supplementary units for measuring angles in SI metrics. The radian relates to a plane angle; the steradian to a solid angle.

Surface Measure The measurement of a part, object, mechanism, or other physical mass that has length and height. A surface is "measured" when its length and height (in the same unit of measure) are multiplied. The product is the *area* in square units.

Symbols A simplified way of identifying and working with quantities, units of measure, and mathematical processes and of communicating information.

Symmetrical The corresponding points of an object are equidistant from an axis.

Take-home Pay The actual money a worker receives after all deductions for federal, state, and local taxes, personal benefits, and other payments are made.

Terms Parts in a mathematical expression that are separated by such signs as (+) and (−). *Like terms* relates to the different terms in an expression that have the same literal factor. When the literal factors are different or unlike, they are called *unlike terms*.

Terms (Proportion) The two outside terms of a proportion are the *extremes*. The two inner terms of a proportion are the *means*.

Theorem A mathematical truth that can be proven.

Transposing Terms A method of moving all known terms to one side of an equation and all unknown terms to the other side. The sign of each transposed quantity is changed.

Transversal A line that intersects two or more lines.

Trapezoid A four-sided figure in which two of the sides (called *bases*) are parallel. The area = the sum of the two bases $\times \frac{1}{2}$ altitude.

Triangle Three straight line segments joined at the ends to form a closed flat shape.

Trigonometric Functions An equation expressing the ratio between two sides. The six common ratios are sine (sin), cosine (cos), tangent (tan), cotangent (cot), secant (sec), and cosecant (csc).

Trigonometry A branch of mathematics that deals with the measurement of angles, triangles, and distances.

True-Credit Balance (Calculator) The visual display on a calculator that indicates whether a resulting numerical value is positive (+) or negative (−).

Variable A letter or other symbol in an equation that represents a quantity that may have more than one value.

Variable Expense An expense item that may be changed.

Vector A line segment that represents direction (+ or −) and magnitude (size or quantity).

Vernier Caliper Micrometer An instrument on which linear measurements are determined by adding the reading on a graduated beam and a graduation on the vernier scale of a movable leg.

Volt The difference of electrical potential between two joints of a conductor carrying a constant current of one ampere. The power dissipated between these two points equals one watt.

INDEX